Solar Water Heating Systems

Solar Water Heating Systems

Proceedings of the Workshop on
Solar Water Heating Systems
New Delhi, India
6-10 May, 1985

edited by

H. P. Garg

Professor of Solar Energy, Centre of Energy Studies,
Indian Institute of Technology,
New Delhi, India

D. Reidel Publishing Company

A MEMBER OF THE KLUWER ACADEMIC PUBLISHERS GROUP

Dordrecht / Boston / Lancaster / Tokyo

Library of Congress Cataloging in Publication Data

CIP

Workshop on Solar Water Heating Systems (1985 : New Delhi, India)
Solar water heating systems.

Includes indexes.
1. Solar water heaters–Congresses. I. Garg, H. P. II. Title.
TH6561.7.W67 1985 697'.78 85-19386
ISBN 90-277-2136-X

Published by D. Reidel Publishing Company,
P.O. Box 17, 3300 AA Dordrecht, Holland

Sold and distributed in the U.S.A. and Canada
by Kluwer Academic Publishers,
190 Old Derby Street, Hingham, MA 02043, U.S.A.

In all other countries, sold and distributed
by Kluwer Academic Publishers Group,
P.O. Box 322, 3300 AH Dordrecht, Holland

Printed in The Netherlands

Dedicated to my parents
who always believed
in
Honesty, Sincerity and Hardwork

CONTENTS

PREFACE

H.P. Garg
Centre of Energy Studies
Indian Institute of Technology
Hauz Khas, New Delhi 110 016
India

Heating of water using solar energy is not new and by using a little science and technology in it, the solar energy can be utilized more effectively and economically for heating the water both for domestic and industrial applications. Solar Water Heaters are popular for the last three decades in countries like USA, Australia, Israel, Japan, India. This is the only solar energy application which is commercially, technically and economically viable and has been studied for more than 30 years in many countries.

Technical advances in solar water heating have been very rapid in the last 30 years. These are becoming popular not only for domestic use but for large establishments like hostels, hotels, hospitals, industries such as Textile, Paper and Food Processing and even in heating of swimming pools in winter. In few instances the cost of solar water heating systems may be higher than those operated by electricity, gas or other fuel but over a period of time this is more than recovered by the savings in the cost of operations and maintenance.

In some cold countries about 20% of the total energy demand of a family is required for heating of water, may be for taking bath, cleaning of clothes and utensils, and heating of space. In Developing as well as in Developed countries, it has been estimated that more than 20% of all the industrial heat is directly used in processes at temperatures below 100°C. In the remaining 80%, where high temperatures are required, considerable amount of heat can be supplied to pre-heat the water up to 100°C. There are many small industries where 100% of the heat is required up to only 140°C, which can easily be supplied by using solar energy employing simple solar energy collection devices like flat-plate collectors, solar ponds, evacuated collectors and linear concentrators, where the technology is sufficiently advanced.

The editor has visited many countries of the world and found that there is a large variation in the design, operation, testing and evaluation procedures of water heating systems. Even in India, where more than 1000 large solar water heaters have been installed, there is no standard procedure for their design or evaluation. A few of these solar water heaters are working satisfactorily but few have failed due to leakage, poor quality materials, poor maintenance, poor design procedures, etc.

The need of such a national effort to sit together and discuss at a common platform and discuss a single topic like that of Solar Water Heating was therefore long felt since this application has got an immediate market and if used in industries, considerable amount of energy can be saved. Even in a small country like Cyprus, where only domestic solar water heaters are used on a large scale, the country is able to save about 4% of the total energy consumed in the country.

The National Workshop on Solar Water Heating Systems was organised by me with support from DNES, to discuss various aspects of solar water heating systems and to give training to the engineers from industries manufacturing the solar water heating systems, officers from nodal agencies involved in the extension programmes, scientists from national laboratories/organisations, academicians from universities/institutions, and officers/planners from Government organisations.

In order to guarantee excellent quality, we had selected and invited more than 25 speakers to speak on the topics suggested to them falling in their field of specializations. More than 200 experts from different industries, nodal agencies, universities, IITs, Government organisations, etc., participated in this training workshop. There were 10 technical sessions on policies and incentives, solar energy materials, selective coatings and paints, solar radiation measurement and computation, solar water heating components, glazing materials, insulation materials, heat exchangers, controls, welding technology, pumps and control, collector plate configurations and domestic and large industrial systems. Some typical case studies, collector testing procedure and economics of solar/hybrid heating systems are also discussed.

I hope the present compilation of papers will be useful to industries, extension workers, designers, policy makers, students and teachers.

ACKNOWLEDGEMENTS

H.P. Garg
Centre of Energy Studies
Indian Institute of Technology
Hauz Khas, New Delhi 110 016
India

The Organising Secretary, who is also the editor of this book, is grateful to Shri Maheshwar Dayal, Secretary, Department of Non-Conventional Energy Sources (Government of India), New Delhi, who has given the idea of convening such a 'National Training Workshop on Solar Water Heating Systems' at IIT Delhi. The Workshop is fully sponsored by Department of Non-Conventional Energy Sources, New Delhi.

I am also grateful to Prof. M.S. Sodha, IIT Delhi, and Dr.G.D.Sootha, Department of Non-Conventional Energy Sources, New Delhi, for helping me in organising the Workshop and for conducting the technical sessions.

I wish to convey my gratitude to the Non-Conventional Energy Development Agency (NEDA), U.P., B-46 Mahanagar Extension, Lucknow (U.P.), India for the financial grant for preparing the manuscript of this Proceeding. I am particularly grateful to Shri D.K. Mittal, Director, Non-Conventional Energy Development Agency, U.P., who has whole-heartedly supported the idea to convene this Workshop and provided the financial support for preparing the manuscript of the Proceeding.

I wish to thank Dr. V.K. Sharma, who has assisted me in organising the above Workshop and in other routine matters while preparing the manuscript of the Proceeding. I am also grateful to Shri M.P. Joseph, who has typed the manuscript efficiently and in a record time. I am also thankful to Shri Kirpal Singh and Shri Sushil who have prepared the art-work of this Proceeding. I also thank the Speakers for submitting the papers for including in the Proceeding and giving permission for their publication. The participants of the Workshop also deserve appreciation for supporting the idea of holding the Workshop and for useful discussions and recommendations.

I would like to express my appreciation to my wife Kusum Garg, daughters Minu, Neelu and Naina, and son Darpan for their patience and moral support during the conduct of the Workshop and preparation of the manuscript.

INAUGURAL ADDRESS

Maheshwar Dayal
Secretary
Department of Non-Conventional Energy Sources
(Government of India)
New Delhi 110 003
India

Mr. Chairman, Ladies and Gentlemen,

Let me first of all thank the organisers of the Workshop and the Indian Institute of Technology Delhi for arranging this workshop on solar hot water systems and for asking me to inaugurate the same. The Department had been feeling a need of arranging such a workshop since long so that the manufacturing agencies and the implementing agencies could be given proper guidance and directions so that properly designed/ engineered systems are installed and materials of desirable properties are used in various solar hot water systems. This has been possible with the help of the Indian Institute of Technology Delhi. I hope all the participants will be benefited from the delebrations of the workshop and will make use of the suggestions made by various learned speakers during the course of the workshop. I find from the programme sheet that almost all the important areas relating to solar hot water technology would be discussed during the coming five days. These discussions would help in deciding the best quality materials for solar hot water systems. I congratulate Prof. H.P. Garg of IIT Delhi for organising this workshop at IIT Delhi.

The Commission for Additional Sources of Energy had initiated a demonstration scheme relating to the utilisation of solar thermal devices in early 1981. Under this scheme a large number of solar thermal systems in general and solar hot water systems in particular were installed in various parts of the country. The primary objective of this programme was to propagate the use of solar thermal devices and to collect data on their performance on actual field conditions. The programme has been very successful since as a result of this programme over 100 manufacturers are now registered with us who are either manufacturing or likely to manufacture low grade solar thermal devices and other components in the country.

Keeping in view the enormous response from the manufacturers and users the Department of Non-Conventional Energy Sources launched an extension programme for the large scale utilisation of low grade solar thermal devices which included solar hot water systems, solar air heaters, solar dryers, solar wood seasoning systems and solar desalinations in April 1984. The scheme relating to the sale of solar cookers

H. P. Garg (ed.), Solar Water Heating Systems, 1–3.

under the subsidy scheme was initiated by the Department in early 1982. Besides, the Department have been providing funds for R&D activities as well as demonstration programmes on various new and renewable sources of energy including solar thermal energy. Under the new scheme of the Department nodal agencies are supposed to implement the scheme in their respective States. During the last about a year now over 200 industrial solar hot water systems have been installed in the country. The capacity of these systems varies from 1500 litres per day to over 60,000 litres per day in the temperature range of 60°C to over 80°C. There is an overwhelming response from various user agencies for the use of solar hot water systems. Over 600 solar hot water projects are under installation in different parts of the country. Besides a large number of other low grade solar thermal systems such as air heaters, dryers, stills, etc. are also under installation. Although great interest has been generated for the use of solar hot water systems in the country, a lot more is yet to be done.

Renewable energy systems are supposed to be the energy devices for the common man. Therefore, our efforts should be to reduce their cost and increase the efficiency and also their life. However, it has been observed that the present day cost of solar hot water systems and other low grade solar thermal devices is still quite high. Although the cost of hot water systems has been reduced from about Rs.80 (US$7) per litre in early 1980 to Rs.40 per litre in 1984-85 for industrial and about Rs.70 per litre for domestic systems, still the cost is quite high as the consumer in India is not so rich that he can afford to spend Rs.6000-Rs.8000 on a solar hot water system of capacity 100 litres per day. The Government is supporting the R&D projects relating to the activities such as reduction of cost, improvement in efficiency, etc.; the manufacturers are also expected to work in that direction so that better systems with lower cost are available to the common man for his day to day use.

The policy of the Government with regard to the utilisation of new and renewable sources of energy is quite clear. In the approach paper to the Seventh Five-Year Plan it has been made clear that we will have to prepare ourselves for a smooth transition from the dependence on oil to coal and electricity and finally to renewable energy. Thus the Government has given its mind for the utilisation of renewable sources of energy not just for tomorrow but for many years to come. It is the responsibility of the scientific community as well as those who are manufacturing these items in the country to work with a view to reduce the cost and increase the efficiency so that large number of these devices are used by common public.

Besides solar thermal energy systems, the Department has been promoting the development and utilisation of various other renewable sources of energy also. These include solar photovoltaic systems, wind energy systems, biogas and biomass including agro and city waste, improved wood burning stoves, energy plantation, etc. During the financial year 1984-85, the Department installed over 1,80,000 family size biogas plants and 8,00,000 improved wood burning stoves/chulhas. A large number of villages have been provided with integrated energy systems, streetlight from photovoltatic systems, water lifting pumps operated on either photovoltaic solar cells or wind energy, etc. These and similar other programmes are now being expanded for implementation in various parts of the country.

The Department has established a Solar Energy Centre which is supposed to test and standardise the solar thermal products being marketed in the country. Besides,the Centre is supposed to give training and also prepare literature for the use of common man so that the new and renewable sources of energy are popularised in various parts of the country including rural areas. The Centre has brought out a booklet on solar hot water system which has already been distributed to the participants in this gathering. The document has been prepared with a view to educate the manufacturers as well as the users. In future the Centre will be bringing out similar documents on other solar thermal devices and also on other renewable sources of energy. It is suggested that all the manufacturers send their products for getting them tested to the Centre in future. It has been informed that some of the manufacturers are not following the guidelines which have been circulated by the Department. These guidelines have been formulated for the benefit of the manufacturers as well as users. It is suggested that in futurre the guidelines and suggestions made by the Department for better utilisation of solar thermal devices are followed by all. Being a new activity we will have to work together in such a manner that new energy devices become popular even in remote rural parts of the country just like a gas stove or a pressure cooker has become a common man's household requirement.

Ladies and gentlemen, the programme of the workshop looks to me very attractive, but unfortunately it will not be possible for me to attend all the sessions. But I am sure you will have a stimulating and profitable meeting which will help in designing an efficient low cost solar water heater. I now open this Training Workshop on Solar Water Heating Systems.

SOLAR HOT WATER TECHNOLOGY AND APPROACH TO POPULARISE THE SAME

G.D. Sootha
Solar Energy Centre
Department of Non-Conventional Energy Sources
New Delhi
India

1. INTRODUCTION

Indian scientists had realised the importance of solar energy just after Independence when a beginning to develop solar thermal devices was made at the National Physical Laboratory in early 1950s. At that time only a few nations of the world were engaged in the R&D activities in this direction. However, today it is being advocated all over the world that various renewable sources of energy in general and solar energy in particular are the future sources of energy as the reserves of fossil fuels are disappearing fast and may not last more than 100 years; some may be exhausted much earlier. Tremendous changes and advances are now taking place in different parts of the world in the fields of co-generation, total energy, waste heat recovery/utilisation, generation of energy from biomass and waste materials utilising different methods like incineration, pyrolysis, gassification etc., and renewable sources of energy. The primary objectives of all these technological developments are to save fossil fuels and to develop sources which may be long lasting, pollution free, easily available and cost effective.

Among the various sources of renewable energy, solar energy is most versatile as it can be converted into thermal, mechanical, electrical, chemical, etc., forms of energy, is available all over the world and also is pollution free. On an average 60×10^{13} megawatt hours of energy is received annually by the land mass of the earth. In India the density of solar energy varies in the range of 4.5 to 7 KWh per square meter per day on a clear sunny day. This is received in the form of visible (about 51%) and infrared thermal (about 47%) radiation. The actual availability, however, depends upon sky conditions, location of the place, time of the day and season of the year.

Conversion of solar radiant energy into heat is the most efficient and well known method of utilising solar energy. A conversion efficiency of over 60% has been achieved by various workers. The thermal energy thus generated is used for variety of applications like cooking of food, heating of water or any other fluid, drying of industrial and agricultural products, generation of power and refrigeration, purifying

H. P. Garg (ed.), Solar Water Heating Systems, 5–14.

brakish water, creation of comfort conditions in office/residential buildings etc. Since India is a vast country having different climatic conditions and energy requirements of varied nature, solar thermal route of utilising solar energy is the best way of supplementing the energy needs of our people. This route is capable of fulfilling the energy requirements of the village poor as well as of big industries.

2. APPROACH OF THE GOVERNMENT

Government of India have given utmost importance to the field of new energy sources in general and solar energy in particular. Keeping in view the importance of renewable sources of energy a Commission, namely the Commission for Additional Sources of Energy (CASE), was set up at Department of Science and Technology (DST) in early 1981 to give a push to the programmes on the development of renewable energy sources. In September, 1982 a new Department, namely the Department of Non-Conventional Energy Sources (DNES) with CASE attached to it, was created to look after the various activities relating to this important area. Recently the Government have further reaffirmed its support for the development and utilisation of new energy sources and devices. The Approach Paper to the Seventh Five Year Plan has recognised the urgent need to promote Non-Conventional Energy Sources. "Conservation of Energy and promotion of Non-Conventional Energy Sources" and "Ecological and Environmental Conservation", are two of the twelve important features listed in the Approach Paper. The paper also observes that "a more ambitious programme of implementation" is called for in Biogas, municipal and industrial waste processing systems, draught animal power and solar energy in its various forms. The Paper further suggests that "It is necessary to plan for a gradual transition from the present dependence on oil, coal and electricity in the medium term and renewable energy in the longer term. Even though the current contribution of renewable energy sources to energy supply is insignificant and many of them are not commercially competitive at the moment, these technologies have to be developed now so that the transition in future will be smooth." The paper thus speaks not only for today and tomorrow, it speaks for several years from hence also.

In order to promote the development and utilisation of various renewable energy sources and devices, Government have been giving several incentives. Some of these incentives are listed below:

(1) Enhanced depreciation allowance on renewable energy devices and systems and also on machinery or plant installed for manufacturing these systems.

(2) Exemption from Central Excise duty on renewable energy devices and systems.

(3) Exemption from Sales Tax in several States.

(4) This industry is taken under the "Priority Sector" and is eligible for loans on soft terms from financial institutions. It is also eligible benefits under the DRI and Deferred Payment Scheme of financial institutions.

(5) The manufacture of renewable energy equipment has been

delicensed subject to certain conditions. Companies registered under the MRTP act are also eligible for setting up of new capacities in this industry.

(6) Import of several renewable energy devices is allowed under O.G.L. In certain cases the duty has been removed/reduced.

(7) Several subsidy schemes are in operation under DNES. The subsidy on various solar thermal devices is given below:

Funding/incentives by Central Government for installation/use of solar thermal systems/devices including community solar cookers during 1984-85

Sl. No.	Details	Rate of Central Govt. Share	Remarks
1.	Funding in respect of solar thermal systems on Central Government own land/buildings where cost of conventional fuel replaced otherwise would have been met out of the consolidated fund of India	100%	-
2.	Funding in respect of solar thermal systems on State Government land/buildings where the cost of conventional fuel replaced otherwise would have been met out of the consolidated fund of the State	75%	-
3.	Funding of Institutions/Public Trusts and bodies like IITs, Universities, etc.	100%	-
4.	Funding in respect of Cooperative Societies for milk, tea and other similar activities in respect of water heating/air heating/desalination systems/timber kilns/crop dryers/community solar cookers	75%	The crop dryers may be subsidised after these are tested and evaluated by STEC and NIDC
5.	Domestic water heating system	50%	subject to a maximum or Rs.3,000/- to individual beneficiaries (This will be reviewed after two years)
6.	Private House-Building Societies - In respect of centralised solar water heating systems/community solar cookers	50% of the cost of the system	This will be reviewed after two years
7.	Government Public Sector Enterprises- Water heating/air heating/desalination timber kilns/dryers/community solar cookers	50%	-

8. Autonomous bodies, e.g. Municipal Corporations etc. - Water heating/air heating/desalination/timber kilns/dryers/ community solar cookers	75%	-
9. Private Sector Industrial/ Commercial Units - Water heating/air heating desalination/timber kilns/dryers/ community solar cookers	33.3%	-
10. Educational Institutions/Agricultural Universities/Colleges/Schools/ Anganwaries/Balwaries/Charitable and Religious bodies - Water heating/air heating/desalination/timber kilns/dryers/community solar cookers	100%	-
11. Solar desalination systems in villages/backward areas	100%	This programme will be drawn up in consultation with Ministry of Works and Housing and Department of Social Welfare.
12. Solar dryers for agricultural produce for agricultural farms/ rice/dal mills and individual farmers	50%	The proposal will be processed in consultation with Ministry of Agriculture.

3. SOLAR WATER HEATER

Heating of water utilising solar energy is a well known and established technology in many parts of the world including India. Several types of solar water heaters have been developed either under the R&D programme of the DNES or by private industries. In order to popularise the use of solar water heating technology and similar other low grade solar thermal technologies, the CASE had initiated, in early 1981, a demonstration programme. Under this programme several solar water heating systems were installed for domestic as well as industrial use. As a result of this intensive demonstration scheme of the CASE over 100 manufacturers are today registered with the Department who are either manufacturing or planning to manufacture solar thermal devices. As these manufacturers are new in this business and many of them do not have any R&D facilities of their own they need proper education and guidance in respect of material specifications, fabrication and systems design technology and use of meteorological parameters while designing a system.

A solar water heater consists of mainly three components, namely (1) Solar collector, (2) Storage Tank and (3) Connecting Pipes. The other components which may be required are (1) Heat Exchanger,

(2) Circulating pump, (3) measuring instruments like flow meter, energy meter, temperature indicator etc., and (4) electronic/electric control instruments. The heart of any solar thermal device is the solar collector. In a solar hot water system flat-plate collector is generally used. It consists of the following components:

(i) front glazing,
(ii) blackened absorber sheet with built-in-channels,
(iii) back and side insulation,
(iv) outer box, and
(v) sealing materials/gaskets.

The performance of any solar thermal system will depend upon the properties of each of the above mentioned components. For example, the front glazing has the function of allowing the solar radiation to pass through it and act as an opaque curtain to the radiation emitted by the heated absorber. The transmission of solar radiation through the glazing should therefore be as high as possible. The glazing should also have long life and should not degrade due to exposure to varying atmospheric conditions.

Ordinary window glass of thickness 4 to 5.5 mm is being used as glazing material in the solar hot water systems. Best quality window glass available in the country has a transmittance of not more than 85%. Low iron or water white glass having a transmittance of over 90% is quite common in other countries. In some countries like Japan, use of fibre glass has been recommended. However, in most of the cases the transmittance of fibre glass is 30-40% less than glass. U.V. stabilized plastics are also being suggested. But Indian experience with plastics has not been very encouraging.

The blackened absorber plate has two functions to play. It abosrbs solar radiation and converts it into heat. It also acts as a heat exchanger. Therefore the absorber has to be very good energy converter and also heat exchanger. The energy conversion properties of the absorber will depend upon the thermo-optical properties of the black coating material. The solar absorptance of the black coating must be as high as possible but the thermal emittance in the infrared at higher temperatures should be low so that heat is not lost due to radiative losses from the coated surface.

The heat transfer properties of the absorber will depend upon the type of material used to construct the system, configuration of the fluid carrying channels or tubes and type of contact of the tubes with the absorber sheet. In order to extract maximum heat from the absorber one will therefore have to optimise the spacing between the two adjacent channels/tubes, diameter of these channels, thickness of the absorber sheet etc. While selecting the best configuration the cost effectiveness of the system will also have to be seen.

Copper is the most acceptable and common tubing material in direct systems. Designs with corrugated sheets of galvanised steel pressed or welded together have also been developed by some organisations. But such systems are most exposed to corrosion and suffer more from energy loss. Bonduct panels of aluminium have also failed due to corrosive nature of the metal. Absorbers of Cu-Al are quite common in many countries including India. However, with these systems there is a problem of proper thermal contact between the Cu-tubes and Al-sheet as they are joined together only mechanically. Plastics are also becoming popular as absorber plate material. The major advantage with plastics is absence of corrosion effects and reduction in weight. The main disadvantages are that they do not withstand high pressures and repairs are possible only with special tools (and also experience). These materials are quite common for swimming pool heating applications in other countries.

For applications below about 60-65°C, absorber system is generally coated with simple black paint. It has been observed that in most of the cases proper painting procedures are not being followed as a result of which the paint starts chipping off after some time, and the performance of the system starts deteriorating. For temperatures above 60-65°C, selectively coated absorber system is used. In most of the cases maxorb foil is being used for this purpose. Theoretically selective coatings deposited on the absorber sheet should work better than the externally applied foil. Besides, it is observed that air bubbles are left while fixing the foil on to the absorber system. These air bubbles expand when temperature increases due to exposure to sunlight. Since the thermal contact is reduced due to the presence of these air bubbles, the performance of the system gets affected and also the life of the system is reduced.

Back and side thermal insulating material is another important component of a solar collector. Various types of insulating materials are being sold in the market. Detailed experiments conducted at the Solar Energy Centre have proved that only a few materials are suitable for solar energy applications. Most of the resin bonded materials are unstable at higher temperatures and hence should not be used in solar systems. Some fibrous materials which are being sold as glass fibre insulating materials are also not suitable for such applications since their fibres break even on applying slight pressure.

In order to seal the collector gaskets of neopren/rubber are being used by almost all the collector manufacturers. It has been observed that none of these materials is stable at higher temperatures. Some oily material is found coming out of these sealing gaskets which gets deposited on the inner side of the front glazing and reduces transmission of solar radiation through the glazing and hence affects the overall performance.

4. SYSTEM DESIGN

There are several concepts of solar water heating systems which can be illustrated in a "genealogical" diagram, at the top of which we find the two basic concepts, namely integrated and separated systems, respectively.

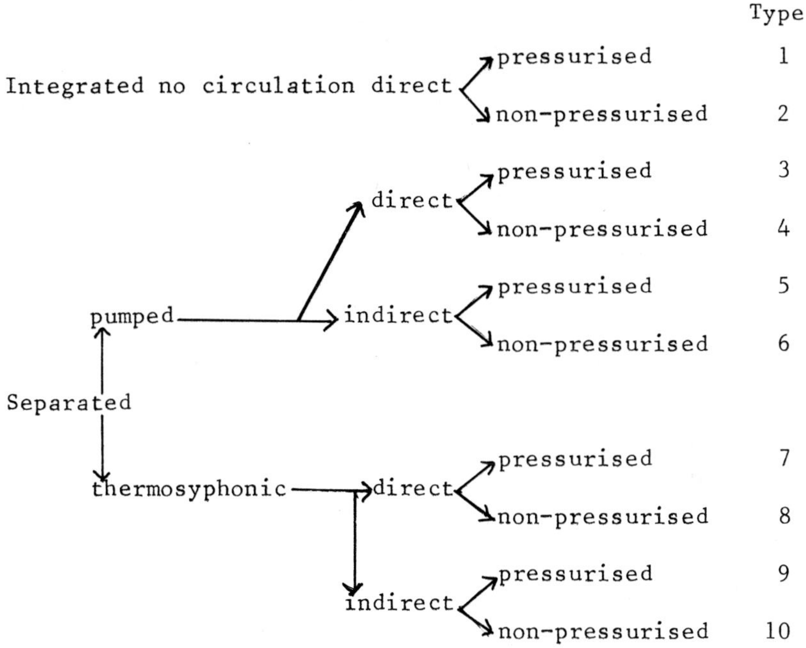

Integrated	The collector and the hot water tank are integrated into the same unit.
Separated	The collector and the hot water tank are two separate units.
Pumped	Circulation is forced by an electric pump.
Thermosyphonic	Natural circulation of water through the system due to gravity.
Direct	The water being heated in the collector is the same as that later used.
Indirect	The heat absorbing fluid circulates through the collector in a closed cycle. The heat is transferred to another cycle of water via a heat exchanger.
Pressurised	The water pressure in the systems is above normal atmosphere, e.g. due to direct connection to the public water distribution network.
Non-pressurised	The water pressure in the system is the same as normal atmospheric.

Any of these concepts can be utilised while designing a solar water heating system. However, the design will depend on the local conditions. Therefore, the designer will have to see the local conditions before a particular concept is selected.

Solar collectors in a large hot water system are connected in series-parallel combination. However, only a definite number of collectors can be connected in series or parallel. The number will depend upon several factors such as design of the collector, flow pattern, temperature required, etc. If proper care is not taken while connecting solar collectors in series-parallel combination some collectors in the system may not have any utility or there may not be any fluid flowing through collectors, if large number of collectors are connected in parallel without optimising the pipe size, flow rate required for such an arrangement, etc. Such designs will also increase the resistance in the flow direction which in turn will affect the capacity of the circulating pump. Large number of pipe bends, elbows, etc., in the system will also increase the flow resistance. The design should therefore be such that the pipe length and bends in the flow path are minimum.

In most of the large solar hot water systems, heat exchanger is being used. There are different types of heat exchangers which can be used for such applications. Heat exchangers having highest effectiveness and low cost should be used. This, however, does not mean that one should use inferior quality materials.

5. SOLAR INSOLATION AND SIZING

The annual average solar intensity per day will decide the potential energy output per collector area. The sizing and consequently the cost of the system is therefore related to this factor. The daily, monthly and seasonal variation of solar intensity will affect the sizing of the collector area and the choice of collector type. It may, therefore, be kept in mind while designing a solar hot water system that collector numbers required for a system of fixed capacity in a place like Delhi will not be the same for another place like Madras or Srinagar.

The energy required to raise the temperature of 1 litre of water by $1^{o}C$ is $4.19x10^{3}$ watts/second. If the capacity of the system is C litres and the temperature of water is to be raised by$\Delta T^{o}C$, the energy required should be

$$E = 4.19x10^{3}xCx\,\Delta T \text{ W sec}$$

$$= 1.16x10^{-3}x\text{Capacity in litres x Temperature KWh}$$

,if I is the solar insolation per day in KWH received per square meter area of the collector and η% is the efficiency of the collector, the collector area required would be

$$\text{Area} = \frac{1.16x10^{-3}xCx\,\Delta T}{Ix\,\eta\%/100}\ m^{2}$$

It may be noted that in the above discussions the thermal losses from storage tank, pumping system and connecting pipes have not been taken into account. Therefore, about 5 to 10% more collector area will have to be used depending upon the system size and system design to compensate these losses.

From the above discussion it is clear that the total collector area required for a fixed capacity hot water system at a particular place will depend upon the efficiency of the individual collectors. Therefore, the collector has to be as perfect as possible in every respect. A higher efficiency of a collector will make it possible to decrease the collector area and the amount of materials used. The collector will be smaller and easier to install and transport. However, these benefits will have to be compared to the additional cost of achieving the higher efficiency. It can now be concluded that designing a solar water heating system is not as easy as many people feel. It is certainly not a job of a village craftsman. It is more a matter of optimization and cost-effectiveness than simply maximizing efficiency. The use of simple materials usually cuts the price of a system quite substantially and consequently makes the system more easily affordable. At the same time it usually decreases the economic lifetime as well. A cheaper system may therefore not necessarily improve the profitability over the system's life time.

6. TESTING AND STANDARDISATION

Manufacturers of solar collectors and hot water systems are required to get their products tested at the Solar Energy Centre (SEC), New Delhi, India. However, only a few manufacturers have sent their products so far. The importance of getting the products tested is clear from the discussion on sizing. Unless and until the efficiency of the product is known proper sizing cannot be done and in the absence of sizing details proper designing of the system cannot be done. It is obvious that most of the systems installed so far have been oversized. It is being taken for granted that one square metre of collector will give 50 litres of water per day at 60-65°C at all places which is absurd.

7. ACHIEVEMENTS

The Department of Non-Conventional Energy Sources has been funding research, development, demonstration and extension programmes on solar thermal energy conversion and utilisation. Over 50 R&D projects are being funded to various R&D centres such as National and Regional Laboratories, Universities, Colleges, IITs, Public Sector Organisations etc. These projects include materials development, systems development, prime-movers development, collector development etc. Technical know-how developed under the R&D programme of DNES for some products materials like cookers, solar kilns, flat-plate collectors, selective coatings etc. have already been passed on to the industry. A bread-box type domestic hot water system and a new design of hot box solar cooker are likely to be released to industry soon. These have been developed at SEC.

Under the subsidy-extension programme, solar cookers, hot water systems, air heaters, dryers, wood seasoning kilns and desalination systems are being installed. So far 31000 solar cookers have been sold. Besides solar water heaters with collector area over 15000 square metres of capacity in the range of 100 litres to over 60,000 litres per day in the temperature range of 60°C to over 80°C have been installed. In addition, 82 other solar systems are installed in different parts of the country and 610 are under installation covering a collector area of about 37,000 square metres.

Under the demonstration scheme, cold storages, steam generators, power generating systems are being installed. A cold storage of storage capacity of 200 MT of potatoes is in operation in a potato seed farm near Gorakhpur. A steam generating plant using line focusing collectors is under installation in a silk factory in Mysore. A 22 Kg power plant using point focusing collectors is nearing completion in a village in Andhra Pradesh. A 50 KW power plant is likely to be installed in a village near Delhi. Another 10 KW power plant is installed at an R&D Centre in Baroda for detailed study and component development.

SOLAR RADIATION

Anna Mani
Indian Institute of Tropical Meteorology
Field Unit
Bangalore

1. INTRODUCTION

The electromagnetic radiation emitted by the sun, covers a very large range of wavelengths from radiowaves through the visible to X-rays and γ-rays. But 99 percent of this energy is contained in the region 0.2 to 4μ, in the near ultraviolet, visible and near infrared regions of the solar spectrum, with a maximum about 0.5μ. Roughly one half of this radiation lies in the visible region between 0.38 and 0.77μ and the remainder in the ultraviolet and infrared regions. The radiation received from the sun, on an average, on a surface of unit area exposed normal to the sun's rays outside the atmosphere at the mean distance of the earth from the sun is called the solar constant. It has a value of roughly 2 $cal/cm^2/min$ or 1.36 kW/m^2. In its passage through the atmosphere, solar radiation is attenuated by scattering and absorption by air molecules, watervapour, dust and aerosols and clouds, and reduced by geometric factors, so that the maximum intensity received at the earth's surface is only about 1.0 kW/m^2, even with very clear, dry air and a cloudless sky.

In a place like Bangalore, the integrated daily energy from the sun and the sky is as high as 6-8 $kWh/m^2/day$ in April-May. The annual incidence of solar energy on a square kilometre of the earth's surface is about 1.5×10^9kWh.

2. MEASUREMENT OF SOLAR RADIATION

The most direct method of assessing the solar energy potential at a place is obviously measurement, using instruments and methods of the required accuracy and reliability.

The radiation parameters that are required to be measured for solar energy utilization programmes are:

1) Global solar radiation or shortwave radiation from the sun and the whole sky, received on a horizontal surface, G. This is usually measured with a thermoelectric pyranometer coupled to a stripchart recorder or integrator printer. Pyranometers using silicon solar cells

H. P. Garg (ed.), Solar Water Heating Systems, 15–35.

or bimetallic strips are not reliable and are not recommended for use, because of the limited spectral sensitivity of the former and the large temperature, azimuth and cosine errors and long response time of the latter. Thermoelectric pyranometers are also used to measure solar radiation on vertical and sloping surfaces with appropriate corrections, and with filters, to measure G in broad and narrow spectral regions.

2) Diffuse solar or sky radiation or shortwave radiation from the sky received on a horizontal surface, D, is measured with a thermoelectric pyranometer provided with a suitable shading ring. Diffuse radiation forms about 17 percent of the global solar radiation with a clear sky and can be as high as 35 percent with hazy skies and 80 percent during the monsoon months.

3) Direct solar radiation at normal incidence, $_oI$, is measured normal to the sun's rays with a pyrheliometer, the Ångstrom compensated type for very accurate measurements and thermoelectric pyrheliometers for routine requirements. The internationally accepted standard for direct solar radiation is the absolute cavity radiometer which has as detector a black body in the form of a receptacle. The thermoelectric pyrheliometers use thermopiles as sensors, the difference between the pyranometers and pyrheliometers being only in their angles of view, pyranometers having a view angle of 2π steradians, while pyrheliometers have an angle of view limited to the sun and 5° of the circumsolar sky. I can also be computed from global and diffuse solar radiation from $I = G\text{-}d/\sin h$, but actual measurements are preferable.

4) Reflected solar radiation or the albedo is measured with an albedometer or a thermoelectric pyranometer mounted with the sensors facing downwards.

5) Longwave terrestrial radiation $L\uparrow$ and atmospheric radiation $L\downarrow$ are measured with a pyrgeometer. Two types are commercially available. Lowcost, accurate, versatile instruments are, however, still to be designed.

6) The spectral distribution of global and direct solar radiation is studied using filter radiometers or spectrographs and monochromators.

7) Atmospheric turbidity is measured using sun photometers or by using pyrheliometers with broad band pass filters.

8) The duration of sunshine, SS, is normally measured with Campbell-Stokes sunshine recorders and electrical sunshine recorders, if found suitable, are acceptable.

The important point to remember in radiation measurement, if not in all measurements,is that wrong measurements are worse than no measurements. After choosing an instrument having the necessary accuracy and reliability and having its calibration factor determined at the National Radiation Centre, Pune, or a similar laboratory authorised or equipped to carry out such calibration, it is necessary to install it correctly and maintain it carefully, if reliable data are to be obtained. Periodic intercomparison of the instrument with travelling standards once a year is essential, if the data are to be considered dependable. The National Radiation Centre at Pune is also one of the two Regional Centres for Asia, the other being Tokyo, and maintains standards that are periodically intercompared with the World standards maintained at the World Radiation Centre in Davos, in Switzerland. With the increase in recent years in the number of radiation instruments in use in India, it has become necessary to evolve a system of verification and quality control of radiation measurements, in order that the instruments used for radiation measurement are installed correctly and calibrated periodically and the data processed with the accuracy required and also published, so that all data will be available to all current and potential users of solar energy in the country.

All radiation measurements are now referred to the World Radiometric Reference WRR, internationally accepted and brought into use from July 1, 1980. The WRR is maintained by means of 4 absolute cavity radiometers designated as World standards and capable of measuring direct solar radiation to an accuracy of 0.1%. The temperature of the cavity is read by a heat flux meter, which is electrically calibrated. Solar radiation measurements is done by exposing the receiver to radiant energy and comparing the temperature field with that caused by heating from an electrical source. They are self-calibrating and in this sense they are absolute.

3. SOLAR RADIATION DATA REQUIREMENTS

There are three main groups of solar technology users, who require solar and terrestrial radiation data in various forms, (1) for the engineering design of collectors and storage systems, (2) for the evaluation of collector efficiency or solar energy system calibration, and (3) in research and for the short and long term predictions of solar radiation.

The first group requires solar radiation data for choosing a suitable site from among several locations and for designing the most efficient collector and storage systems after the site is chosen. Climatological data from national networks maintained by meteorological services will provide in part of information to this group, particularly the natural variability of radiation. Storage systems require information on the persistence of solar radiation greater than a fixed threshold value. Often the network data must be supplemented by site-specific data and actual measurements at the site will be required. Predictions of future energy receipt are also essential for the design engineer.

The second group requires solar radiation measurements immediately next to an operating solar collector, from which its efficiency may be evaluated. The meteorological parameters required to be measured in addition to G, D and I, are longwave radiation $L\uparrow$ and $L\downarrow$, temperature T, windspeed W, humidity h, turbidity β, cloud amount N, sunshine duration SS and precipitation P. The information is required hourly or every 15 minutes or oftener. Their frequency distribution and spectral distribution is also necessary.

There are at least two types of a scientific requirements for solar radiation data related to a comprehensive solar energy research programme. First, to understand the atmospheric and environmental properties which attenuate and determine the solar radiation received at the ground. The accuracy demands are very high for such measurements but this work would be undertaken by specialists at only a few places.

The second is for the elucidation of long term trends in the transmission of solar radiation through the atmosphere. The instrumental accuracy demands will again be very great.

Finally, predictions of future solar radiation will be needed on two time scales, first, for one to several days in advance, for operational purposes and second, on a time scale of years or longer, to assess their effect on solar energy systems.

The requirements of radiation and meteorological data for solar energy technology are summarised below:

1.	Water heating and space heating	: detailed information on G, frequency distribution f(G), temperature T and windspeed, W and cross correlation of G and T or W.
2.	Passive house heating systems	: information on L↓ and L↑, besides G, T, W and h.
3.	Solar cooling systems	: G, L↓ , L↑ , T, W, f(G),f(L)
4.	Concentrating systems	: I, D, f(I,D)
5.	Solar heat engines (water pumps) and solar thermodynamic generators	: I, D, f(I,D), I, W
6.	Photovoltaic power generators	: G, spectral distribution of G, $\frac{\partial G}{\partial \lambda}$, T, W.
7.	Solar energy conversion based on photosynthesis	: G, $\frac{\partial G}{\partial \lambda}$, L↓ , L↑ , T, W, h, p, soil temperature T_{soil}.

The inclination of the collector and albedo have to be taken into account for flat plate collectors and in view of the temporal behaviour of the collection and storage system, the frequency distribution of G. Environmental parameters such as T and W are required for calculating the performance of the system, in view of the heat exchange with the environment. In the passive house heating systems and for the solar cooling systems, L↓ and L↑ are as required besides G, T, W and h, because again of the heat exchange with the environment.

For concentrating systems, with concentration factors between 2 and 100, using linear focusing collectors, parabolic and spherical collectors and heliostats, I and D and their frequency distribution are required, I for high concentrators and D for mild concentrators, which reflect diffuse radiation to a certain part. High concentrators, which are very sensitive to temporal variations, need high temporal resolution (1 minute) of measurements.

Power generators based on thermodynamic processes in the power range 1 to 100 kW have typical operation temperatures of 300°C, the working fluid being oil and need information on I, D, f(D), T and W.

The photovoltaic power generator reacts with a high response to global radiation and its sensitivity strongly depends on the spectral distribution of radiation. Hence information on G, $\partial G/\partial \lambda$ as well as T and W are required, the latter as the efficiency is slightly dependent on the operating temperature.

Biomass represents a stable form of energy which can be easily stored, used for combustion in domestic applications as well as in heat engines and which can be converted by microbiological action into liquid or gaseous fuel. In addition to G, L↓ and L↑ information on spectral distribution is vital as well as agromet data on T, W, h, p and T_{soil}.

In all these measurements great accuracy is not required. Though meteorological radiation measurements aim at an accuracy of 5%, in view of the natural variability of radiation due to changing cloudiness, turbidity and ground albedo (even 5-year monthly averages are highly variable) and depending on engineering needs an accuracy of 15-25% is considered adequate for engineering design needs (Ref.1).

For the evaluation of collector efficiency much higher accuracies of the order of 5-15% or less are required and in addition continuous monitoring of the solarradiation immediately next to an operating solar collector. It may be advantageous to use a 'standard year' as in the U.S.A. to avoid incompatible results of testing of collectors from one year to another, since one type of collector might prove better than others in one year and worse in another.

4. COMPUTATION OF SOLAR RADIATION DATA

The India Meteorological Department maintains a network of 16 stations with thermoelectric pyranometers for the measurement of global solar radiation G, 14 stations for the measurement of diffuse solar radiation D and 13 stations for the measurement of direct solar radiation I. Some of these stations measure albedo and longwave radiation as well. Data from all these stations for periods extending from 2-22 years have been processed and published in the Handbook for Solar Radiation Data for India (Ref.2). Computed data for a number of sloped surfaces of varying azimuths have also been included in the volume.

But this network is quite inadequate for a country as large as India. While a spatial density of 500 km is considered adequate for homogeneous regions, in regions with strong gradients, such as coastal and mountainous areas, a closer network is necessary. Therefore, using two different techniques, daily values of global, diffuse and direct solar radiation were calculated for 145 stations in India, where no radiation measurements are now made. The results have been published in Solar Radiation for India (Ref.3) (1981).

4.1 Computation of Radiation from Sunshine and Cloudiness

The first method is empirical using regression techniques correlating duration of bright sunshine and cloudiness with global solar radiation. Over 100 such formulae have been proposed, since Ångstrom first formulated the empirical relationship relating sunshine duration to cloudiness. The formula most commonly used is

$$\frac{G}{ETR} = a + b\left(\frac{n}{N}\right) \qquad (1)$$

where n and N are the daily and maximum possible duration of sunshine, ETR is the daily sum of the extra-terrestrial solar radiation on a horizontal surface and a and b are regression parameters for each location.

Following Hay (Ref.4), G' was used in place of G, to allow for multiple reflections between the ground, atmosphere and clouds and N' for N, to allow for the fact the sun burns sunshine recorder cards only when it is 5° or more above the horizon. So the relationship becomes

$$\frac{G'}{ETR} = a + b \left(\frac{n}{N'}\right) \tag{2}$$

A similar empirical relationship was found for G', D' and ETR given by

$$\frac{D'}{G'} = c + d \left(\frac{G'}{ETR}\right) \tag{3}$$

where c and d are again regression parameters. For the 14 stations where both G and D and n measurements were available, the regression parameters a, b, c and d were calculated, using the above formulae. Tables 1 and 2 give their value for some of the stations. The very high correlation coefficient and the small standard error of estimation indicates the rigorousness of the relationship.

TABLE 1 - Regression Analysis Between G'/ETR and n/N'

Station	No.of months	a	b	r	r^2	S.E.
Ahmadabad	202	0.252	0.448	0.960	0.921	0.030
Bhavnagar	141	0.259	0.433	0.985	0.971	0.020
Bombay	100	0.283	0.397	0.920	0.847	0.046
Calcutta	248	0.271	0.378	0.897	0.807	0.040
Goa	175	0.270	0.423	0.914	0.835	0.048
Jodhpur	132	0.313	0.401	0.874	0.763	0.042
Kodaikanal	194	0.273	0.521	0.916	0.0839	0.044
Madras	255	0.242	0.436	0.913	0.834	0.032
Mangalore	115	0.214	0.458	0.944	0.892	0.041
Nagpur	218	0.227	0.459	0.975	0.951	0.024
New Delhi	255	0.256	0.454	0.874	0.763	0.042
Port Blair	114	0.156	0.472	0.860	0.739	0.062
Pune	254	0.269	0.417	0.944	0.891	0.037
Shillong	140	0.225	0.463	0.920	0.847	0.044
Trivandrum	215	0.309	0.408	0.912	0.831	0.028
Vishakhapatnam	210	0.247	0.435	0.968	0.938	0.025

The values of the regression parameters when plotted on a map were found to follow an organised spatial pattern over the country and so, from maps drawn for each parameter, values of a, b, c and d were interpolated for 105 stations where sunshine data are available for periods ranging from 8-20 years. Using these interpolated values of a, b, c and

TABLE 2 - Regression Analysis Between D'/G' and G'/ETR

Station	No.of months	c	d	r	r^2	S.E.
Ahmedabad	134	1.329	-1.701	-0.960	0.921	0.057
Bhavnagar	141	1.406	-1.809	-0.979	0.959	0.044
Bombay	44	1.306	-1.769	-0.593	0.909	0.063
Calcutta	147	1.124	-1.386	-0.860	0.740	0.076
Goa	142	1.285	-1.659	-0.941	0.885	0.072
Jodhpur	58	1.325	-1.703	-0.883	0.781	0.071
Madras	193	1.108	-1.351	-0.903	0.816	0.048
Nagpur	123	1.312	-1.761	-0.975	0.951	0.042
New Delhi	237	1.137	-1.373	-0.893	0.797	0.059
Port Blair	26	1.354	-1.801	-0.935	0.875	0.067
Pune	234	1.469	-1.963	-0.959	0.920	0.065
Shillong	116	1.211	-1.651	-0.915	0.837	0.050
Trivandrum	157	1.271	-1.632	-0.915	0.837	0.050
Vishakhapatnam	122	1.229	-1.587	-0.939	0.881	0.056

d mean monthly values of G and D were calculated for all 121 stations from SS. A comparison between computed and observed values at the 16 stations, where measurements are available, showed an agreement within 3-5% for G and 5-10% for D, indicating that if reliable records of global and diffuse solar radiation and sunshine duration are available at a well-distributed network of stations in a region, for a minimum period of at least 5 to 6 years, it is possible to compute daily values for other years for the same stations or for at an extended network of stations for any period.

There were still gaps in the data coverage over the country. Therefore 24 stations, where cloud observations were available, were chosen as gap filler stations, making a total of 145 stations, to provide adequate representation for all major climatic zones in the country. G and D were computed for these 24 stations from cloud cover observations, using the inverse relationship between sunshine and cloud cover $C = 1 - \frac{n}{N}$. An improved regression formula derived to suit Indian data was used to compute n from C. The general agreement between cloud-derived values of G and D and those derived from sunshine shows that where no sunshine data are available, G and D can be calculated with a considerable degree of confidence from cloud cover observations.

For solar energy applications it is also necessary to obtain hourly values of G and D. Following Liu and Jordan's (Ref.5) work, a regression relationship was derived which gives hourly values of G and D from daily values, using hourly and daily values of G and D actually measured at 14 stations. For almost all stations, the hourly/daily ratios were found to be smooth functions only of the given hour angle and the sunset hour angle. A single set of curves could therefore be drawn for evaluating hourly values of G and another set for hourly values of D. Figs. 1 and 2 show the two sets of curves. A random verification showed that an accuracy of 2-3% is possible for G and 4-6% for D, in the determination of the mean hourly values from mean daily values using these curves.

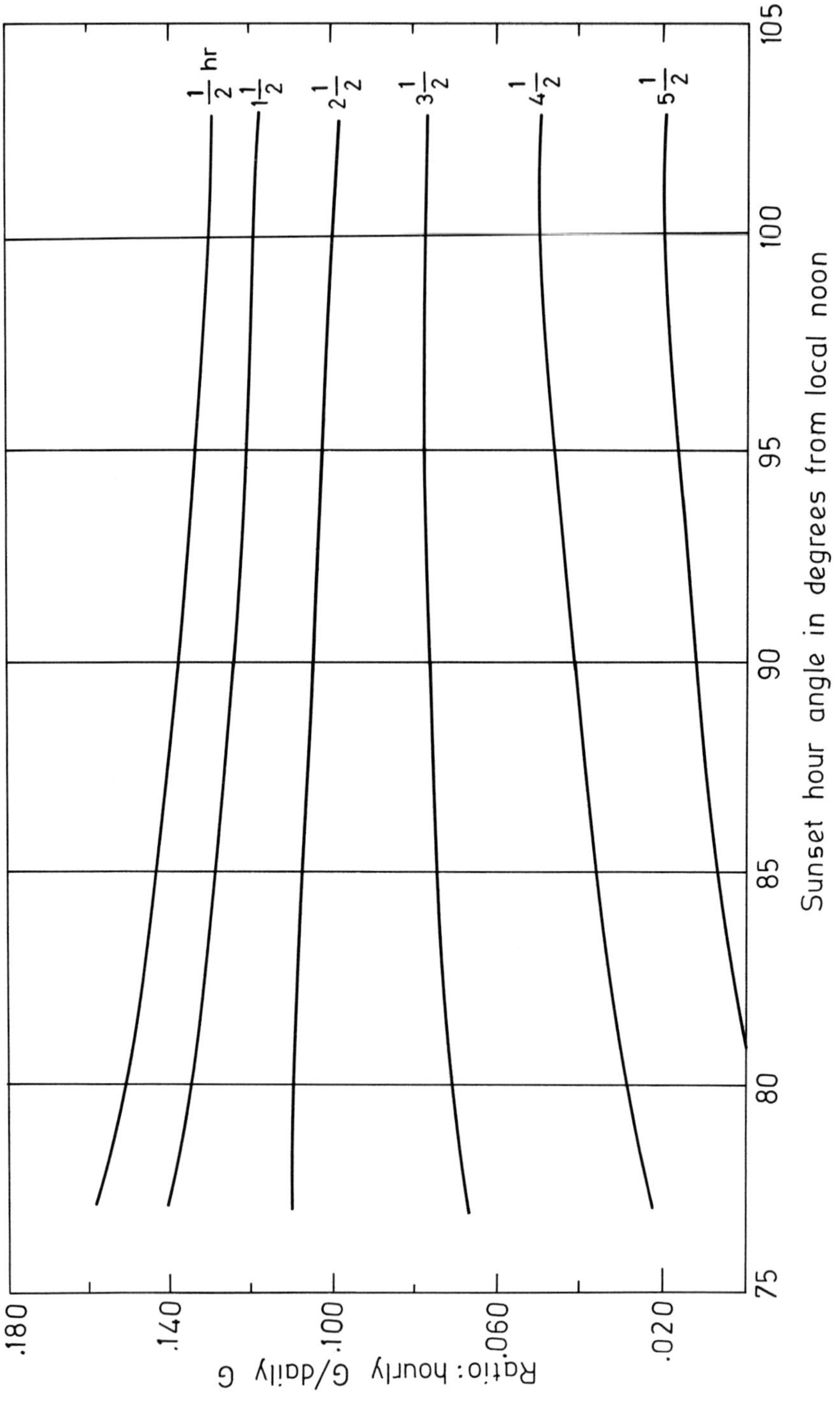

FIG.1. EXPERIMENTALLY DETERMINED RATIOS OF HOURLY GLOBAL TO DAILY GLOBAL SOLAR RADIATION.

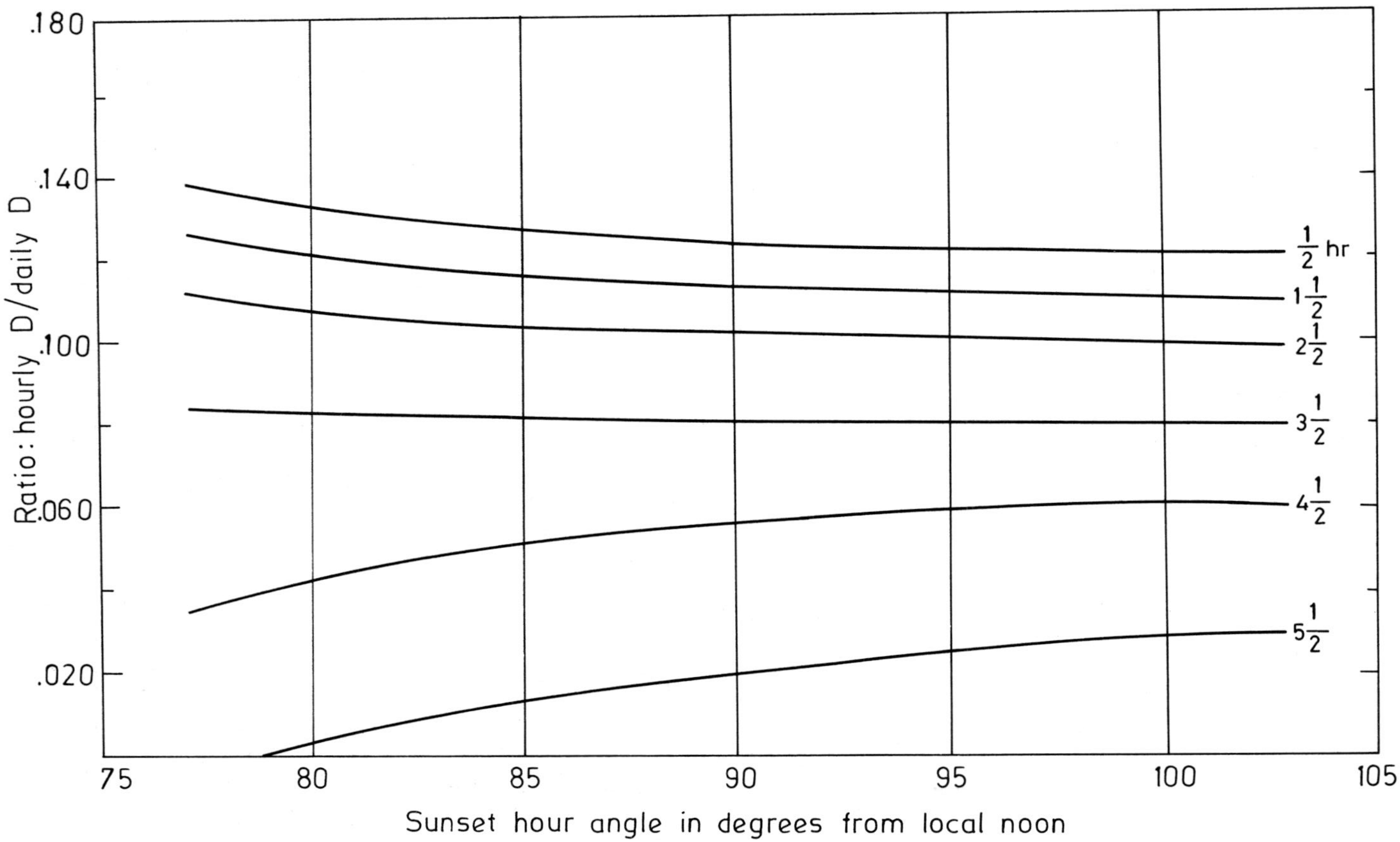

FIG. 2. EXPERIMENTALLY DETERMINED RATIOS OF HOURLY DIFFUSE TO DAILY DIFFUSE SOLAR RADIATION.

The main reason for the success of the empirical regression techniques to provide hourly values of G and D with a high degree of accuracy from sunshine and cloudcover data, should be ascribed to the availability of accurate measured values of G, D and n, for periods extending from 8 to 21 years, at a widely distributed network of stations in the country using carefully calibrated and well maintained instruments of the required quality.

5. COMPUTATION OF SOLAR RADIATION FROM SOLAR CONSTANT

The other method used to compute solar radiation at the ground was from extra-terrestrial solar radiation, allowing for its depletion by absorption and scattering by atmospheric gases, dust aerosols and clouds. The theoretical model used was by Hoyt (Ref.6). While the depletion of solar radiation due to absorption by atmospheric gases and scattering by air molecules can be calculated with precision, absorption by watervapour and scattering and absorption by dust have to be computed from measurements of precipitable water and atmospheric turbidity made at a number of stations in India.

Precipitable water content in the atmosphere over India was first calculated for 19 stations from radiosonde measurements of atmospheric water vapour content. These values, supplemented by data on surface water vapour mixing ratios at 100 stations were then used to derive interpolated values of precipitable water at 145 stations.

Atmospheric rubidity coefficient β was computed using a new technique from the ratio D/I, as the turbidity coefficients normally measured at field stations were found to be too low in the premonsoon summer months when dust scattering is almost spectrally neutral in character.

Using interpolated values of precipitable water content and atmospheric turbidity, simple calculations of global and diffuse radiation on clear days were made for 145 stations. The computed values were compared with observed values at 16 stations and were found to agree within 3-5% as seen from Table 3. Of the different variables that enter into the computation of clear sky noon and day values, the dust scattering function is the one that is known with the least reliability and it is this uncertainly that puts a limit to the accuracy of the computed data.

The ratio $\frac{G}{ETR}$ on clear days is a measure of the monthly mean transmission factor for daily global solar radiation, since $\frac{G}{ETR} = a + b\ (\frac{n}{N})$. When n = N on clear days $\frac{G}{ETR} = a + b$. The values of $\frac{G}{ETR}$ and $(a + b)$ are tabulated in Table 4 and the agreement between the two sets of values for all the stations establishes the soundness of the regression technique on one hand and the reliability of the theoretical model for computing clear sky radiation on the other.

5.1 Computation of Solar Radiation from Satellite Measurements

The third method for mesoscale mapping solar radiation at the earth's surface is by the use of satellites. Simple physical models have been used by a number of workers to estimate incident solar radiation at the earth's surface from visible brightness measurements from geostationary satellites. Cloud coverage and density as a function of time of day and

TABLE 3(a) - Comparison of Computed and Observed Values of Mean Daily Global Solar Radiation

Station		Jan	Feb	Mar	Apr	May	Jun	Jul	Aug	Sep	Oct	Nov	Dec
	Calc.	5.01	5.88	6.24	6.97	7.45	6.59	5.02	4.67	5.52	5.95	5.31	4.82
Ahmadabad	Obs.	4.90	5.79	6.73	7.33	7.61	6.39	4.85	4.52	5.55	5.77	5.00	4.59
	% Diff.	2.3	1.6	7.3	4.9	2.1	3.1	3.5	3.2	0.5	3.1	6.2	5.0
	Calc.	5.19	5.91	6.49	7.12	7.67	6.20	4.67	4.46	5.28	5.98	5.48	4.86
Bhavnagar	Obs.	5.14	5.96	6.87	7.27	7.60	6.03	4.62	4.31	5.46	5.91	5.22	4.78
	% Diff.	1.0	0.8	5.5	2.0	0.9	2.7	1.1	3.5	3.3	1.2	5.0	1.7
	Calc.	5.12	5.83	6.31	6.81	6.91	5.49	4.40	4.30	5.05	5.66	5.33	4.95
Bombay	Obs.	5.03	5.75	6.45	6.99	7.26	5.18	4.06	3.98	4.88	5.44	5.07	4.79
	% Diff.	1.8	1.3	2.1	2.6	4.8	5.9	8.3	8.0	3.5	4.0	5.0	3.3
	Calc.	4.36	5.09	5.58	6.21	6.21	4.87	4.74	4.60	4.59	4.69	4.54	4.17
Calcutta	Obs.	4.22	5.03	5.79	6.32	6.53	4.97	4.64	4.47	4.47	4.56	4.37	4.10
	% Diff.	3.3	1.2	3.6	1.7	4.9	2.0	2.2	2.9	2.7	2.9	3.9	1.7
	Calc.	5.62	6.33	6.57	6.90	6.73	4.94	4.29	4.93	5.37	5.67	5.72	5.37
Goa	Obs.	5.69	6.38	6.76	6.93	6.74	4.79	3.99	4.76	5.31	5.61	5.61	5.39
	% Diff.	1.2	0.8	2.8	0.3	0.1	3.1	5.0	3.6	1.1	1.1	2.0	0.4
	Calc.	4.67	5.61	6.19	7.09	0.52	7.19	6.25	6.05	6.24	5.95	5.10	4.51
Jodhpur	Obs.	4.71	5.57	6.55	7.23	7.55	7.07	5.98	5.54	6.10	5.83	4.90	4.43
	% Diff.	0.9	0.7	5.5	1.9	0.4	1.7	4.5	9.2	2.3	2.1	4.1	1.8
	Calc.	6.03	6.78	6.96	6.64	6.02	5.04	4.49	4.95	5.13	4.88	4.86	5.19
Kodaikanal	Obs.	6.28	6.71	6.87	6.43	5.84	5.28	4.54	4.70	4.80	4.44	4.74	5.22
	% Diff.	4.0	1.0	1.3	3.3	3.1	4.5	1.1	5.3	6.9	9.9	2.5	0.6
	Calc.	5.37	6.29	6.59	6.72	6.32	5.52	5.03	5.28	5.41	5.10	4.56	4.57
Madras	Obs.	5.23	6.29	6.89	6.92	6.53	5.82	5.40	5.58	5.71	4.90	4.34	4.30
	% Diff.	2.7	0.0	4.4	2.9	3.2	5.2	6.8	5.4	5.3	4.1	4.7	6.2

Unit : $kWh.m^{-2}.day^{-1}$

TABLE 3(a) (cont'd)		Jan	Feb	Mar	Apr	May	Jun	Jul	Aug	Sep	Oct	Nov	Dec
Mangalore	Calc.	5.82	6.37	6.37	6.45	6.02	4.24	3.53	4.33	4.99	5.26	5.33	5.36
	Obs.	5.33	6.02	6.38	6.42	5.73	3.86	3.21	4.07	4.96	4.93	5.17	5.10
	% Diff.	9.2	5.8	0.2	0.5	5.1	9.8	10.2	6.3	0.6	6.7	7.0	5.1
Nagpur	Calc.	5.02	5.84	6.18	6.69	6.89	5.51	4.33	4.30	4.91	5.58	5.27	4.81
	Obs.	4.91	5.68	6.31	6.78	6.91	5.64	4.39	4.13	5.11	5.61	5.13	4.67
	% Diff.	2.2	2.8	2.1	1.3	0.3	2.3	1.4	4.1	3.9	0.5	2.7	3.0
New Delhi	Calc.	4.06	5.07	5.60	6.50	6.61	6.22	5.61	5.53	5.58	5.48	4.71	3.95
	Obs.	3.99	5.00	6.14	6.93	7.29	6.54	5.33	5.05	5.60	5.35	4.52	3.84
	% Diff.	1.7	1.4	8.8	6.2	9.3	4.9	5.3	9.5	0.4	2.4	4.2	2.9
Port Blair	Calc.	4.94	5.75	5.90	5.79	4.23	3.47	3.48	3.57	3.69	4.34	4.39	4.64
	Obs.	4.87	5.65	6.07	5.95	4.05	3.34	3.59	3.49	3.91	4.20	4.01	4.31
	% Diff.	1.4	1.8	2.8	2.7	4.4	3.9	3.1	2.3	5.6	3.3	9.7	7.7
Pune	Calc.	5.32	6.10	6.53	6.91	6.99	5.74	4.57	4.68	5.19	5.60	5.35	5.03
	Obs.	5.30	6.17	6.81	7.16	7.30	5.88	4.54	4.58	5.31	5.67	5.24	4.93
	% Diff.	0.4	1.1	4.1	3.5	4.3	2.4	0.7	2.2	2.3	1.2	2.1	2.0
Shillong	Calc.	3.64	4.75	5.34	5.69	5.56	4.26	4.18	4.14	4.18	4.26	3.78	3.96
	Obs.	4.01	5.08	5.63	5.70	5.23	4.11	4.26	4.21	3.87	4.02	3.94	4.04
	% Diff.	9.2	6.5	5.2	0.2	6.3	3.6	1.9	1.7	8.0	6.0	4.1	2.1
Trivandrum	Calc.	6.08	6.55	6.69	6.24	5.78	5.00	5.06	5.64	5.85	5.52	5.33	5.50
	Obs.	5.93	6.36	6.68	6.17	5.49	5.25	5.03	5.56	5.94	5.23	4.95	5.13
	% Diff.	2.5	3.0	0.0	1.1	5.3	4.8	0.6	1.4	1.5	5.5	7.7	7.2
Vishakha-patnam	Calc.	5.34	6.11	6.24	6.62	6.63	5.04	4.59	5.12	5.18	5.42	5.22	4.99
	Obs.	5.37	6.08	6.51	6.66	6.68	5.17	4.67	4.95	5.16	5.28	5.16	5.07
	% Diff.	0.6	0.5	4.1	0.6	0.6	2.5	1.7	3.4	0.4	2.7	1.2	1.2

Unit : $kWh.m^{-2}.day^{-1}$

TABLE 3(b) - Comparison of Computed and Observed Values of Mean Daily Diffuse Solar Radiation

Station		Jan	Feb	Mar	Apr	May	Jun	Jul	Aug	Sep	Oct	Nov	Dec
Ahmadabad	Calc.	1.10	1.26	1.99	2.15	2.13	2.89	3.32	3.21	2.72	1.60	1.18	1.00
	Obs.	1.17	1.43	1.74	2.18	2.33	3.16	3.48	3.24	2.53	1.43	1.12	1.05
	% Diff.	6.6	11.7	14.6	1.6	8.5	8.8	4.6	0.8	7.4	11.3	4.9	4.8
Bhavnagar	Calc.	1.10	1.39	1.89	2.09	1.92	3.18	3.47	3.36	2.97	1.73	1.11	1.17
	Obs.	1.17	1.38	1.66	2.17	2.23	3.24	3.55	3.37	2.61	2.51	1.15	1.09
	% Diff.	5.9	0.9	13.7	4.1	14.0	1.8	2.4	0.4	14.0	14.3	3.6	7.0
Bombay	Calc.	1.18	1.35	1.77	1.92	2.09	2.97	3.09	3.03	2.73	1.82	1.24	1.10
	Obs.	1.28	1.47	1.78	2.19	2.51	2.97	3.07	3.17	2.86	1.98	1.19	1.11
	% Diff.	7.8	8.1	0.6	12.3	16.8	0.0	0.7	4.1	4.5	8.0	4.2	0.5
Calcutta	Calc.	1.56	1.81	2.23	2.44	2.69	2.97	2.94	2.84	2.64	2.21	1.67	1.49
	Obs.	1.50	1.74	2.16	2.66	3.09	3.20	3.17	2.92	2.60	2.03	1.54	1.33
	% Diff.	4.1	4.0	3.3	8.1	13.3	7.4	7.5	2.7	1.3	8.5	8.1	11.6
Goa	Calc.	1.39	1.50	2.01	2.18	2.43	3.12	3.12	3.08	2.84	2.28	1.54	1.41
	Obs.	1.15	1.34	1.88	2.32	2.69	3.04	3.03	3.34	2.94	2.11	1.47	1.12
	% Diff.	20.7	12.1	6.8	6.3	9.7	2.6	2.7	7.6	3.2	7.9	4.6	26.1
Jodhpur	Calc.	0.99	1.15	1.80	1.94	2.10	2.57	3.07	2.89	2.14	1.25	0.86	0.83
	Obs.	1.14	1.36	1.77	2.34	2.68	3.06	3.39	3.21	1.88	1.22	0.91	0.93
	% Diff.	13.5	15.9	1.6	16.9	21.6	16.0	9.4	10.1	14.1	2.5	4.9	10.5
Madras	Calc.	1.87	1.83	2.14	2.32	2.54	2.76	2.82	2.80	2.69	2.50	2.23	2.05
	Obs.	1.87	1.73	1.87	2.12	2.57	2.86	3.11	3.01	2.58	2.26	2.07	2.04
	% Diff.	0.4	6.1	16.0	9.5	1.2	3.5	9.3	7.1	4.4	10.8	7.4	0.4
Nagpur	Calc.	1.06	1.17	1.80	2.03	2.18	3.00	3.09	3.01	2.73	1.76	1.09	1.00
	Obs.	1.21	1.43	1.76	2.16	2.49	2.99	3.22	3.02	2.53	1.64	1.25	1.07
	% Diff.	13.0	17.9	2.5	5.7	12.7	0.3	4.0	0.4	7.8	7.6	12.7	6.3

Unit: $kWh.m^{-2}.day^{-1}$

TABLE 3(b) (cont'd)

Station		Jan	Feb	Mar	Apr	May	Jun	Jul	Aug	Sep	Oct	Nov	Dec
	Calc.	1.35	1.56	2.16	2.45	2.81	3.14	3.12	2.92	2.45	1.65	1.17	1.20
New Delhi	Obs.	1.24	1.47	1.86	2.47	2.92	3.54	3.13	2.73	2.15	1.47	1.14	1.12
	% Diff.	9.2	5.8	16.2	1.1	3.7	11.2	0.5	7.1	13.9	12.5	2.8	7.1
	Calc.	1.23	1.32	1.83	2.11	2.29	3.21	3.43	3.37	3.02	2.18	1.51	1.28
Pune	Obs.	1.11	1.19	1.59	2.07	2.24	3.32	3.47	3.49	2.93	1.73	1.22	1.14
	% Diff.	10.2	10.6	15.1	2.0	2.4	3.3	1.1	3.6	3.2	25.5	23.1	12.5
	Calc.	2.09	2.05	2.40	2.65	3.03	2.86	2.87	2.91	2.90	2.78	2.38	2.12
Port Blair	Obs.	2.00	1.90	2.10	2.00	2.80	2.80	2.96	2.90	3.00	2.60	2.30	1.80
	% Diff.	4.5	7.7	14.4	33.7	8.0	2.3	3.1	0.2	3.3	6.9	3.3	18.0
	Calc.	1.52	1.57	1.99	2.40	2.74	2.91	2.87	2.77	2.53	2.07	1.65	1.13
Shillong	Obs.	1.21	1.37	1.87	2.36	2.83	3.14	3.20	2.91	2.49	1.75	1.30	0.93
	% Diff.	25.8	14.2	6.3	1.6	3.1	7.4	10.1	4.9	1.6	18.3	26.4	21.5
	Calc.	1.79	1.91	2.22	2.64	2.77	2.92	2.93	2.84	2.75	2.69	2.43	2.08
Trivandrum	Obs.	1.77	1.91	2.20	2.62	2.83	2.97	3.13	3.13	2.77	2.56	2.26	2.03
	% Diff.	0.9	0.0	1.2	0.8	2.1	1.8	6.6	9.4	0.5	5.0	7.2	2.4
	Calc.	1.30	1.41	2.05	2.25	2.40	3.00	3.00	2.92	2.71	2.18	1.64	1.39
Vishakha-	Obs.	1.42	1.49	1.86	2.17	2.64	2.97	3.10	3.05	2.55	1.83	1.55	1.40
patnam	% Diff.	8.0	5.6	9.9	3.6	8.9	1.3	3.1	3.9	6.4	19.1	5.5	0.5

Unit: $kWh.m^{-2}.day^{-1}$

TABLE 4 - Mean Annual Transmission Factor for G Compared with (a+b)

Station	G(c)/ETR	a+b	Percentage difference
Ahmadabad	0.707	0.749	4.5
Bhavnagar	0.709	0.731	3.1
Bombay	0.700	0.720	2.9
Calcutta	0.696	0.687	1.3
Goa	0.706	0.733	3.8
Jodhpur	0.709	0.754	6.3
Kodaikanal	0.817	0.844	3.3
Madras	0.710	0.716	0.8
Nagpur	0.707	0.725	2.5
New Delhi	0.708	0.749	5.8
Pune	0.720	0.724	0.5
Shillong	0.721	0.731	1.4
Trivandrum	0.710	0.760	7.0
Vishakhapatnam	0.698	0.719	3.0

season are evaluated through the use of digital data processing techniques and seasonal geographic distributions of cloud cover and sunshine converted to solar radiation at the earth's surface, through relationships developed from long-term measurements of these two parameters at widely distributed stations.

5.2 Computation of Solar Radiation on Sloped Surfaces

A number of methods have been proposed by various workers for the computation of global, diffuse and direct solar radiation on sloped surfaces. Measurements are difficult to make and difficult to interpret and subject to many errors, arising from the changes in calibration factor with the tilt angle, the difficulty in accurate positioning of the instrument and the determination of the influence and representativeness of the surrounding ground cover or neighbouring obstacles. Assuming diffuse radiation to be isotropic in one case and anisotropic in the other, tilt factors for the conversion of global solar radiation on horizontal surfaces to that on sloped surfaces have been calculated by many workers. Data for 145 stations in the country for five slopes for south facing surfaces and for vertical surfaces facing east, north and west have been published in Solar Radiation over India (Ref.3). In the earlier volume (Ref.2) tilt factors for 14 stations had been given for five tilt angles and seven azimuths.

6. RADIATION CLIMATOLOGY OF INDIA

Considerable data on solar radiation have been collected by many national meteorological services since the IGY in 1957 when global observations of different components of short and longwave radiation were organised at a large number of stations in the world. Based on these data and on computed values from other meteorological parameters, maps showing the distribution of global, diffuse and direct solar radiation and terrestrial radiation have been prepared for the world and many countries.

6.1 Sunshine

A study of the distribution of sunshine duration for the whole year over the Indian subcontinent shows that 3300-3700 hours of bright sunshine are received in a year in the northwest and west central zones of the subcontinent and 2900 hours over the central peninsula.

During the cold season December to March, the duration is highest over central India and least over the north and northwest, as a result of the passage of winter disturbances and the decrease in length of the day with latitude. With the advance of the year, the region of clear skies and maximum sunshine shifts from the central to the northern parts. With the onset of the monsoon in June over the peninsula, the duration is highest over Pakistan and northwest India. With the withdrawal of the monsoon, skies began to clear in October in north and central India, and by November the region with clearest skies lies again over Central India.

6.2 Global Solar Radiation (Figs. 3 and 4)

Since cloudiness is the main parameter influencing sunshine and radiation, the distribution of global solar radiation is similar to that of sunshine. A gradient exists in radiation intensities on clear days in winter with low values in the north and high in the south. A reversal occurs during May-July with the apparent northward movement of the sun and the onset of the monsoon over the peninsula. During April the radiation received is high and fairly uniform over the whole subcontinent, Gujarat and adjoining areas receiving over 210 kWh/m^2/month. The highest values are thus reached in April-June in the arid and semiarid regions of the subcontinent. Similarly it is low and fairly uniform again in August (140 kWh/m^2/month) when the monsoon has extended over the whole country. The lowest values are in July-August and November-February.

On the whole, global solar radiation of the order of 7.5 kWh/m^2 is received daily or 225 kWh/m^2 per month on a horizontal surface for the major portion of the year over the arid and semiarid regions in the north and northwest of the subcontinent. During the monsoon and winter months the global solar radiation falls to about 3-6 kWh/m^2/day but it is still possible to collect over the year 80% of the available solar energy. Assuming 6 kWh/m^2 are received daily with an efficiency of collection of even 50%, 1 MWh/m^2 can be collected from each square metre of horizontal surface.

6.3 Diffuse Solar Radiation (Figs. 5 and 6)

The pattern, as expected, closely resembles the distribution of cloudiness. On a day with average clouding diffuse radiation contributes

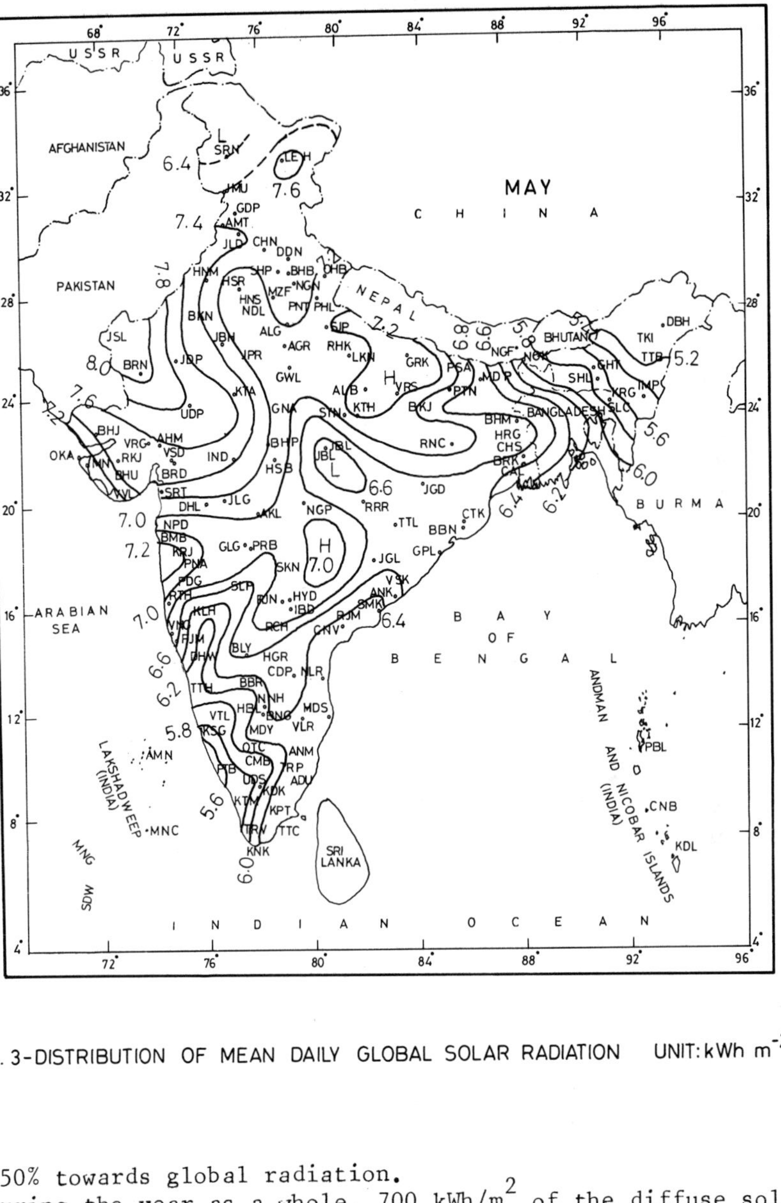

FIG. 3-DISTRIBUTION OF MEAN DAILY GLOBAL SOLAR RADIATION UNIT: kWh m^{-2} day^{-1}

40 to 50% towards global radiation.

During the year as a whole, 700 kWh/m^2 of the diffuse solar radiation are received over the country, highest in the extreme southwest and northeast regions, where the maximum clouding occurs. About 50% of the diffuse radiation is received in the monsoon months and 30% in the dusty cloud-free premonsoon summer months.

The proportion of diffuse radiation to global solar radiation varies widely, depending on the cloudiness of the sky, the maximum 77% occurring during June-August and the minimum 17% during November-February. This means that as much as 80% of the shortwave radiation received is scattered sky radiation during the monsoon months. In the

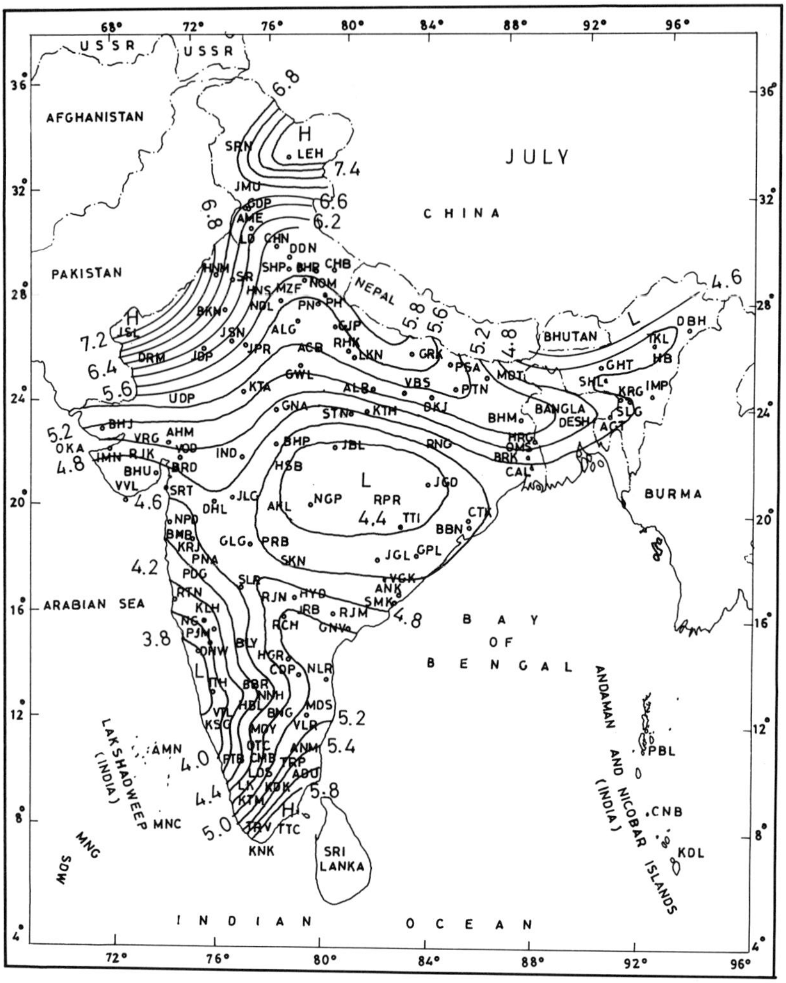

FIG.4. DISTRIBUTION OF MEAN DAILY GLOBAL SOLAR RADIATION UNIT: $kWh.m^{-2}.day^{-1}$.

arid and semi-arid regions 20-25% of the global radiation is diffuse even with clear skies. It increases to 35% in June. Thus Delhi receives about 3.5 kWh/m^2 as diffuse radiation in a day in June and Jodhpur 2.8 kWh/m^2.

6.4 Direct Solar Radiation

Direct solar radiation is highest in winter and lowest in the summer and monsoon months, for the same optical air masses and solar elevations. The reduced values in the premonsoon summer months is to be ascribed to scattering and absorption by dust and aerosols. It is also

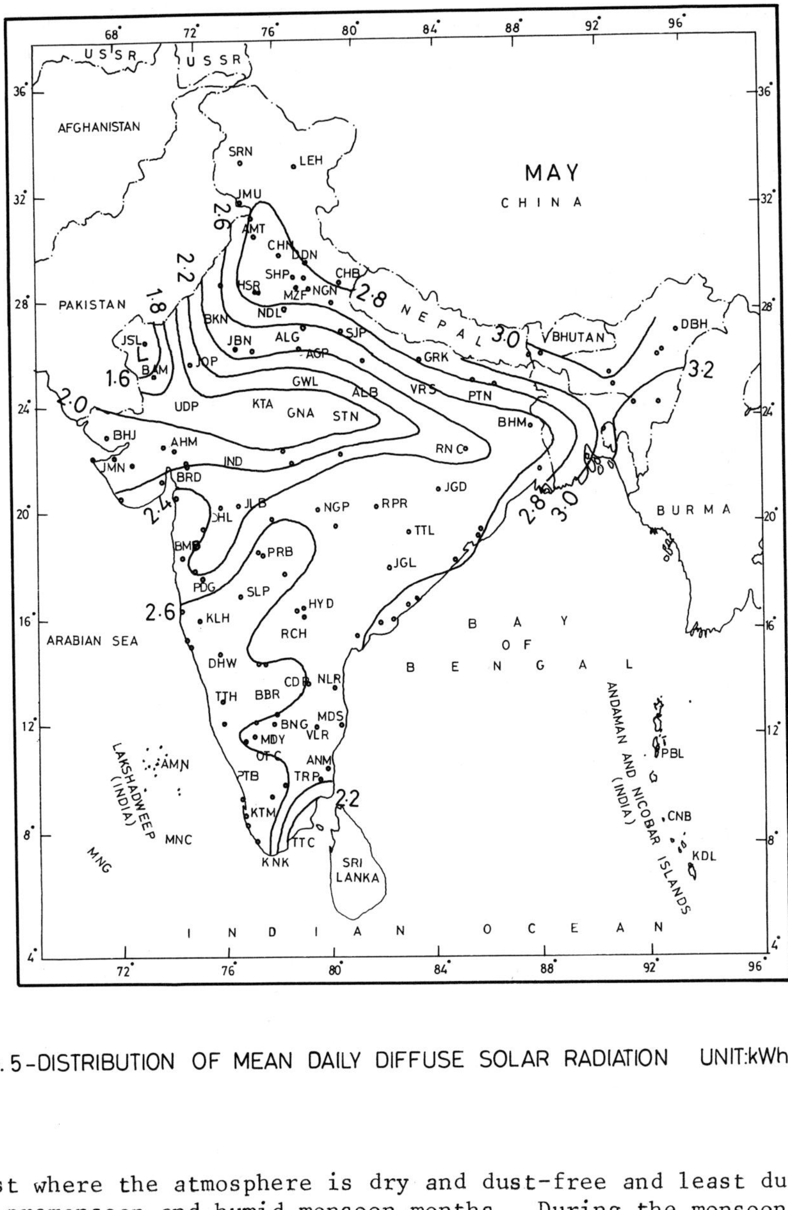

FIG. 5-DISTRIBUTION OF MEAN DAILY DIFFUSE SOLAR RADIATION UNIT: $kWh.m^{-2}day^{-1}$

highest where the atmosphere is dry and dust-free and least during the dusty premonsoon and humid monsoon months. During the monsoon months the dust content is reduced by rainout and washout but the increased watervapour content causes a reduction in the intensity of direct solar radiation.

7. DATA STORAGE, ANALYSIS AND PUBLICATION

Solar energy applications require data summaries other than the standard monthly mean maps and tables that are normally provided by climatologists.

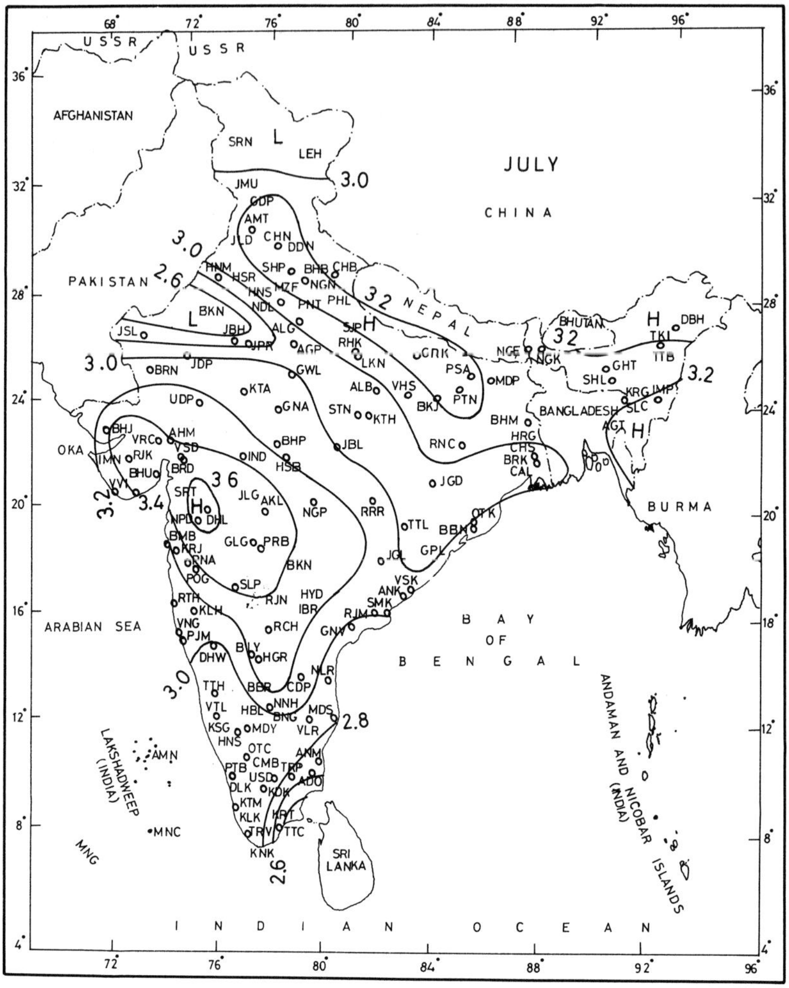

FIG. 6-DISTRIBUTION OF MEAN DAILY DIFFUSE SOLAR RADIATION UNIT: $kWh.m^{-2} day^{-1}$

Various statistical analyses are required. These involve median values, percentage frequency distribution and cumulative percentage frequency distribution and run lengths of global solar radiation in hours or days above or below given energy levels and finally skewness and kurtosis. Such statistical analysis for global solar radiation for 16 stations in India have been published (Ref.3).

8. CONCLUDING REMARKS

To meet the requirements of radiation and meteorological data for solar energy applications -

i) an improved national network of radiation stations in the country is essential to provide temporal and spatial representativeness. New stations should be started immediately, as the utility of the data is directly related to the length of the record. If 5% of the total amount of money spent in the country on the development of solar energy is allocated to the national weather service for starting and maintaining this network, it will bring in returns manifold.

ii) in order to ensure the accuracy and comparability of the data, the National Radiation Centre at Pune should be strengthened so that all radiation instruments used in the country can be periodically checked and calibrated and observers trained in the operation and maintenance of radiation instruments and the processing of data.

iii) it is also necessary to establish a number of selected baseline research stations with a high frequency of very precise measurements for studies of climate change, variability and long term trends.

iv) the radiation and other data collected at all the stations should be archived and published, periodically so that the information will be available to all solar energy users in the country in the form they need.

REFERENCES

1. L. Machta, (1979), 'Monitoring solar radiation for solar energy', UNESCO/WMO Solar Energy Symposium, World Meteorological Organisation, Geneva.
2. A. Mani, (1981), Handbook of Solar Radiation for India, Allied Publishers, New Delhi.
3. A. Mani and S. Rangarajan, (1982), Solar Radiation Over India, Allied Publishers, New Delhi.
4. J.E. Hay, (1979), 'Study of shortwave radiation on non-horizontal surfaces', Report No.79-12, Atmospheric Environment Service, Canada.
5. B.Y.H. Liu and R.C. Jordan, (1960), 'The interrelationship and characteristic distribution of direct, diffuse and total solar radiation', Solar Energy, 4(3), 1-19.
6. D.U. Hoyt, (1978), 'A model for the calculation of solar global radiation', Solar Energy, 21, 27-35.

COMPONENTS OF SOLAR WATER HEATERS

H.P. Garg
Centre for Energy Studies
Indian Institute of Technology
Hauz Khas, New Delhi 110 016
India

1. INTRODUCTION

The first recorded use of solar energy was made about 2500 years ago in the Greek where people not only worshipped the God 'Sun' but built their homes facing south, with porticos (or gaps in the roof) so as to heat the indoors during winter. These were the first solar homes, and solar architecture including many other passive features were taken up much later. The great Roman baths were heated by the sun and the invention of glass led to the discovery of the green house. The sun also had more adventurous and romantic roles. In 212 B.C., Archimedes used mirrors to destroy the ships of Roman invaders. Apart from this several peaceful applications of solar energy were also made in the past. However, as 'Modern' fuel sources became popular in the middle ages, solar power was almost a forgotten art. It was only another energy crisis (1972) that turned man once again to the sun.

Solar energy can be put to variety of applications like water heating, crop drying, desalination of water, heating and cooling of space, refrigeration, mechanical and electrical power production, direct conversion of electricity, etc. Heating of water using solar energy is the only commercial application which is economically viable (Ref.1). For more than 90 years work on solar water heaters is in progress. It is estimated that about 30000 to 50000 solar water heaters were installed in USA by 1950, but about that time their popularity began to decline due to readily available cheap energy from fossil fuels (Ref.2).

2. TYPES OF SOLAR WATER HEATER

Technical advances in solar water heating have been very rapid in the last 30 years, and the obvious benefits to the householders can no longer be overlooked where the climate is ideally suited for application of solar energy for water heating, particularly in the present situation of acute energy shortage. Solar water heaters find wide applications in large establishments like hostels, hotels, hospitals, industries such as textile, paper, and food processing, domestic uses and even in heating

H. P. Garg (ed.), Solar Water Heating Systems, 37–59.

swimming pools in winter. The initial cost of a solar water heating system is higher than that operated either by electricity, gas or other fuel but over a period of time, this is more than recovered by the savings in the cost of operation and maintenance. The application, which is perhaps the easiest to adopt, is to use solar energy for heating water. Solar water heaters offer a number of advantages such as:

- o Simple to construct and install,
- o Almost no maintenance and running cost,
- o Save time and energy,
- o Retrofittable to existing buildings,
- o Economically competitive with electrical and gas water heaters,

and o Required temperature easily achieved.

For water heating it is not economical to concentrate the radiation because the mechanism required to follow the sun throughout the year is both costly and complex. Developmental work on solar water heating has therefore in the main, been done on flat plate collectors utilizing solar radiation as the low grade heat source (Ref.3). Almost all solar water heaters are based on the principle of the flat plate collector. Over the past 60 years many different types of collector have been designed and built, and the pioneering work has been mainly done in USA (Ref.4), the U.K. (Ref.5), Australia (Ref.6), South Africa (Ref.7), Israel (Ref.8), and India (Ref.9). The main objective in these studies have been to convert as much solar radiation as possible into useful heat, at the highest attainable temperature, for the lowest possible capital investment.

Solar water heaters employing flat-plate collectors can be divided into the following four types according to their main applications, temperature of operation, and capacities:

i) Swimming pool water heater, where a cheap plastic collector can be used even without any cover and insulation. A high flow rate is maintained to limit the temperature rise to less than 2°C.

ii) Built-in-storage type solar water heater where all the three functions/components, i.e. collection, storage, and control are combined into a single unit. The early solar water heaters in (1910-1920) Southern California were of this type, in which shallow, blackened metal water tanks are enclosed in pine boxes, covered with glass, and exposed to the sun. Hot water (up to 60°C only) from such water heaters had to be used during the days, otherwise the heat would be lost during the night.

iii) Domestic solar water heaters where the maximum temperature required is no more than 80°C. Here the collector and storage functions are separate. The control function is still accomplished through the use of natural principles and this technology is frequently employed for domestic systems in the form of rather well known (thermosiphon systems). While this technology requires no manual interruption and is largely self-controlling, the maximum system capacity is usually limited because of the rather weak nature of the thermal gradient forces employed for fluid flow.

iv) Large size solar water heaters designed for community use and industrial use. Since the requirement of hot water is large, a large number of collector banks are employed with a storage tank along with a control system. A sophisticated control arrangement may control the outlet water temperature, fluid flow rate, and auxiliary energy supply.

3. SWIMMING POOL HEATING

One of the most practical and simple applications of solar energy is the heating of swimming pool with solar energy. Desired pool temperatures are 24°C to 27°C. In a swimming pool since very lage amount of water is to be heated, heating with fossil fuels or natural gas or electricity is very expensive and in some countries not allowed. Solar energy can be more profitably used for heating both indoor and outdoor swimming pools. There are two basic methods of heating swimming pools by solar energy:

1. Avoiding waste of the solar radiation, incident on the swimming pool by using pool covers, and
2. Heating of water in simple and inexpensive solar collectors located outside the swimming pool.

Heat loss by evaporation from the open swimming pool constitute the largest heat loss which can be prevented by covering the pool water surface by a transparent thin plastic sheet. Ideally the cover should transmit solar radiations (0.2-2.5 microns), should not transmit thermal radiations (4-40 microns), reduce conduction and convection loss, and should be easy to put on when heating is required and take off for swimming. Research on type of pool covers and their suitability for swimming pool heating has been carried out in USA (Ref.10), Australia (Ref.11,12), England (Ref.13) etc. In case of large swimming pools, the use of mechanical roller system is recommended which help in the removal and replacement of plastic pool cover. A double layer plastic cover with an air space between the layers or bubles of different shapes enclosed in two layers with air enclosed are becoming popular as covers for pool heating (Ref.14). Since such covers considerably reduce the convection, conduction and radiation loss to the outside, it has been observed that a temperature rise of 5 to 10°C in the pool water can be obtained by using such cover compared to a pool without cover.

Since large amount of water is to be heated through a small temperature difference for swimming pool heating, low cost solar collectors even without any glazing and back insulation, making the system most cost effective, are used. A few typical commercially available rubber and plastic panel solar collectors (Ref.15) which are used in USA for swimming pool heating are shown in Fig.1. The solar energy collectors in a typical solar pool system is inserted into the circulation line between the filter and the pool. When adequate sunshine is available, the filtered pool water is circulated through the collector tubes, where it is heated by solar radiation and then return to the pool. Circulation (of all the water in the pool), through the filter about once every 5 to 10 hours is recommended. The collector area required depends on many variables; the size of the pool, the type of collector used, the general weather pattern of the location, the actual temperature to which the pool is to be heated, the time of the year when it is to be heated, and the micro-climate. Rigorous methods are available for sizing collectors

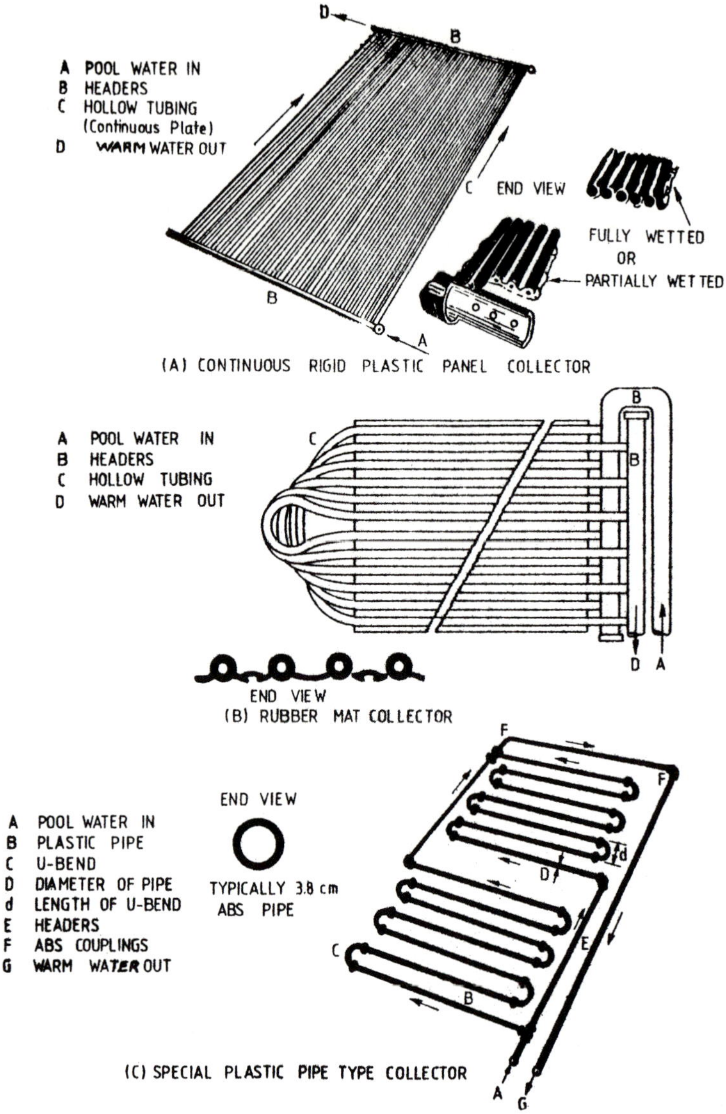

FIG.1 - SIMPLE COMMERCIAL SWIMMING POOL HEATING COLLECTORS

for solar pool heating, but the rules of thumb is generally employed which says that the solar collector area should be about one half the pool area (Ref.16).

The solar collector and the pool cover can also be suitably combined which can float on the swimming pool water. Such floating heaters are now being marketed. These floating heaters will reduce the evaporation, convection, and radiation heat loss and heat the pool water sufficiently.

4. BUILT-IN STORAGE TYPE SOLAR WATER HEATERS

As the name suggests, there is no separate storage tank in these solar heaters, and the absorber (collector) performs the dual function of absorbing the solar radiation and storing the solar heated water. Therefore, these solar water heaters are simple in design, low in cost, easy in operation, low maintenance cost and easy to install. These water heaters are efficient to provide hot water during day time only and are divided into the following three types depending on the design details:

i) Shallow solar pond water heater
ii) Built-in storage solar water heater
iii) Other simple low cost solar water heater

The above types of solar water heaters are studied by scientists of many countries including USA (Ref.1,17,18), Japan (19), India (20,21), South Africa (22), Ceylon (23), Israel (24) etc. A few popular types of these water heaters are schematically shown in Fig.2.

The shallow solar pond water heater as shown in Fig.2(E) is essentially of plastic water bag having clear plastic at the top and a black plastic in the bottom. The depth of the water within the plastic bag is normally in the range of 4-15 cm. This plastic bag may have a built-in insulation in the bottom or can be placed on an insulated platform. Solar radiation directly falls on to the water through upper transparent film and heats the water. The solar energy collection efficiency is directly proportional to water depth. In Japan several thousands of such plastic pillow type solar water heaters are in use particularly in rural areas. In USA large size plastic shallow solar pond water heaters are made, and hot water supplied to four process industries, for process heat applications (Ref.17,18).

Typical peak temperatures for a shallow solar pond range from 60°C in the summer down to 40°C in the winter. With a nominal 7.5-10.0 cm operating depth, the annual efficiency of the shallow solar pond system is about 50 percent.

The above shallow solar ponds made up of plastic or open type solar water heaters (a few of them shown in Fig.2) can only be placed horizontally thus giving poor performance in winter season (due to a low altitude of sun) and at high altitude stations. Therefore a completely closed type solar water heater (as shown in Fig.2B and 2D), made of either rectangular metal sheet or several closed large diameter pipes made of metal or plastic or glass are preferred which would be inclined at a desired angle to capture more solar radiation. Such closed rectangular box type solar water heaters which can be inclined to a desired level are studied extensively in India by the author (Ref.20,21).

The optimized (from thermal performance view point) configuration of a built-in storage water heater as developed by the author is described briefly here (Ref.21). The water heater is schematically shown in Fig.3. It consists of a rectangular 20 gauze galvanized iron tank measuring 112 x 80 x 10 cm with a capacity of 90 litres of water. It is contained in a mild steel sheet box with 5 cm fibre glass insulation below it and one glass cover on the top. The front face (exposed side) of the tank is blackened by ordinary black board paint. The hot water is taken out from the heater outlet pipe at the top by opening the gate valve from

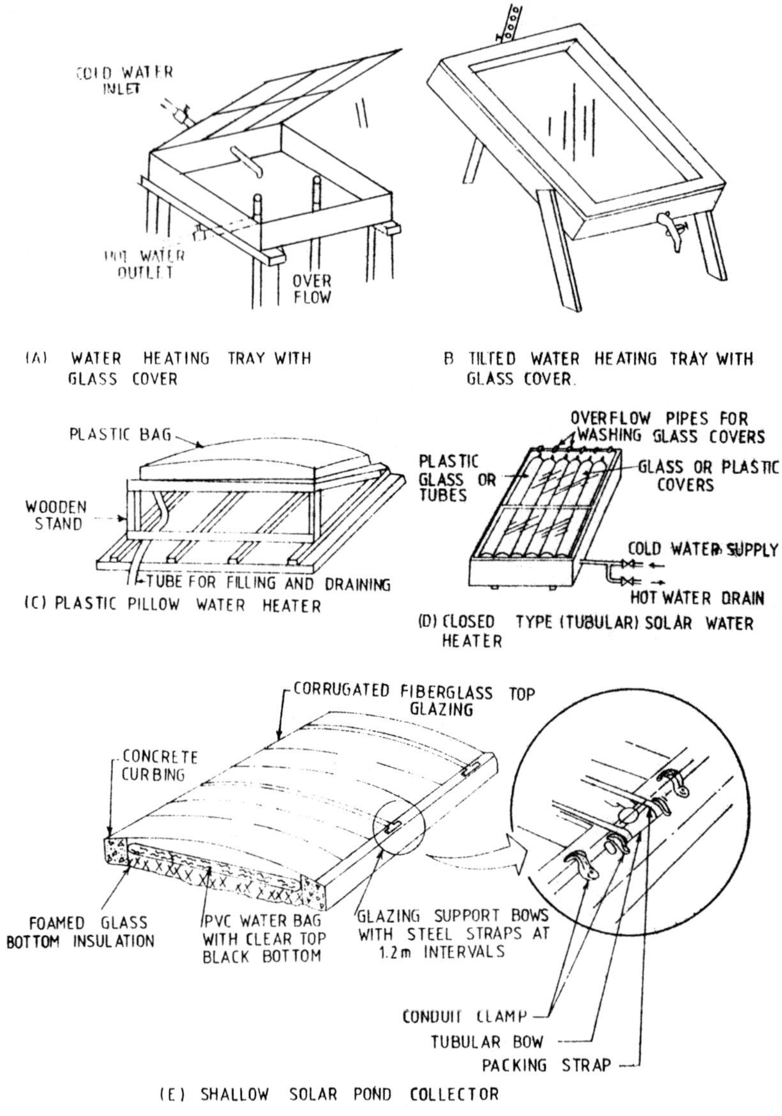

FIG.2 - SIMPLE BUILT-IN-STORAGE SOLAR WATER HEATERS

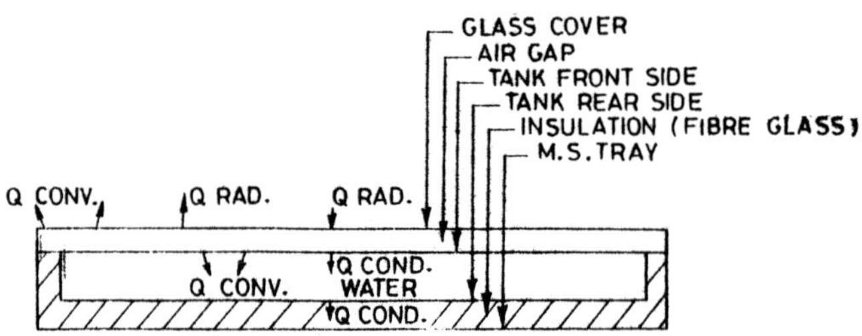

FIG.3 - SCHEMATIC DIAGRAM OF BUILT-IN-STORAGE TYPE SOLAR WATER HEATER

the inlet pipe side of the heater fixed at the bottom. For maximum collection of solar radiation, the heater is tilted with a tilt equal to latitude of place plus 15° from horizontal and facing south. For rural use, where there is no central water supply arrangement, a large funnel can be fixed at the top of the heater and then connected to the inlet tube to the collector. Hot water can be taken out immediately by putting the same amount of cold water in the funnel. The diurnal variation of water temperature in the solar water heater, ambient temperature, and a water temperature is shown in Fig.4. The maximum water temperature

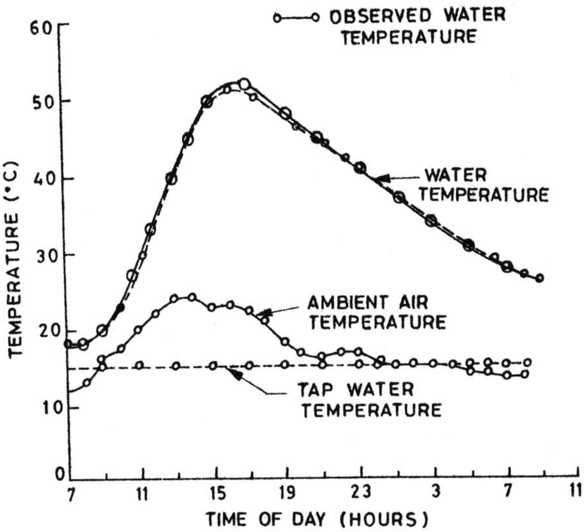

FIG.4 - DIURNAL VARIATION OF WATER TEMPERATURE IN BUILT-IN-STORAGE TYPE WATER HEATER

reaches at around 5 p.m. and then starts decreasing because of heat loss to the surroundings. Therefore, this type of solar water heater is suitable only for day time use. Hot water can be stored in the heater for next day morning use by incorporating any of the following features (Ref.25):

1. Covering the water heater with about 5 cm thick insulation blanket at around 5 p.m. till next day morning.
2. Draining the entire hot water from the water heater into a separate insulated hot water storage tank.
3. Providing an insulated (about 2.5 cm thick) baffle plate in the rectangular tank.
4. Providing honeycomb structure between the absorption plate and the glass cover which reduces the convection losses to the outside considerably.

Other very simple low cost collector-cum-storage type solar water heaters are also studied extensively at IIT Delhi (Ref.26) where either plastic or metal pipes are embedded in the ground or in a concrete block or sand block, the front side of which is blackened and glazed. This concrete block gets heated due to absorption of solar radiation and transfers its heat to the water flowing in the tubes. Such systems have great potential but are not adopted and used anywhere.

5. COMPONENTS OF SOLAR WATER HEATER

The combined collector and storage type solar water heaters are generally not preferred and more than 90 percent of the solar water heaters installed in USA, Israel, Australia, India and other countries have the collectors and storage tanks as separate units. A basic Domestic Conventional water heater is shown in Fig.5. Here the hot water tank

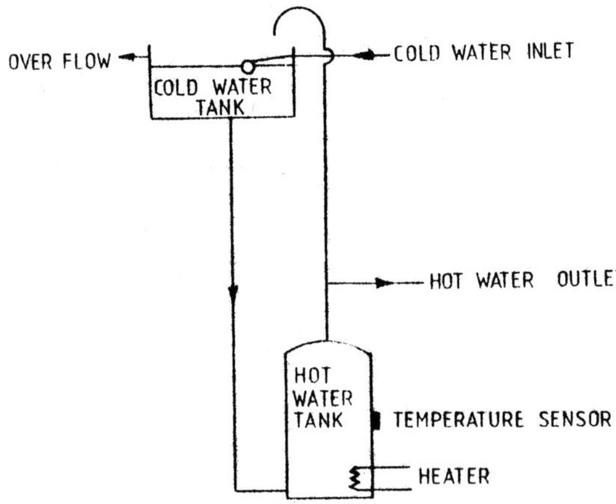

FIG.5 - CONVENTIONAL DOMESTIC HOT WATER HEATER

is fitted with a heater in the bottom and a temperature sensor in the middle which also controls the water temperature. Heating of water by solar energy can be accomplished in several ways. The working principle of solar hot water system is schematically shown in Fig.6.

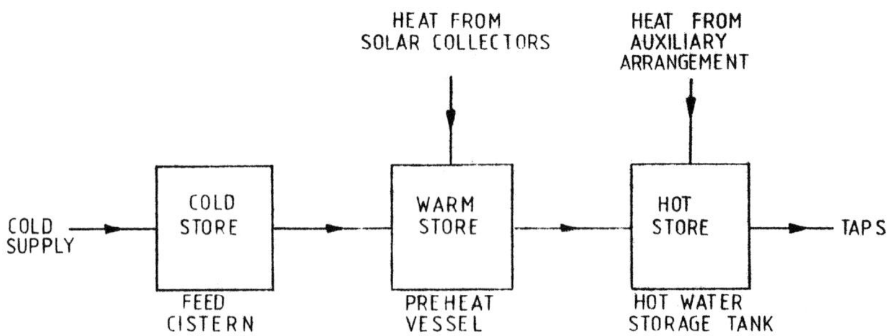

FIG.6 - WORKING PRINCIPLE OF SOLAR HOT WATER SYSTEM

Water from the cold supply is supplied to the cold water storage tank from where it goes to the pre-heat vessel, forming, the 'Warm Store', where it is heated by solar energy. The warm water then goes to the hot water storage tank where it is heated by a boiler or electric heater. Generally the auxiliary heating arrangement is made within water store itself, thus saving one cylinder. In hot weather or tropical countries, the auxiliary heating arrangement is not required, and solar energy alone may be sufficient to heat the water to desired draw-off temperature. In cold climates or in countries where the solar insolation is low, the water cannot be heated to the required draw-off temperature at all times of the day by solar energy alone, and therefore arrangements for providing heat by other means such as a solid fuel, gas or oil fired or an electric commercial heater are made.

Many different designs of solar water heaters are possible and they may be classified in many ways. Each type has its own advantages and disadvantages, and depending on the situation a particular design is recommended. Some of the solar water heating configurations are schematically shown in Figs.7-12. These systems are as follows (Ref.27):

- o A direct natural circulation solar water heater (Fig.7).
- o An indirect natural circulation solar water heater (Fig.8).
- o An indirect forced circular solar water heater (Fig.9).
- o A direct forced circulation solar water heater (Fig.10).
- o A single cylinder indirect forced circulation solar water heater (Fig.11).
- o An indirect system with air heating collectors (Fig.12).

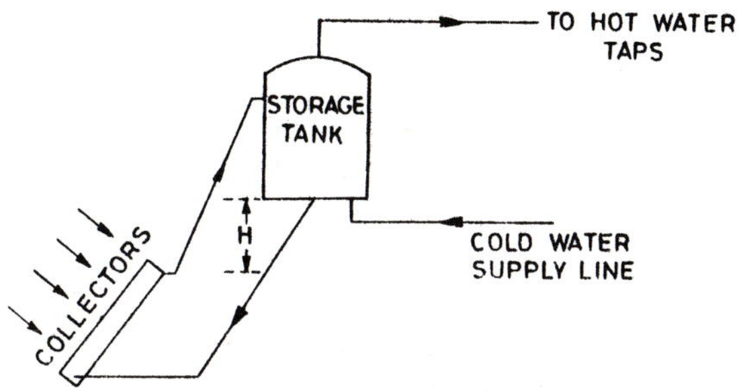

FIG.7 - DIRECT NATURAL CIRCULATION SOLAR WATER HEATER

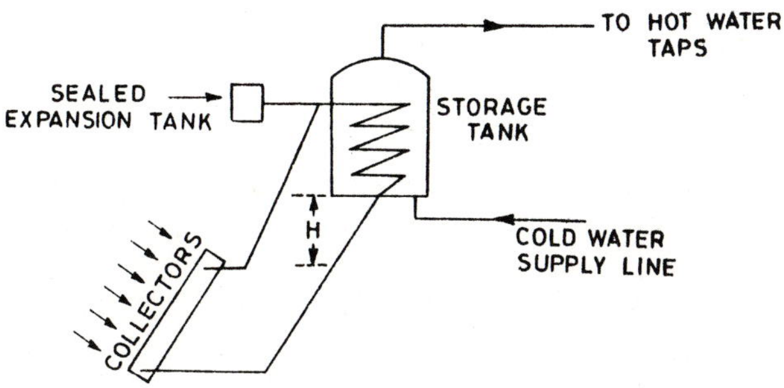

FIG.8 - INDIRECT NATURAL CIRCULATION SOLAR WATER HEATER

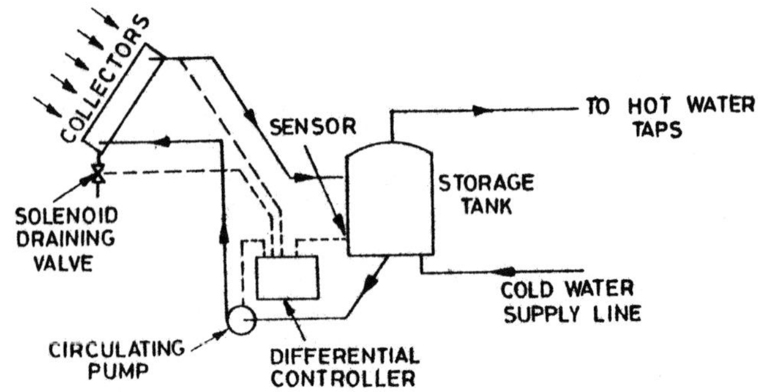

FIG.9 - DIRECT FORCED CIRCULATION SOLAR WATER HEATER WITH FROST PROTECTION ARRANGEMENT

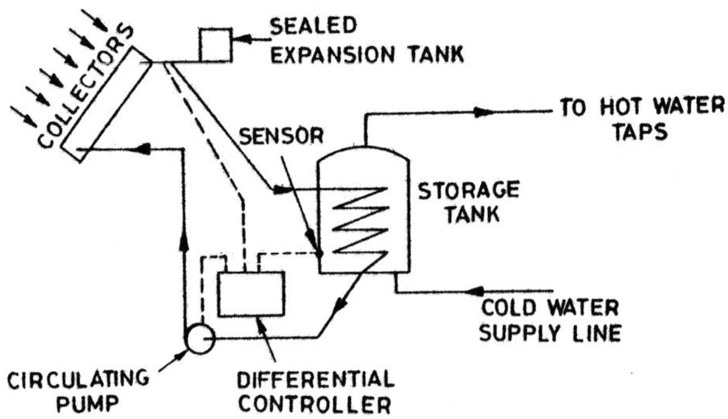

FIG.10 - INDIRECT FORCED CIRCULATION SOLAR WATER HEATER

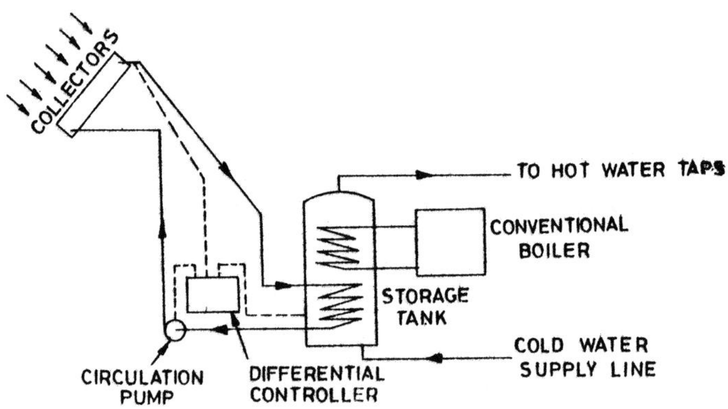

FIG.12 - SCHEMATIC OF INDIRECT SOLAR WATER HEATER WITH AIR HEATING COLLECTOR

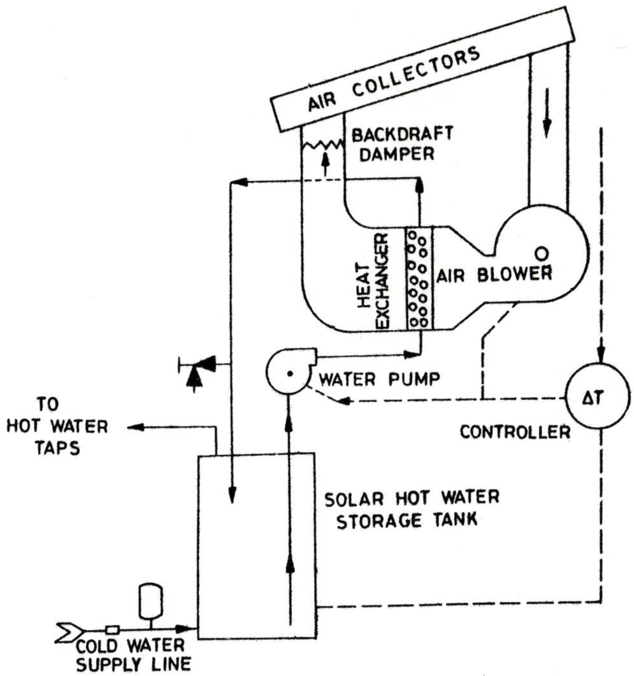

FIG.12 - SCHEMATIC OF INDIRECT SOLAR WATER HEATER WITH AIR HEATING COLLECTOR

As is seen in these figures, a solar water heating system consists of the following components:

- o flat plate collectors
- o storage tank
- o heat exchanger
- o automatic controls
- o pumps, pipe work, valves and fittings

5.1 Flat Plate Collector

The main purpose of the collector is to absorb the sun's energy and transfer this energy efficiently to the liquid flowing in it. There is a great variety of flat plate collectors, but a tube in plate type of collector, as is shown schematically in Fig.13 is used. The collector

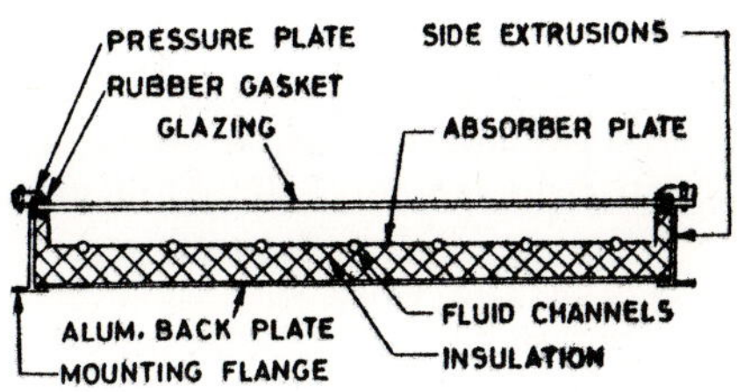

FIG.13 - CONVENTIONAL TYPICAL METALLIC PLATE FLAT PLATE COLLECTOR WITH SINGLE GLAZING

can be all metallic or plastic, single glazed or double glazed, selectively coated or ordinary black painted depending on the temperature of operation and outside climatic conditions. A list of flat plate collector materials is given in Table 1.

TABLE 1 - List of Collector Materials

Component	Materials
1. Collector plate (painted with ordinary black paint or with selective coating)	Galvanized iron, steel, aluminium, copper, stainless steel, plastic
2. Fluid pipe (circular or rectangular)	Galvanized iron, steel, aluminium, copper, stainless steel, plastic
3. Working fluid	Water, air, oils, silicone fluids, antifreeze solutions
4. Cover (single or double)	Glass, FRP, tedlar, teflon, fibre-glass
5. Seals	Rubber, butyle, neoprene, silicons, tar, putty
6. Tray	Galvanized iron, mild steel,aluminium, plastic, AC sheet, wood
7. Insulation	Fibreglass, glass wool, wood shave, styrofoam, urethane, polyimide

As is seen earlier, a flat plate collector has the following components:

(a) a blackened or selectively coated flat-absorbing plate, normally metallic, which absorbs the incident solar radiation, convert it into heat and conducts the heat to the fluid passages.

(b) tubes, channels or passages attached to the collector absorber plate to circulate the fluid required to remove the thermal energy from the plate.

(c) insulation material provided at the back and sides of the absorber plate whose principal function is to reduce heat loss from the back and sides of the absorber plate.

(d) a transparent or translucent cover or covers whose principal functions are to reduce the upward heat losses and to provide weather proofing.

(e) an enclosing box whose principal functions are to hold the other components of the collector in their correct relative positions and to protect the collector plate and insulation material from the weather.

Collectors generally available in the market, although conforming to the above general design, have some differences between them. The components most often changed are the absorber plate configuration, the black coating on the absorber plate, and the glazing.

5.2 THE STORAGE TANK

The storage tank stores the heat collected during the day for use when needed. For the storage of hot water, copper, steel, galvanized iron, aluminium, concrete, plastic, and sometimes wooden tanks are used. The tank should be sized to hold between 1.5 and 2 days supply of hot water. The auxiliary heating arrangements may be electric or gas booster and thermostat should be fitted in the central part of the tank, and not in the bottom of the tank. For domestic purposes, the thermostat setting is done between 50-60°C. There are many variations in the tank design and a few are listed below:

- o Vertical or horizontal type
- o Pressure or non-pressure type
- o Gas, electric or solid fuel booster, off-peak or continuous tariff, or
- o Internally or externally mounted

There is very little information available on the difference the above options have on system performance. A few storage tanks with electric booster and gas booster are schematically shown in Fig.14.

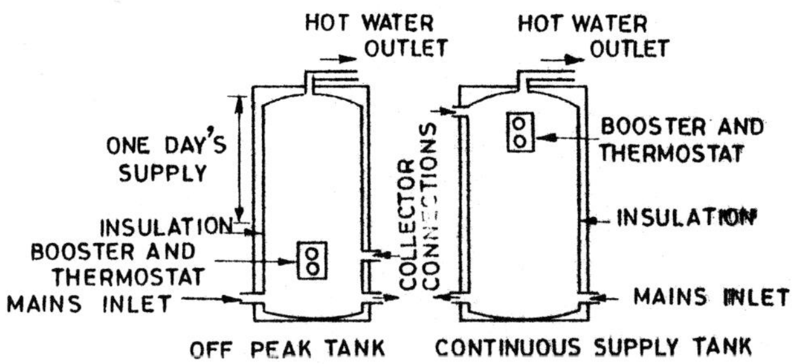

(a) SOLAR STORAGE TANKS WITH ELECTRIC BOOSTING.

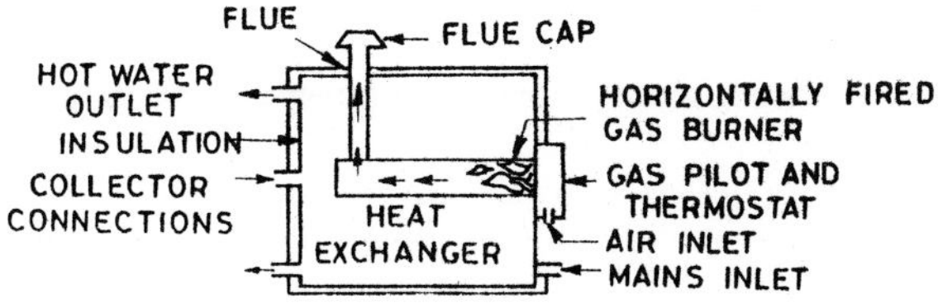

(b) SOLAR STORAGE TANK WITH GAS BOOSTING.

FIG.14 - SOLAR STORAGE TANKS

The storage tank should have the following pipe connections:

- o The cold water supply provided at the bottom of the tank.
- o Hot water outlet pipe provided at the top of the tank.
- o Flow pipe (inlet pipe) to the absorber provided near the bottom of the tank, and

o Return pipe from the absorber, the position of which depends on the use pattern. Generally for domestic applications it is provided at the height of two-third of the total height of the tank.

The hot water storage tank should be heavily insulated. Generally a thickness of around 7-10 cm fibreglass insulation or cork insulation is used.

5.3 Heat Exchanger

A heat exchanger is a device for transferring thermal energy from one fluid to another. In some solar systems, a heat exchanger is required in the collection loop where a fluid is circulated through the collector and the storage medium or in a storage-delivering loop where the fluid is circulated from the storage to the distribution point. In certain solar heating systems heat is required to be transferred from air to liquid based solar heating systems, heat exchanger are used between the collector and the storage for freeze protection and or to reduce the pressure on storage tanks. The main requirement of a heat exchanger is that it should transfer the heat efficiently which needs a largeer heat exchanger area and large fluid flow rates. Whatever type or position of a heat exchanger is used, some penalty is paid for this. Some commonly used heat exchangers for liquid to liquid heat transfer are shown in Fig.15. Complete analyses of the effectiveness of heat exchangers are discussed elsewhere (Ref.28,29).

5.4. Automatic Controls

In a natural circulation or thermosyphon type solar water heaters, where pump is not used, there is no need for a control system. In a forced circulation or large solar water heater a control is required whose primary purpose is to 'turn on' the pump only when the heat can be gained and 'turn off' the pump when heat cannot or should not be collected. Several types of control are proposed and used; these are: time switch control, temperature set or radiation set control, differential control, proportional control, etc.

An electronic differential controller is generally preferred which operates the 'turn on' and 'turn off' the pump at a predetermined temperature difference between solar hot water storage tank and collector outlet temperature. Circuit diagrams of electronic differential controlleters are now available in textbooks (Ref. 30). The main components of the controller are:

i) One temperature sensor (thermistor) which senses collector outlet temperature and one sensor (thermistor) which senses tank bottom temperature. They provide the signal for operating the controls. When fluid temperature at the collector outlet reaches 8-10°C warmer than one water in the bottom of the storage tank the control circuit switches the pump on which starts circulating the fluid between storage tank and collectors. When the temperature differential drops below 1-2°C, the thermostat automatically turns the pump off.

ii) An electronic circuit (bridge) which compares the signals received

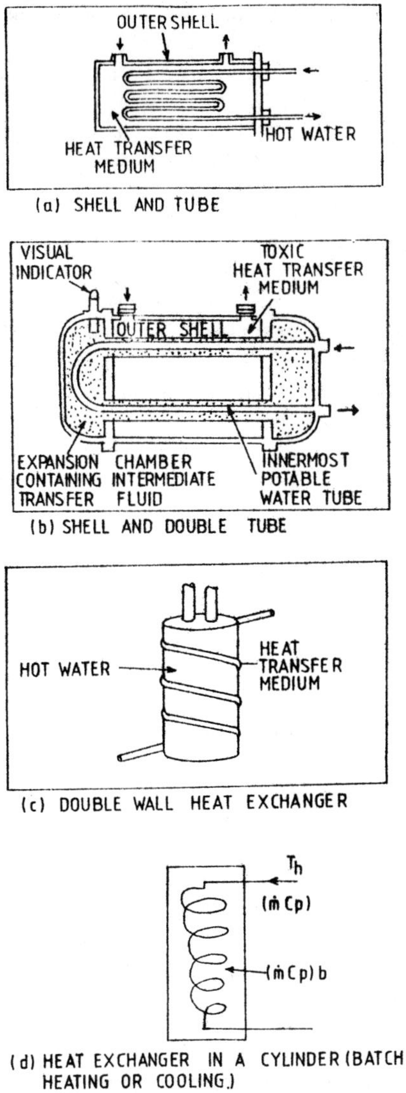

FIG.15 - SOME COMMON LIQUID TO LIQUID HEAT EXCHANGER DESIGNS FOR SOLAR ENERGY USE

from the sensors and to operate the pump switch.

iii) An electrically operated switch to control the supply to the pump.

iv) A power source to provide supplies at the required voltages.

5.5 Pumps, Pipe Work, Valves and Fittings

Small centrifugal circulating pumps are employed in domestic or industrial or large solar water heaters for circulating water through storage tank and collectors. These are generally 40 watt or 60 watt rating for domestic applications and 90 watt or more for large systems. The length of all the pipes in a solar water heater should be as small as possible of appropriate diameter and should be insulated to reduce heat loss to the environment.

All the valves should be of proper specifications and of standard make otherwise it may later give trouble like fluid leaking and may create pressure loss. Similarly all the fittings should be properly sized. All the hot water pipes should be insulated.

6. NATURAL CIRCULATION TYPE SYSTEMS

In a natural circulation type or thermosyphon type solar water heater, a storage tank is placed at a certain height (30 to 60 cm) related to the top of the collector to prevent the reverse circulation during off sunshine hours (Ref.31,32). An open loop and closed loop thermosyphon type solar water heaters are schematically shown in Fig.7 and 8 respectively. In the morning, the sun heats the collectors, the hot water inside rises by natural convection and reaches to the storage tank and the colder storage tank water leaving from its bottom, flows into the collectors. Thus the circulation loop is automatically established whenever there is a sufficient sunshine, and circulation automatically stopped during insufficient insolation when the upward buoyancy force is unable to overcome the fluid friction losses inside the pipe. In freezing weather a closed loop solar water heater as shown in Fig. 9 is used, wherein the collector loop anti-freeze solution is used and the storage tank contains the fresh water.

7. FORCED CIRCULATION OR PUMPED SYSTEMS

A forced circulation or pumped system may either be direct type or indirect type (Ref.33). In a direct system, the potable or service water is directly circulated between the solar hot water storage tank and solar collectors (Ref. 34). In an indirect system a fluid (such as anti-freeze, air, distilled water, or an ortanic heat transfer fluid) other than the potable or service water is circulated in the solar collectors. The main advantage of a forced circulation system is that the storage tank can be placed at a convenient location like in attic space or bathroom or ground floor etc. while the collectors are exposed. Some of the forced circulation system with its components are shown in Fig.9,10,11 and 12.

8. COLLECTOR ORIENTATION AND TILT

Collectors should always face the equator, i.e. true south in the northern hemisphere, such as in India and true north in southern hemisphere such as in Australia. Variantions of 20 degrees east or west of south (in northern hemisphere) or north (in southern hemisphere) are acceptable (Ref.35). The optimum tilt of the collectors should be determined from the period of use of hot water. If the water heater is to be used in winter season then the collector tilt from the horizontal should be equal to latitude (L) of the place plus 15 degrees (tilt = $(L+15)^o$). For summer use and year round use of hot water, the collector tilts from horizontal can be approximately $(L = 15)^o$ and 0.9 L^o respectively (Ref.36). A few degrees of variation in the tilt will not affect the performance significantly.

9. SIZING SOLAR HEATING SYSTEM

The size of the solar heating system depends on the daily demand of hot water, the local weather conditions, the hot water use pattern, the type of collectors, etc. The amount of hot water required per person per day varies from place to place. It may be as low as 20 litres of hot water per person per day in a developing country to 100 litres of hot water per person per day in America or Japan. Therefore, the demand of the hot water varies drastically with individual living habits. A good rule of thumb in a developing country like India is 30 lites per person per day. If hot water in a family is required for other purposes also like in the kitchen or in the washing machine then the demand of hot water will increase accordingly. After assessing the daily requirement of the hot water for a family, the size of collector can be approximately determinted by using a thumb rule that 50-70 litres of water will be heated up to 55-60°C by 1 m^2 collector area on a sunny day. There are sophisticated computer simulation programmes such as TRANSYS (Ref.37) are available in the literature which can be used for performance prediction and design of solar water heating systems under a given situation. Simple design procedure like f-chart method, $\bar{\phi}$ -f chart method, etc. are also available which can be used using simple calculators or log tables for designing and performance prediction of solar water heating systems. A review of design methods of Solar Water Heating System is recently prepared by Garg (Ref.40).

10. HOT WATER, SPACE HEATING AND AIR CONDITIONING SYSTEMS

The large solar water heating systems can also be used for space heating and cooling purposes. Several solar heated and cooled houses with hot water supply arrangements have been made and are in use in countries like USA, Germany, Japan, etc. In the space heating and hot water supply system, flat plate collectors are used to heat the water which delivers its heat if sufficiently hot directly to the space either through radiant panels or liquid to air heat exchangers. If the hot water is not at the required temperature, additional heat is supplied by the auxiliary heating arrangement and then supplied to the space. A solar water supply and heating system as developed by Furukawa Electric Co., Japan, is schematically shown in Fig.16. This system is fully automatic and calls the energy directly either from the collector or from the storage tank of auxiliary heating arrangement, depending on the requirements.

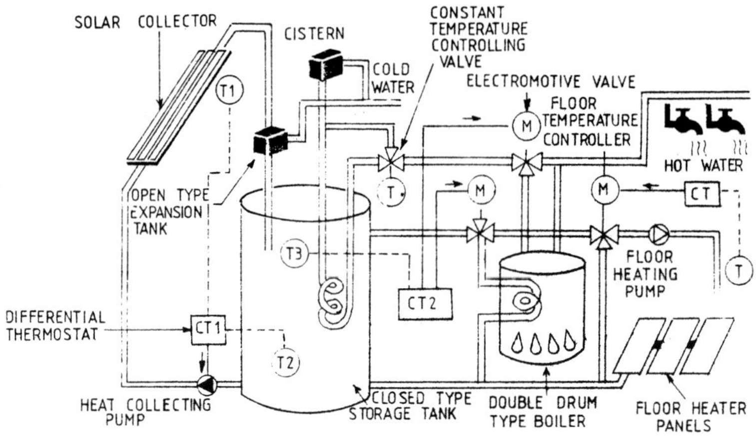

FIG.16 - SCHEMATIC DIAGRAM OF SOLAR HOT WATER SUPPLY AND HEATING SYSTEM (ADOPTED FROM FURUKAWA ELECTRIC CO., JAPAN)

A more sophisticated but dependable system is the one which can supply the requirement of the hot water and can heat and cool the space as and when required (Fig.17). The Furukawa Electric Co., Japan, has also developed such a system which can supply the heating and cooling load of the space and also the hot water requirement. In this system an absorption type chiller which can operate with hot water around 90°C is used. Hot water from the high temperature storage operates this chiller to produce cold water which is stored in the low temperature storage. This cold water is transferred to the fancoil unit as and when required. For heating,both the high and low temperature storages are utilized together.

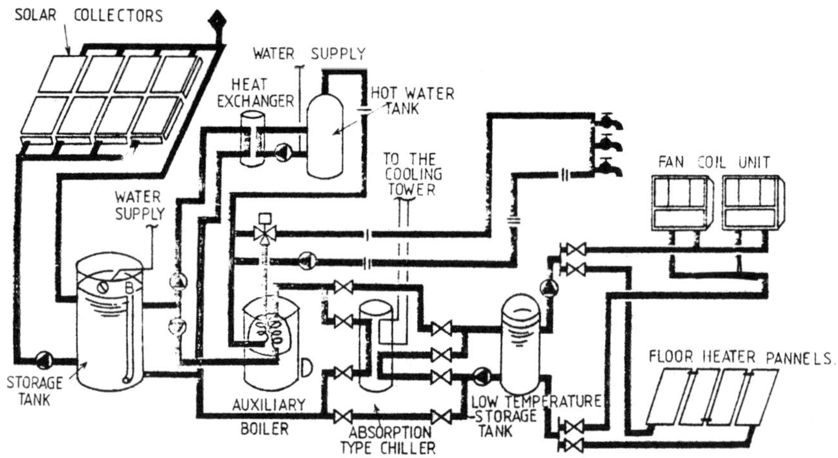

FIG.17 - SCHEMATIC DIAGRAM OF SOLAR HOT WATER SUPPLY AND AIR CONDITIONING SYSTEM (ADOPTED FROM FURUKAWA ELECTRIC CO., JAPAN)

11. PERATION AND MAINTENANCE OF SOLAR HOT WATER SYSTEMS

Generally it is assumed that solar systems do not need any maintenance and will work trouble free. This is not correct. It is always desirable to understand the working of each and every component before it is purchased. A schematic diagram of system will always help in locating the fault. Since the system is basically a simple one, there are relatively few problems which can easily be diagnosed through the use of common sense and system understanding. Routine maintenance is a necessity as in the case of any other mechanical device. Efficient use of a solar water heater pays good dividends. It is possible to solve simple problems with relative ease, using the troube shooting checklist as shown in Table 2 (Ref.15).

TABLE 2 - Solar Water Heating Systems - Homeowner Troubleshooting Checklist (Ref.15)

Symptoms	Component	Possible Cause	Remedy / Check
Not enough hot water	Collectors	1.Shading	1.Remove or reduce shading
		2.Improper orientation	2.Check orientation-face south $\pm$ 20°
		3.Improper tilt	3.Check tilt-set tilt equal to latitude + 15°
		4.Insufficient area	4.Check sizing - approx. 50-70 litres tank per sq.m. of collector
		5.Improperly plumbed	5.Compare with system schematic
		6.Dirty glazing	6.Clean periodically - with mild soap and water when cool
	Controller	1.Improper operation (cycling, late turn on)	1.Check sensor placement. Adjust for good thermal contact and insulation from surrounding air.
		2.Faulty sensors and/ or faulty controller	2.Check by placing sensors in hot and cold water and watching pump function
		3.Improper wiring and/or loose connections.	3.Compare with system schematic

TABLE 2 (cont'd)

Symptoms	Component	Possible Cause	Remedy / Check
	Mixing Valve	1. Improper adjustment	1. Adjust to desired hot water temperature
		2. Faulty valve	2. Check water temperature at house faucet for various valve settings. Replace if necessary
	Piping	1. High heat losses	1. Check insulation
		2. Night convection	2. Check for pump running at night and amount of hot water in morning
		3. Improperly plumbed	3. Compare with system diagram
		4. Isolation valves	4. Open valves
		5. Flow blockage	5. Flush system
	Pump	1. Little or no flow	1. Feel pipes for high temperature difference
		2. Runs continuously	2. Check sensor wiring for breaks or shorts
		3. Faulty pump	3. Listen for irregular noises in pump operation
	Tank	1. Insufficient size	1. Consult sizing formulae
		2. High storage losses	2. Check insulation and location of tank
		3. Improper backup wiring	3. Check service connection
		4. Breaker or timer off on cloudy day	4. Turn on backup power
		5. High limit switch activated	5. Reset to enable heating element
		6. Thermostat set too low	6. Check setting and adjust to desired temperature
No hot water in morning	Check Valve	1. Pump cycle on at night (Nighttime thermosiphoning)	1. Contact installer
Noisy system	Pump	1. Bearing need lubrication	1. Oil per manufacturer's recommendation

TABLE 2 (cont'd)

Symptoms	Component	Possible Cause	Remedy / Check
	Piping	1.Entrapped air	1.Purge system by running water up return pipe and out drain on supply line (isolation valves closed)
Does not drain	Vacuum Breaker Air Vent	1.Dirty or stuck seat	1.Disassemble and clean
	Piping	1.Insufficient slope	1.Check for vertical drop in all horizontal runs
		2.Air pockets	2.Check for high points in collector piping

*Adapted from Booklet, Florida Solar Energy Centre, Florida (USA), Report No.FSEC-GP-80-1.

REFERENCES

1. K. Butti and J. Perlin (1977), 'Solar water heaters in California 1890-1930', The Co-Evaluation Quarterly, 4-13, Published by the Whole Earth Catalog.
2. F.A. Brooks (1936), 'Solar energy and its uses for heating water in California', Bull. Calif. Agric. Exp. Sta., No.602.
3. H.P. Garg, (1982), Treatise on Solar Energy, Vol. 1 : Fundamentals of Solar Energy, John Wiley and Sons, England.
4. H.C. Hottel and B.B. Woertz, (1942), 'The performance of flat-plate solar heat collectors', Trans. ASME, 64, 91-104.
5. H. Heywood, (1954), 'Solar energy for water and space heating', J. Inst. Fuel, 27, 334-347.
6. R.N. Morse, (1956), 'Solar water heaters', Proc. World Symp. on Applied Solar Energy, Stanford Research Inst., University of Arizona, Phoenix, Arizona, 191-202.
7. D.N.W. Chinnery, (1967), 'Solar water heating in South Africa', National Building Research Institute, Bulletin 44, CSIR Research Report 248, Pretoria, South Africa.
8. H. Tabor, (1955), 'Solar energy collector design', Bull. Res. Coun., 5C, No.1, Israel.
9. K.N. Mathur, M.L. Khanna, T.N. Davey and S.P. Suri, (1959), 'Domestic solar water heater', J. of Scientific and Industrial Research, 18A(2), 51-58.
10. F.A. Brooks, (1955), Solar Energy Research, Edited by F. Daniels and J.A. Duffie, Madison, p. 76.
11. J.T. Czarnecki, (1963), 'Methods of heating swimming pools by solar energy', Solar Energy, 7(1), 3-7.

12. J.L.A. Francey, P. Golding and R. Clarke, (1980), 'Low-cost solar heating of community pools using pool covers', Solar Energy, 25, 407-416.
13. L.M.T. Bell, (1961), 'The prevention of heat losses from open air swimming pools', Heating Air Cond. Vent. Insulation, 23 (187), 229.
14. J.L.A. Francy and P. Golding, (1981), 'The optical characteristics of solar pool covers used for direct solar heating', Solar Energy, 26, 259-263.
15. Collen Mc. Cann Kettles and Hilary Krom Kiell (1980), 'A comprehensive consumer guide to solar water heating', Report No.FSEC-GP-80-1, Florida Solar Energy Centre, Florida, USA.
16. R. Talwar, (1980), 'Swimming pool heating', Solar Energy Technology Handbook, Part B (Edited W.C. Dickinson and P.N. Cheremisinoff), Marcell Dekker, Inc., New York, pp. 15-31.
17. W.C. Dickinson, A.F. Clark, J.A. Day and L.E. Wovters (1976), 'The shallow solar pond energy conversion system', Solar Energy, 18, 3-10.
18. A.B. Casamajor and R.E. Parsons, (1979), 'Design guide for shallow solar ponds', Lawrence Lewermore Laboratory Report No.UCRI-52385, Rev.1, University of California, Lewermore, California.
19. I. Tanishita, (1970), 'Present situation of commercial solar water heaters in Japan', 1970 Int. Solar Energy Soc. Conf., Melbourne, Australia, March 2-6, 1970, Paper No.2/73.
20. H.P. Garg, (1972), 'Design and performance prediction of low cost solar water heater', Research and Industry, 17.
21. H.P. Garg, (1975), 'Year round performance studies on a built-in-storage type solar water heater at Jodhpur, India', Solar Energy, 17, 167-172.
22. S.J. Richards and D.N.W. Chinnery, (1967), 'A solar water heater for low cost housing', NBRI Bull 41, CSIR Research Report 237, Pretoria, South Africa.
23. J.C.V. Chinnappa and K. Gnanalingam, (1973), 'Performance of Colombo, Ceylon, of a pressurised solar water heater of the combined collector and storage type', Solar Energy, 15, 195-204.
24. A.I. Kudish and D. Wolf, (1978), 'A compact shallow solar pond hot water heater', Solar Energy, 21, 317-322.
25. H.P. Garg and Usha Rani, (1982), 'Theoretical and experimental studies on collector/storage type solar water heater-, Solar Energy, 29(6), 467-478.
26. M.S. Sodha, G.N. Tiwari, S.C. Kaushik and M.A.S. Malik, (1979), 'Analysis of an inexpensive solar collector/storage system', Proc. Indian National Science Academy, 45(2), 172-180.
27. J.C. Mcveigh, (1977), Sun Power: An Introduction to the Applications of Solar Energy, Pergamon Press, Oxford.
28. W.M. Keys and A.L. London (1964), Compact Heat Exchanger, II Edition, McGraw Hill Book Co., Inc., New York.
29. F. De Winter, (1975), 'Heat exchanger penalities in double loop solar water heating systems', Solar Energy, 17(6), 335-337.
30. Donald Rapp, (1981), Solar Energy, Prentice Hall, USA, Chapter 6: Domestic Hot Water Systems, 157-196.

31. C.L. Gupta and H.P. Garg, (1968), 'System design in solar water heaters with natural circulation', Solar Energy, 12, 163-182.
32. C.L. Gupta, H.P. Garg and R. Ganguli, (1967), 'Solar water heater', Bulletin of Central Building Research Institute, Roorkee, India.
33. H.P. Garg, (1973), 'Design and performance of a large size solar water heater', Solar Energy, 14(3), 303-312.
34. E.T. Davey, (1970), 'Solar water heating in Australia', Paper No.4/71 1970 International Solar Energy Society Conference, Melbourne, Australia.
35. H.P. Garg and C.L. Gupta, (1980), 'Design data for direct solar utilization devices, Part II - Solar radiation data', Indian J. of Met. and Geophysics, 20(8), 221-228, 1980.
36. H.P. Garg and C.L. Gupta, (1967), 'Optimizing the tilt of flat-plate solar collectors for India', J. of Inst. of Engrs (India), 48(I), 21-28, PT CEL, 1967.
37. S.A. Klein, W.A. Backman and J.A. Duffie, (1976), 'TRNSIS - A transient simulation programme', ASHRAE Trans., 82, Part I, 1976.
38. W.A. Beckman, S.A. Klein and J.A. Duffie (1977), Solar Water Heating Design by the f-chart Method, Wiley-Interscience, New York.
39. S.A. Klein and W.A. Beckman, (1979), 'A general design method for closed loop solar energy systems', Solar Energy, 22, 269-282.
40. H.P. Garg, (1985), 'An overview of design methods for solar water heating systems', Solar and Wind Technology, 2(3).

MATERIALS FOR LOW TEMPERATURE SOLAR THERMAL APPLICATIONS

S.K. Gupta
Solar Energy Centre
Department of Non-Conventional Energy Sources
New Delhi, India

1. INTRODUCTION

It has long been recognised that the widespread use of solar water heating technology is being held back by excessive cost and to some extent by lack of confidence in the long term durability of solar heating systems. Thus the selection of the materials and manufacturing processes are of vital importance to the commercial viability of and user satisfaction with solar heating systems.

The materials used for one component will, in general, affect the possible choice for other components in the system; it is therefore important to consider interaction between various components of one main component of the system and also between various main components. It is also essential to know whether or not they play a major role in solar energy collection and/or heat transfer/heat loss from the system.

The design of solar hot water system is also equally important. Several basic types of designs of solar water heating systems are currently available which are given in Table 1. The most common design consists of flat plate solar collectors mounted on south facing shadow free piece of land or on the roof of a building connected to a heat management and storage system. A heat transfer fluid is made to flow through the solar heating system which collects heat from the solar collector and transfers it either directly to the hot water storage system (direct system) or via a heat exchanger in a hot water tank (indirect system).

A typical solar collector,illustrated in Photograph 1, consists essentially of a blackened absorber plate having channels to transfer fluid. The blackened plate is insulated at the back and is encased in a weather-proof box having transparent front surface facing the sun. The essential component of a solar collector therefore are absorber, front glazing, casing, seals, insulation and fixing devices; the materials aspects of these components are discussed in this paper.

The heat management and storage system consists of the heat transfer fluid itself, the pipe work carrying the fluid, the storage vessels and the control system, which includes temperature and pressure sensors and electronic control unit and associated wiring, valves and vents and a

H. P. Garg (ed.), Solar Water Heating Systems, 61–84.

TABLE 1

HOT WATER SYSTEMS

- SOLAR
  - PARABOLIC TROUGH [@]
  - FLAT PLATE
    - CLOSED [o]
      - FORCED [o]
      - THERMO-SIPHON [o]
    - OPEN [o]
      - FORCED [o]
      - THERMO-SIPHON [o]
  - BREAD BOX [o]
  - EVACUATED TUBE COLLECTOR
    - CLOSED [*]
      - FORCED [*]
      - THERMO-SIPHON [*]
    - OPEN [*]
      - FORCED [*]
      - THERMO-SIPHON [*]
- HEAT PUMP
- CONVENTIONAL
  - ELECTRIC

- CIRCULATION
  - FREEZE PROTECTION
    - DRAIN BACK [*]
    - ANTI FREEZE [*]

- CIRCULATION
  - FREEZE PROTECTION
    - DRAIN BACK
    - ANTI FREEZE

[o] Installed, under monitoring   [*] Under installation   [@] Under fabrication

PHOTOGRAPH 1

pump. The material aspects of these components are also discussed.

The desirable properties of various materials, systems, subsystems and components and also some test results are discussed in this paper.

2. MATERIALS FOR COMPONENTS OF THE COLLECTOR

2.1 Absorber

The absorber is the central component of a solar collection system and as such its function is to absorb solar radiation and transfer heat to a circulating fluid. Metals such as copper, aluminium, steel and stainless steel are commonly used as absorber materials; since these are not strongly absorbing, a coating must be provided which absorbs solar radiation. Plastics such as polypropylene and acrylic on the other hand be pigmented black to absorb solar radiation.Built-in-channels or metallic tubes welded on to the absorber sheet carry the heat transfer fluid.

Heat transfer to the fluid depends on the thermal extract between the sheet and the tubes, conductivity of the plate and tube material, and on the distance between fluid carrying channels/tubes. Thermal conductivities of commonly used materials are given in Table 2.

Adequate heat transfer can be achieved even with low conductivity materials by properly designing the system. In a plate and tube design lower thermal conductivity metals will require greater thickness of the plate for equivalent heat transfer. This leads to weight penalties and to high thermal mass. Light weight and low thermal mass absorbers can

TABLE 2 - Thermal Conductivities of Some Absorber Materials

Material	Thermal Conductivity (W/m^{o}C)
Copper	376
Aluminium	205
Mild Steel	50
Stainless Steel	24
Acrylic	0.20
Polyethylene	0.30-0.44
Polypropylene	0.20
PVC	0.16

PROPERTIES OF SOME SELECTIVE COATINGS

Coating	Substrate	Absorptivity	Emissivity	Durability: Maximum Service Temp. ^{o}C	Durability: Humidity	Method of Application
Black Chrome	Copper	.95-.97	.08-.14	316	Good	Electroplating
	Galvanized Steel	.91-.97	.07-.16	427	Poor	- do -
	Zinc Plated Aluminium	.95-.97	0.12	427	V.Poor	- do -
Black Chrome on Nickel Plate	Copper Steel	.90-.95	.09-.15	427	V.Good	- do -
	Aluminium	.95	.50		V.Good	- do -
Black Nickel on Bright Nickel	Steel,Copper Zinc Plated Aluminium	.87-.96	.07--.1	288	Variable	- do -
Black Nickel	Steel	.89-.96	.07-.17			- do -
Selective Coating Nickel Foil	Any	.95-.96	.08-.11	400	Good	
Selective Paint	Any	.94-.96	.35-.45	250	Good	Spray
Lead Sulphide Paint	Any	.90-.94	.3			Paint

be obtained with polymeric sandwich absorbers.

For plate and tube or tube and fin designs, good thermal contact must be achieved between the tube and plate. The area of contact should be large, which is most easily achieved by profilling the plate and good thermal contact can be achieved by soldering of copper tubes and plates, by mechanical methods preferably in conjunction with use of a heat transfer paste or by integral forming of the tube and fin by extrusion (aluminium).

To maintain its performance over many years, an absorber must be resistant to both internal and external corrosion. External corrosion may occur because of humid conditions in the collector box, because of leaks in the enclosure, allowing ingress of rain water or because of drops of condensation falling from the glazing. Steel collectors are particularly susceptable to external corrosion even when coated.

Internal corrosion may occur due to attack by the heat transfer fluid or due to the presence of dissimilar metals in the system. The heat transfer fluid may contain high levels of oxygen, particularly if the system is an unsealed drain down system and this may cause dioxidation of the less corrosion resistant metals such as steel. Chloride ions in solution can also cause attack of some metals such as copper and certain grades of stainless steel. Solar absorbers are connected to conventional copper pipe work, which causes galvanic corrosion to absorbers made from the less noble metals such as aluminium or steel.

Aluminium absorbers in a copper circuit can be protected from corrosion to some extent by adding inhibitors to the heat transfer fluid but corrosion of the aluminium has been observed whether the heat transfer fluid is water, anti-freeze inhibitor mixture or heat transfer oil. In general the most reliable materials for the absorber are Cu, aluminium and stainless steel.

Specifications of absorber materials

Material	Grade
Copper	Phosphorus dioxidized non-arsenical copper.
Aluminium	Wrought heat treatable aluminium alloy and anodizing of the aluminium after fabrication is preferable to avoid corrosion.
Stainless steel	316, 317 austenitic type, 444 - ferrite type, pitting and crevice corrosion resistant grade containing molybdenum.

Polymeric materials - resistant to dioxidation by water like polypropylene and polyethylene (maximum operating temperature 30°C), chlorinated PVC, ABS, polybutylene, and nylon provide promising low cost alternative to the metals.

2.2 Heat Transfer Fluid

The heat transfer fluid extracts heat from the solar collector and transfers it to the hot water either directly or indirectly via a heat exchanger. For a high rate of heat transfer the fluids should have:

- high thermal conductivity
- high heat capacity for ease of transfer of fluid
- low viscosity
- low density
- low thermal expansion coefficient

Commonly used heat transfer fluid is water but other fluids have also been tried out and the properties of some such liquids are given in Table 3.

2.3 Glazing

The primary function of the glazing of a solar collector is to reduce convective heat losses by restricting air flow. A glazing also protects the absorber from the environment and may reduce radiative heat losses by reflecting thermal radiation emitted by the absorber. The most important property required of a glazing material is high transmittance of solar radiation, and loss in transmittance will lead to a direct reduction in efficiency of the collector.

Optical properties of some glazing materials available in India are given in Table 4 along with photograph of test plate in Photograph 1. The first column gives the time of measurement of transmittance of solar radiation.

2.4 Insulation

The basic requirements of insulation material for application in solar collectors are:

(i) Low thermal conductivity values;
(ii) Stable up to a temperature of 200°C;
(iii) No degassing up to 200°C;
(iv) Self-supporting without tendency to settle;
(v) Should not give rise to or accelerate corrosion;
(vi) Ease of application.

Bearing in mind the above requirements, the thermal insulating materials listed in Table 5 have been tested for suitability for solar application. Photographs 2 to 14 show testing of materials.

2.5 Collector Box

The function of the collector box in a solar collector is to house absorber and insulation and to hold the front glazing and also to prevent the absorber panel from atmospheric pollutants, rain, dust, so as to increase the life of components and systems and to prevent heat losses. It should:

(i) have long life,
(ii) have low cost,
(iii) have low weight,
(iv) be maintenance free,
(v) have good weathering properties,
(vi) be resistant to UV exposure,

TABLE 3 - Properties of Heat Transfer Fluids

Fluids	Sp. Heat at 82°C ($KJm^{-3}K^{-1}$)	Thermal Conductivity at 82°C (W/mK)	Density (Kg/m^3)	Viscosity at 82°C (m^2/sec)	Thermal Expansion Coefficient (/K)	Option
Water	$4.19x10^3$	0.640	1000	$3.5x10^{-3}$	$1.8x10^{-4}$	Optimum
Glycols Ethy-lene-Propylene)	$3.63x10^{-3}$	0.415	1020	$10-20x10^{-3}$	$5.4x10^{-4}$	More pumping power. periodically replacement
Silicone Oils	$1.48x10^3$	0.140	930	$5-10x10^{-3}$	$10.8x10^{-4}$	Higher flow rate. More power corrosion problem. Toxicity degradation by oxidation
Hydrocarbon Oils	$1.77x10^3$	0.121	880	$1-10x10^{-3}$	$5.4-14.4x10^{-4}$	

Water as heat transfer fluid is optimum fluid can be used directly depending upon the water quality.

TABLE 4 - Glazing Transmittance

Material / Time (hr)	10.00	10.30	11.00	11.30	12.00	12.30	13.00	13.30	14.00	14.30	15.00	15.30	16.00
Safex 4.0 mm	.84	.84	.85	.85	.85	.86	.88	.88	.86	.85	.85	.81	.80
Atul (Old) 5.5 mm	.70	.70	.71	.72	.75	.78	.78	.78	.75	.72	.72	.70	.70
Atul (New) 5.0 mm	.80	.80	.82	.82	.84	.84	.84	.84	.82	.82	.80	.80	.80
S.K. Tuf. 5.8 mm	.66	.66	.71	.72	.73	.73	.75	.75	.73	.71	.66	.66	.66
FRP (2 mm) FGP	.78	.78	.80	.80	.82	.82	.82	.80	.80	.77	.75	.73	.70
Vallabh Glass 4.04mm	.80	.81	.82	.84	.84	.85	.85	.84	.82	.82	.81	.80	.80
5.0 mm	.70	.71	.72	.74	.75	.75	.75	.75	.75	.75	.73	.72	.70
5.5 mm	.67	.67	.69	.69	.71	.72	.72	.73	.73	.70	.70	.68	.68
Holland 5.0 mm	-	.93	.94	.94	.94	.95	.95	.95	.95	.95	.94	-	-

TABLE 5 - THERMAL INSULATION FOR SOLAR ENERGY

S. No.	Name of Material	Manufactured by	Thermal conductivity at 200°C (W/m°C)	Density (Kg/m^3)	Out-gassing	Saging	Colour change	Remarks
1.	Crown White Wool	PFG	0.34	48	No	Yes	No	Good, but expensive
2.	Crown Bonded 150	FPG	0.066	48	Yes	No	Yes	Not good
3.	Spintex 300 Industrial	Lloyd	0.075	48	No	No	No	Good, reasonable cost
4.	Glass Wool	U.P. Twiga	0.044	48	No	Yes	Yes	Under testing
5.	Calsium Silicate	Hyderabad Asbestos	0.07	251.60	No	No	No	Good, but component system becomes very heavy
6.	Expanded Polystyrene	Synthetic Foam	0.017	32	Yes	No	Yes	Not good
7.	ISO Cyanurate	Lloyd	0.020	32	No	No	Yes	Under testing
8.	Phenotherm	Bakelite Hylam	0.029	32	Yes	No	Yes	Not good
9.	Thermocole	Lloyd	0.035	16	Yes	No	Yes	Not good
10.	Polyurethene Foam	Synthetic Foam	0.016	32	Yes	No	Yes	Not good
11.	Cellular Foam	Synthetic Foam	0.093	400	Yes	No	Yes	Not good
	PIPE SECTIONS							
12.	Rocklloyd	Lloyd	0.075	48	No	No	No	Good
13.	Isoloyd	Lloyd	0.021	32	No	No	No	Good
14.	Thermocole	Lloyd	0.035	16	No	No	No	Not good
15.	Foam	Synthetic Foam	0.017	32	No	No	No	Good

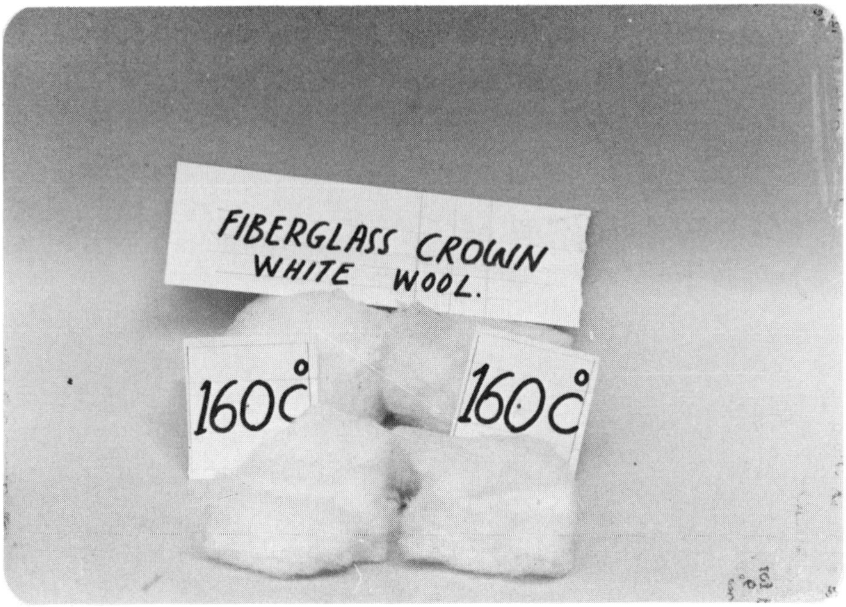

PHOTOGRAPH 2

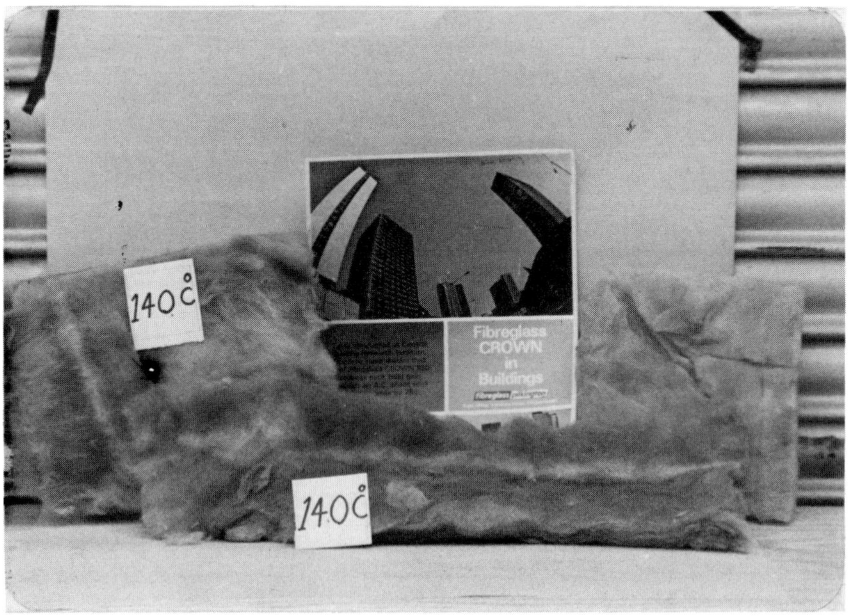

PHOTOGRAPH 3

PHOTOGRAPH 4

PHOTOGRAPH 5

PHOTOGRAPH 6

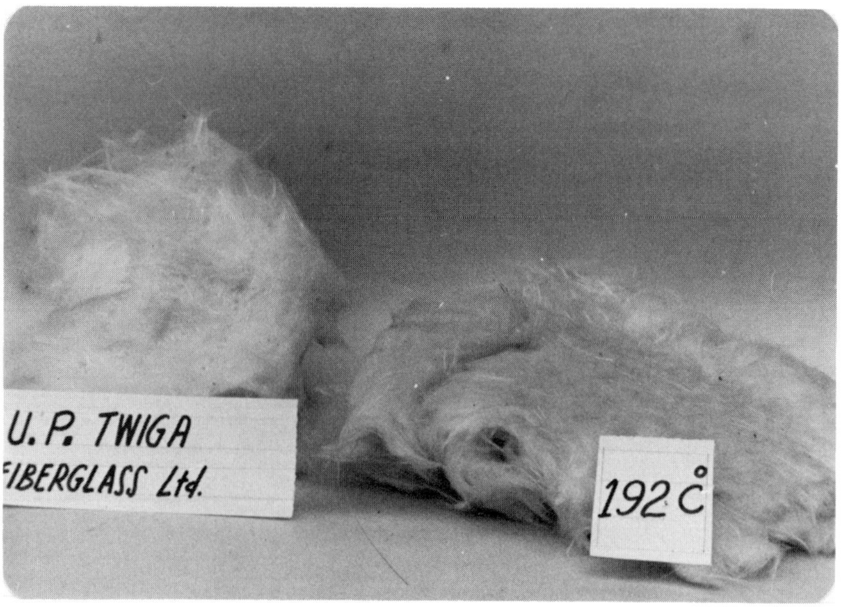

PHOTOGRAPH 7

PHOTOGRAPH 8

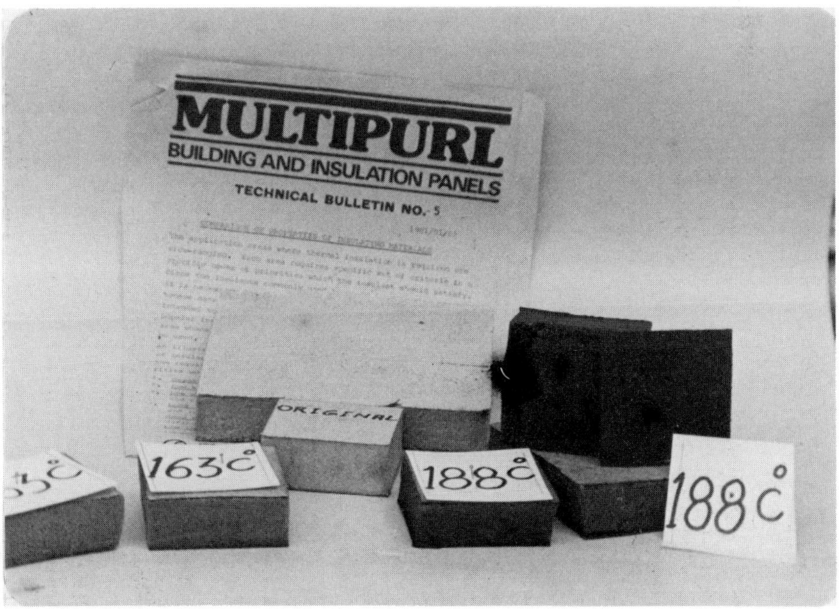

PHOTOGRAPH 9

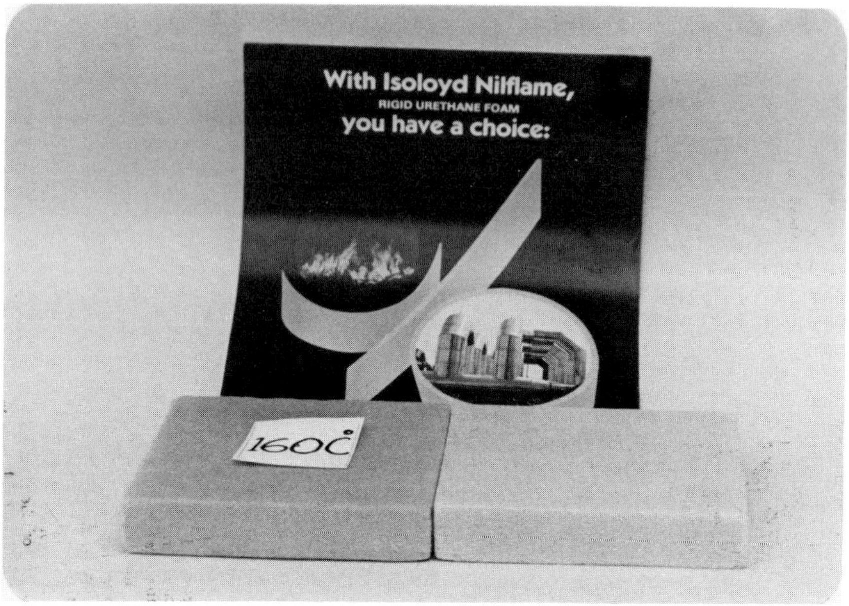

PHOTOGRAPH 10

PHOTOGRAPH 11

PHOTOGRAPH 12

PHOTOGRAPH 13

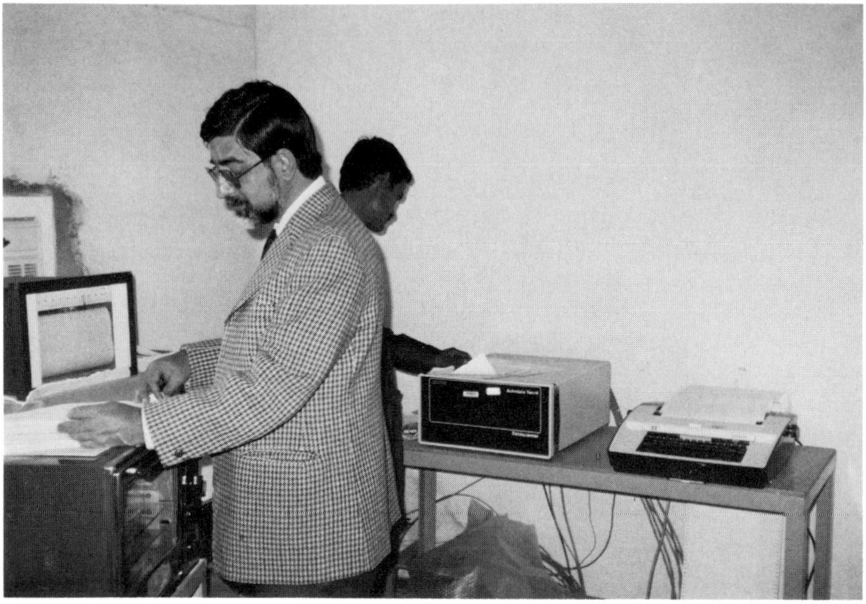

PHOTOGRAPH 14

(vii) have good mechanical strength,
(viii) have low coefficient of expansion and contraction,
(ix) have temperature cycle stability.

Table 6 indicates the possible alternatives for casting materials.

TABLE 6 - Suitable Materials for Collector Box

Material	Fire resistance	Mechanical properties	Cost	Temperature and weathering resistance
Aluminium	Good	Good	High	Not good for humid conditions
Galvanized mild steel	Good	Good	Moderate	Not good
GRP	Poor	Good	High	Good
Stainless steel	Good	Good	High	Good
Polypropylene	Poor	Not V. good	Low	Not bad
P.V.C.	V. Poor	Not V. good	Low	Not bad
ABS	Poor	Not V. good	High	Good
Acrylic	Poor	Good	High	Good
Wood	V.V. poor	Good	Low	V. poor
Extruded aluminium sectios	Good	V. good	Moderate	Excellent

The majority of solar collectors available in India utilize aluminium as the casing material. Other casing materials like FRP, stainless steel are also used. FRP casing used in actual installation is shown in Photograph 15.

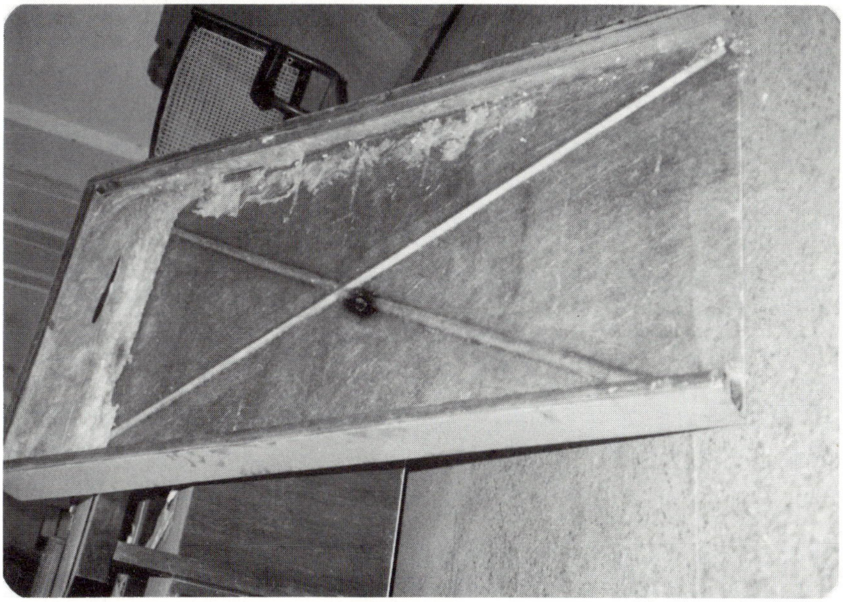

PHOTOGRAPH 15

Cheaper casing materials may be produced by using polymeric materials. Care must be taken to select a material with adequate mechanical strength, temperature resistance and weathering resistance, although the weathering resistance of many polymeric materials is not proved. GRP is a suitable material with acceptable properties in an outdoor environment, although it costs nearly as much as aluminium or stainless steel. Other possibilities are extruded polypropylene or PVC, moulded thermoplastics and vacuum formed thermoplastics such as polypropylene or ABS. Filled polypropylene appears to be the lowest cost option. Some potential problems with polymeric casings are UV degradation (likely to be less severe than in the transparent grades used for glazings), high thermal expansion coefficients and fire properties.

Wood and wood based products are even cheaper alternatives than plastics, but they are not favoured materials for casings because of their need for maintenance in outdoor conditions. Photograph 16 indicates the deterioration of wood after one year of use.

To summarise, aluminium, stainless steel and GRP are all suitable materials for collector casings, and there is little to choose between them in terms of both cost and performance. Coated mild steel and wood based products are not favoured because of the likelihood of the need for maintenance. Polymeric materials are very promising low cost alternatives, although further data are required before their use can be recommended.

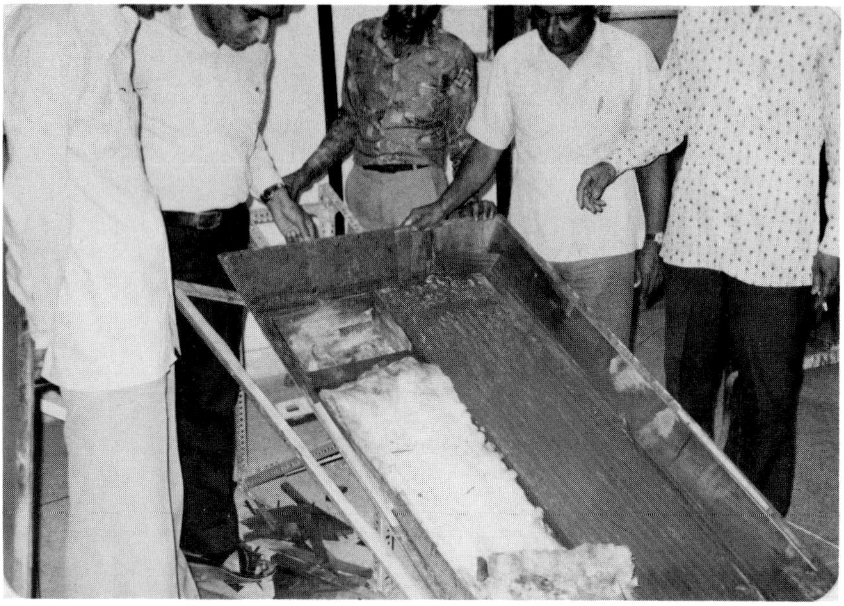

PHOTOGRAPH 16

2.6 Seals

At the junctions between glazing and casing, a seal must be provided to ensure weather-tightness of the enclosure; this is usually effected by a sealing material. The most common materials recommended for this application are EPDM and silicon rubbers while neoprene and butyl rubbers are also used. Property of seals should be:

(i) temperature stability between 30°C and 100°C,
(ii) good mechanical properties,
(iii) good weathering resistance,
(iv) resistance to water,
(v) compression stress,
(vi) percentage elongation to failure.

Properties of some materials are given in Table 7.

Neoprene at elevated temperatures releases organic oil which deposits on the inner surface of the front glazing and reduces the glazing transmittance as a result of which system performance is affected. Photograph 17 shows the degradation of neoprene at elevated temperatures after organic oil has come out. Photograph 18 shows the bush made out of fluoelastomers suitable for solar applications.

To summarise, EPDM, fluoroelastomers, neoprene and silicone have adequate properties for use as seals between solar collector glazings and casings. Performance experience and cost considerations indicate that EPDM is the optimum material and should form a leak-tight seal for many years if care is taken during fitting of the seal. For higher temperature applications (e.g. sealing of an inner glazing in a double

TABLE 7 - Properties of Some Candidate Sealing Materials

Sealing Material	Working Temperature (°C) Minimum	Working Temperature (°C) Maximum	Resistance to compression	Resistance to creep	Resistance to weathering	Resistance to water
Acrylic	-40	130	Good	Fair	Excellent	Poor
Neoprene	-20	130	Good	Fair	Excellent	Fiar
Silicone	-60	230	Good	Good	Excellent	V. Good
Putty	-60	80	Poor	Poor	Good	Good
Araldite	-50	70	Poor	Poor	Poor	Good
Urethanes	-50	100	Poor	Excellent	Good	Poor
Rubber (Butgle or Hypalon)	-50	120	Good	Good	Good	Fair

glazed collector or in a concentrating collector) silicone or fluoro-elastomer should be used as the sealing material.

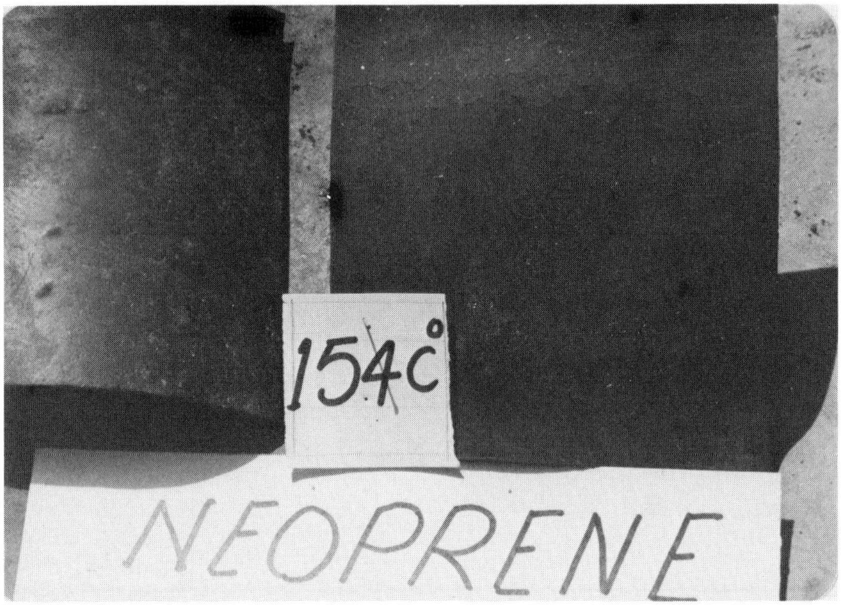

PHOTOGRAPH 17

PHOTOGRAPH 18

2.7 Pipes and Connections

The pipework in a solar heating system must be able to withstand circulating fluid at temperatures up to 100°C without any degradation and corrosion. G.I. Grade 'B' is the commonly used material in India but it has less resistance to corrosion, oxidation rate at elevated temperatures, chloride attack and high oxygen levels. Copper and stainless steel have excellent corrosion resistance properties but are more expensive. Polymeric materials such as neoprene pipes, steam hose pipes, acrylic, chlorinated PVC can also be considered for hot water distribution pipe work. Polymeric materials in general have low thermal conductivity, low density, ease of forming and assembly and flexibility and are cheaper than metallic piping and are therefore promising alternatives. The only consideration while designing the polymeric pipes may be to take care of high coefficient of thermal expansion. tendency to creep and thickness to withstand the system or line pressure.

While selecting materials for connectors, potential galvanic corrosion may also be considered. Use of polymeric materials like teflon is an excellent choice. Photograph 19 indicates the test of acrylic pipe for piping system and may be considered for use. Suitable materials for interconnecting pipes are shown in Table 8.

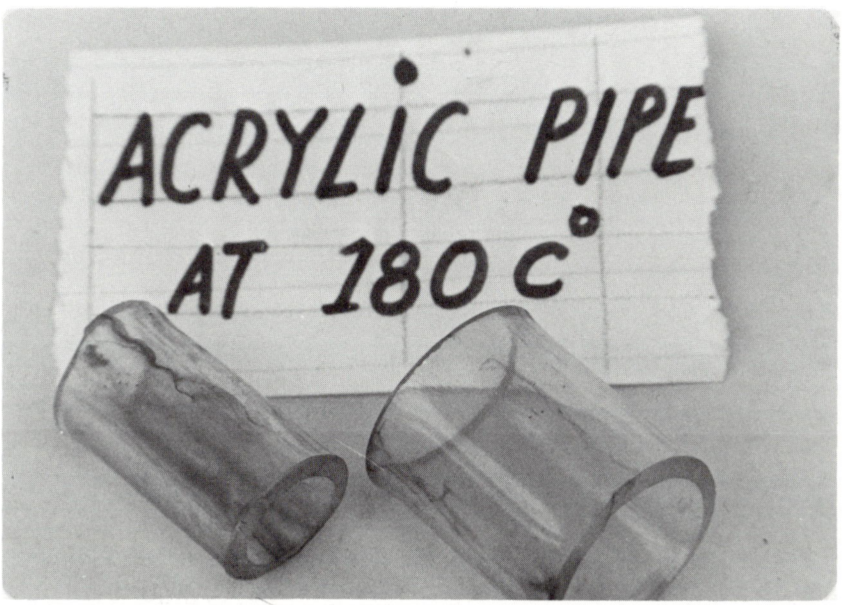

PHOTOGRAPH 19

2.8 Storage Vessels

Storage vessels in solar heating systems must contain the hot heat transfer fluid without degrading or corroding and must withstand the pressure involved. M.S. with epoxy paint is the commonly used material.

TABLE 8 - PIPES AND CONNECTIONS

SYSTEM INTER-CONNECTING PIPES			
Copper	Good corrosion resistance	Sensitive to chloride attack and to high O_2 level	Expensive
Stainless Steel	Good corrosion resistance	Yes, if 316, 317 or 444 grades are not used	Expensive but lower thermal conductivity. Lower thermal coefficient. High strength to mass ratio allow thinner sections.
Mild Steel	Very poor	Poor	Cheap, high oxidation rate
Galvanised Iron	Poor	Poor	Cheap. Zn not stable, so oxidation is problem. Special.
DISTRIBUTION PIPING			
Acrylic	Yes	Yes	Cheap
Polybutylene	Yes	Yes	Cost etc.
Chlorinated PVC	Yes	Yes	Cost etc.
Neoprene	Yes	Yes	Cost etc.
Steam Hose	Yes	Poor	Cost etc.

TABLE 9 - MATERIAL USED FOR STORAGE VESSEL

Material	Corrosion resistance	Cost	Disadvantages/Advantages
Copper	Excellent	Costly	High 'K' and loss of thermal stratification in the tank.As a result systemefficiency goes down. Insulation cost is higher.
Stainless Steel	Excellent	Costly	Low 'K'
Galvanised Steel	Fair	Cheap	Low 'K', Moderate corrosion.
Mild Steel	Poor	Cheap	Low 'K', High corrosion.
FRP	Excellent	Moderate	Reduced insulation cost. Is comparable to stainless steel or copper.
Polypropylene	Excellent	Moderate	Low cost but cannot be used above $60^{o}C$.

Table 9 shows the possible materials with disadvantages of each.

Use of FRP tanks,because of its ease of transportation, weight, ease of handling, repair and maintenance, is recommended.

2.9 Fixing Devices

Solar collectors are generally mounted at an angle. In all cases, some form of fixing framework is therefore required to support the collectors and keep them in inclined position, transmitting the loads (dead weight, wind loads) to structural members. The fixing material should have:

(i) sufficient strength,
(ii) low density,
(iii) good corrosion resistance,
(iv) weathering resistance.

Table 10 shows the possible materials for support structure.

TABLE 10 - Suitable Support Structure Materials

Material	Cost	Mechanical strength	Servicing
M.S. (Mild Steel)	Cheap	Strong	Require frequent maintenance
Stainless Steel	Expensive	Strong	No corrosion
Aluminium	Cheap	Strong	Corrosion resistance
G.R.P.	Lower	Strong	No
G.I.	Cheap	Strong	Yes, corrosion oxidation

M.S. is commonly used for support structure but use of aluminium angle and channels is recommended to avoid galvanic action, rusting, oxidation and on the basis of weight.

2.10 Screws, Nuts and Bolts

Should be corrosion resistance and selected with regard to parent materials. Coated steel will corrode in outdoor service and stainless steel fitments are more appropriate.

Philips screws should be preferred over self-threading screws.

DESIGN OPTIMIZATION OF COLLECTOR PLATES

Arun G. Joshi
A.J. Electronics
Pune 411 004
India

1. INTRODUCTION

There are various types of Collectors used in India, all have their merits and demerits and accordingly, they have their respective applications. The collector plates are primarily consisting of fin section to receive solar insolation. They also have tubes or passages within the fins to carry heating fluids. There are various combinations of fins and tubes, they may be of similar or dissimilar metals. Here specific investigations are carried out for copper tube insert Aluminium Roll Bond Panels. By design optimisation, if efficient collectors are manufactured, less collector area per system will be required, thereby saving not only the cost but also the overall volume of the solar system and the space is saved thereby.

Now though plenty of solar energy is available still it is in too scattered form but one must make use of this abundant energy, though the costs today are prohibitive. However, to offset the high costs at least partly, one has to develop very efficient collector panels at most reasonable price.

2. SOLAR ENERGY COLLECTORS

The principle of collecting solar energy is simple. It is just necessary to use receiving surfaces, which are able to absorb major part of radiation and retain and transfer a maximum, to desired heating medium and lastly give out as little as possible.

There are different types of collectors. They may be classified as tracking and stationary collectors or focusing and non-focusing collectors. Of all these collectors, the most commonly used for water heating is the flat plate collector.

The majority of flat-plate collectors have five major components:

i) A TRANSPARENT COVER: A transparent cover generally a Glass, is useful for protecting the absorber plate from atmosphere and other external foreign bodies. It also offers a good media to transmit solar radiation on to the absorber plate, so that maximum energy is received

H. P. Garg (ed.), Solar Water Heating Systems, 85–91.

by the plate. It also stops longwave radiation emitted by the absorber plate.

ii) TUBES, OR PASSAGES: They are to carry smoothly the heating fluid. They should be of materials to resist corrosive effect of heat carrying agents such as water, fluid or air.

iii) A COLLECTOR PLATE: This is an important part of the collector absorbing heat from solar insolation and converting the heat to the fluid. This absorber plate has to be very efficient in collecting heat and transferring this heat effectively to the fluid flowing in tubes.

iv) INSULATION: The major role of insulation is to stop loosing heat collected by the collector and thereby increase the efficiency of the collector.

v) A CASING OR CONTAINER: A casing has to be rugged, elegant and at the same time of reasonable cost. It has to be also air-tight to prevent heat from within the collector.

3. TYPES OF COLLECTOR PLATES

Some of the variety of liquid absorber designs are illustrated in Fig.1.

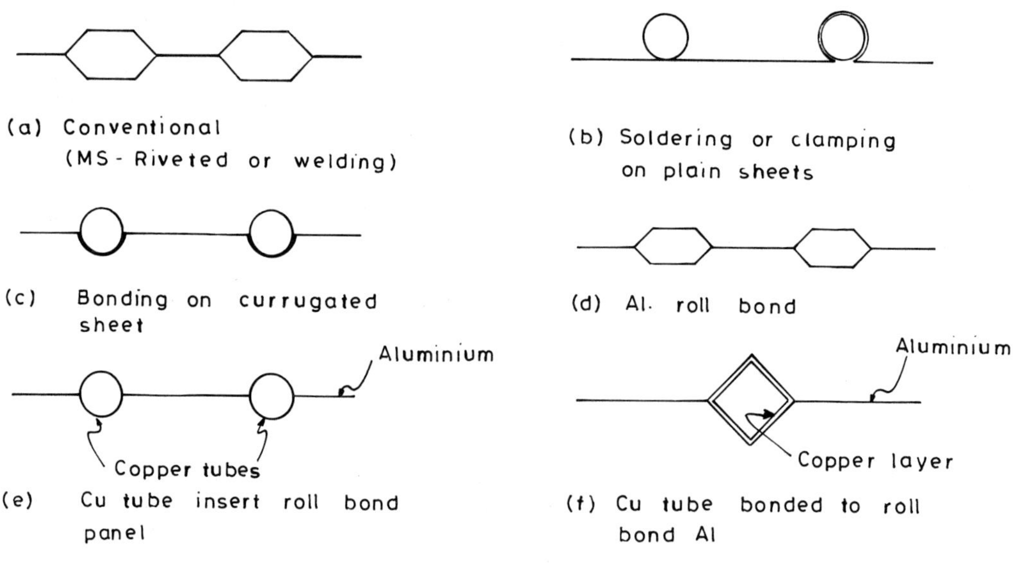

FIG.1 - SECTIONS OF VARIOUS SOLAR PANELS

Fig.1(a) indicates the conventional standard Mild Steel (MS) Radiators reveted at the joints. These are, however, not of longer life and also less efficient. Fig.1(b) and (c) indicate methods of bonding or clamping tubes either copper or GI to flat/corrugated sheets of copper, aluminium or GI. However, the tube-in-plate type absorber which is soldered bond type is expensive and bond may crack due to heating and cooling. There is considerable difficulty in soldering tubes to a thin sheet and the whole frame or plate may have to be discarded if there is a scale formation in the tube. Mechanically riveted or clamped tubes to sheet type absorbers are not efficient since the contact is poor. Copper fin, copper tube combination to be the best but expensive, due to scarcity of copper in many countries including India. Value Engineers suggested, aluminium for fin section, especially when aluminium is indigeneously available in abundance. However, copper was retained for passages.

All Aluminium Roll Bond Panels as shown in Fig.1(d) are the best type of collectors. They are cheap, light weight and most thermally efficient. However, they are very susceptable to corrosion due to mineral content in the water and have to be discarded within few months of their installation. Many experiments were made to overcome the corrosion problem, however no commercial viable solution is available and this type of absorbers have a very big setback in practice.

In Aluminium Roll Bond passages corrosion is the major impediment. Especially if fluid is of high pH values, or with dissolved heavy metals or chloride ions. Hard water and dissolved oxygen increase the risk of corrosion. Flow velocity below 0.9 m/sec causes increase in corrosion rate, while increase in flow rate increases erosion. Field reports indicate occurrence of severe pitting leading to pin holes in channel within 6-12 months with city water.

To avoid corrosion problem in all Aluminium Roll Bond Panels, in Western countries, people started using thin aluminium extruded sections with copper tube inserts. In India thin sheets were not available, so the trials were made successfully to insert copper tube in Roll Bond Panels of thickness 1.25 mm to 1 mm (Fig.1(e)). In short, copper tube insert Aluminium Roll Bond Panels have been developed as the extension of all Aluminium Roll Bond Panel. It is done by inserting copper tubes and mechanically bonding the same, to the Roll Bond Sheets, to offer water passages. The copper tubes offer a good corrosion resistance to the mineral content in the water. This has eliminated the corrosion problem completely.

Further extension of this technology, trials are being made in India to Roll Bonding Copper tubes to Aluminium sheets, during rolling operations only. This ensures a perfect metallurgical bond between copper tube and aluminium sheets (Fig.1(f)). This arrangement is further efficient and of longer life compared to copper tube insert Roll Bond Panel, where the contact between the copper tube and aluminium sheet is of mechanical nature.

For domestic applications, where high temperatures and long life are required, the pipe/fin tube absorber would appear to be the best choice: corrosion resistance is good and low thermal capacity is a definite advantage, giving fast response to changes in radiation intensity in variable cloudy climates, together with good mechanical strength.

4. SOLAR ABSORBER PLATE MATERIALS

The metals, copper, aluminium and steel, which have relatively high heat conductivities are usually used in a tube and fin configuration, since heat collected across the fin can readily flow to the tube and into the working fluid. For a fixed fin width and thermal conductance, the required thickness for Cu, Al and steel increased in the ratio 1:2:8 (Table 1) and the weight per unit area of collector approximately in the ratio of 1:0.6:7; thus the use of steel leads to considerable disadvantage with respect to thermal mass and support structure.

TABLE 1 - Physical Properties, Energy Contents and Comparative Fin Thickness for Equal Thermal Conductance

Material	Thermal conductivity W/°C m	Density gm/cc	Thickness	Mass per unit area Kg/m²	Energy content kWh/Kg
Copper	377	8.69	0.54*	4.69	12.8
Aluminium	206	2.70	1.00*	2.70	27.0
Steel	45	7.85	4.4*	34.40*	12.2

*Theoretical equivalent fin thickness for equal conductance.

The performance of various materials are plotted in Fig.2.

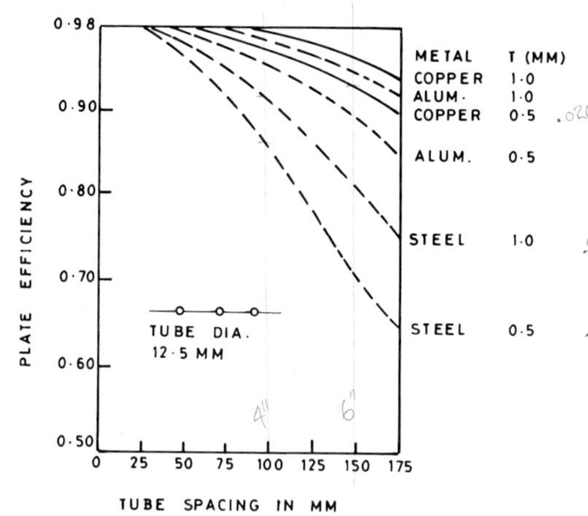

FIG.2 - TUBE SPACING VS. PLATE EFFICIENCY

Mention may be made about polythylene and EPDM materials. EPDM is a synthetic rubber which has several favourable properties for solar absorber. This material is being used in USA extensively. The main advantage of EPDM is age factor, good heat and oxygen resistance and very good low temperature properties. However, in India it is still imported and the life may be doubtful due to typical extreme climate of tropical nature. Again it is useful for low temperature applications such as swimming pools etc. only.

5. SOLAR COLLECTOR EFFICIENCY

Besides the absorber surface, absorber geometry like heat extracting channel system, sheet thickness, channel diameter is of considerable importance in determining the collector efficiency. Normally, tube sheet type absorbers are used. The cross sectional view is shown in Fig.3. The absorber geometry should be so chosen that the energy

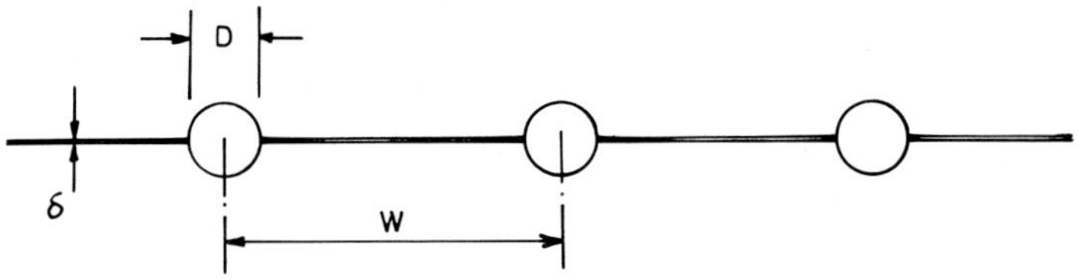

FIG.3 - CROSS SECTION OF TUBE SHEET ABSORBER PANEL

absorbed by the absorber is efficiently transferred to the heat extracting fluid flowing through the channels. The energy is absorbed by the absorber. This energy is conducted through the fin across the tube to the heat extracting fluid. The fin, the tube thickness etc. offer resistance to the flow of heat. The total resistance to the flow of heat depends upon (1) fin efficiency factor, (2) nature of bonding of sheet to the tube, (3) thickness of the tube wall and (4) the heat transfer factor from the tube to fluid. The collector efficiency factor F' can be expressed in terms of physical dimensions and characteristics of the absorber plate, Bond Conductance, heat loss coefficient, fin efficiency factor etc. as given below:

$$F' = \frac{1/U_L}{W\left\{\frac{1/1}{U_L\left[(W-D)F+D\right]} + \frac{1/1}{DK/\delta'} + \frac{1}{C_b} + \frac{1}{\pi Dh}\right\}}$$

F = Fin efficiency factor

U_L = Overall heat loss coefficient from the absorber

C_b = Bond conductance between the tube and the sheet

δ' = Wall thickness of the tube

h = Heat transfer coefficient from the inner surface of the tube to the fluid

k = Thermal conductivity of the absorber

Obviously, for maximisation of collector efficiency factor F', every factor in the above expression need to be optimized. It is instructive to look at the orders of magnitude of the various terms appearing in the denominator.

6. FIN EFFICIENCY FACTOR FOR ALUMINIUM ABSORBER

$U_L = 8\ w/m^2\,^oC$

Sheet thickness	(mm)	2	1.6	0.91	0.56
Fin efficiency factor	F	0.999	0.998	0.997	0.995

It is seen that the variation in F is insignificant for aluminium absorber for thickness varying between 0.56 mm and 2 mm. The corresponding theoretical range of sheet thickness of copper would be 0.3 mm to 1.1 mm for mild steel absorber 2.2 mm to 8 mm. In case of mild steel since the thickness of the absorber is necessarily large, the thermal capacity of the absorber becomes prohibitively large.

For a typical case of copper tube in copper sheet and the copper tube in aluminium, the variation of collector efficiency factor F' as a function of absorber sheet thickness is tried. It was observed that for lower thickness, there is an appreciable difference between the collector efficiency factor for copper tube, copper sheet absorber and copper tube aluminium sheet absorber. As the thickness of the absorber increases, the difference in the collector efficiency factor decreases. It is, therefore, necessary to use 0.10 - 0.20 cm thick aluminium absorber sheet. This is highly cost effective.

7. BOND CONDUCTANCE

The bond conductance term becomes important if the bonding between the absorber sheet and tube is poor. For example, the bond conductance value of pressure or clamped wire bonding is about 5 $W/m^2\,^oC$ as reported in the literature. This increases the $1/C_b$, term by an order of magnitube and reduce the value of F'. For a typical case, if C_b is increased from 3.4 to 35 $W/m^2\,^oC$, F' value increases from 0.6 to 0.78, i.e. by about 30%. However, if C_b = 35 $W/m^2\,^oC$, the effect of bond conductance is negligible. The bond conductance of soldered bonding is given by,

$$C_b = K_b \frac{b}{r}$$

where b is the bond length, r is the thickness of the bond and K_b is the thermal conductivity of the bond. Typically if b = 3 mm, r = 1 mm and K_b = 20 $W/m^2\,^oC$, then C_b = 60 $W/m^2\,^oC$. This is possible by making a proper soldered bond.

In case of aluminium absorber, roll bond process can be effectively used. This process involves pressure rolling of two aluminium sheets leaving channels are blown by using compressed air or water. However, aluminium corrodes quickly if water quality is not suitable. Consequently pure roll bond aluminium collectors are not recommended. One can fix copper tubes mechanically in roll bond aluminium absorbers. This is done by manufacturing first all aluminium roll bond panels with straight passages and then insert copper tubes in the passages. The panel is then passed through the grooved rollers where aluminium passages are literally squeezed on to the copper tubes to have a perfect mechanical bond.

Since copper tube is completely surrounded by aluminium metal surface extended as a fin and with perfect mechanical bond, the contact is so positive that the heat transfer characteristic is considerably high and so also the bond conductance. This is one of the major advantages of this type of plate compared to other conventional ones. Hence the heat transfer efficiency is higher.

A better way is when a copper tube is metallurgically bonded to aluminium sheets, which is being developed in India. There is a relationship between the fin width and fin thickness for a given channel opening.

When water is forced through the channels, there will be a certain pressure drop. It is necessary to know the pressure drop for a given flow rate. It is not necessary to know the bursting pressure of the channel. The copper tube inserted in roll bond panels can easily stand the pressure.

8. CONCLUSION

Design optimization was made of copper tube insert roll bond panels.

Various experiments are carried out for copper tube diameters and tube spacing that is the fin width by comparing the performance of these collectors simultaneously under the same condition of insolation, wind speed and identical collector parameters.

Keeping in view the economy and all above factors the optimum parameters especially from economy and efficiency points of view, the plate geometry most suitable is as under:

Absorber material	-	Aluminium roll bond type.
Roll bond fin thickness	-	1 mm
Water passages	-	Copper tubes 0.5 to 0.7 mm wall thickness. 9 Nos of 10 mm OD/meter width. 6 Nos of 12 mm/meter width.
Fin width per tube	-	150 mm to 165 mm for 1/2" Ø copper tubes. 105 mm for 3/8" Ø copper tubes.

ABSORBER PLATE CONFIGURATION AND OPTIMIZATION

H.P. Garg
Centre of Energy Studies
Indian Institute of Technology
Hauz Khas, New Delhi 110016
India

1. INTRODUCTION

The most important component of a Solar Water Heater is the Solar Energy Collector whose main function is to absorb radiant energy from the sun as much as possible and convert it into heat and transfer it to the fluid flowing in the channel which is welded on the absorber plate. The most important step to utilize solar energy in an effective way for heating of water is to develop an efficient low cost flat plate solar collector. As discussed earlier a solar energy flat plate collector has five components: absorber plate, glazing, insulation, tubes or channel in the absorbing plate, and tray. There are several advantages of using flat-plate collectors for heating water such as (a) it absorbs direct as well as diffuse radiation, (b) simple to construct, (c) no need of sun following arrangement, (d) low in cost, and (e) the collector can form an integral part of the roof. Since the most expensive part of a solar water heater is the flat plate collector, while designing a collector one should consider the following three factors:

1. Thermal performance
2. Cost of construction
3. Life time and reliability

A highly thermally efficient collector if costs high and does not last long will supply thermal energy at a very high cost and therefore the system will not be cost effective.

2. TYPE OF ABSORBING PLATE

A typical Flat Plate Collector which can be used for the supply of hot water up to 80°C is schematically shown in Fig.1. There can be a large variety of absorber plates for liquid heating as shown in Fig.2. The absorbing plate should have good heat transfer properties, high thermal conductivity, high solar energy absorbence, low thermal emittance, and non-corrosive and stable at high temperature. Although there is a large variety of absorbing plate but these can be broadly divided into three basic types as shown in Fig.3. These three basic types of absorbing

H. P. Garg (ed.), Solar Water Heating Systems, 93–111.

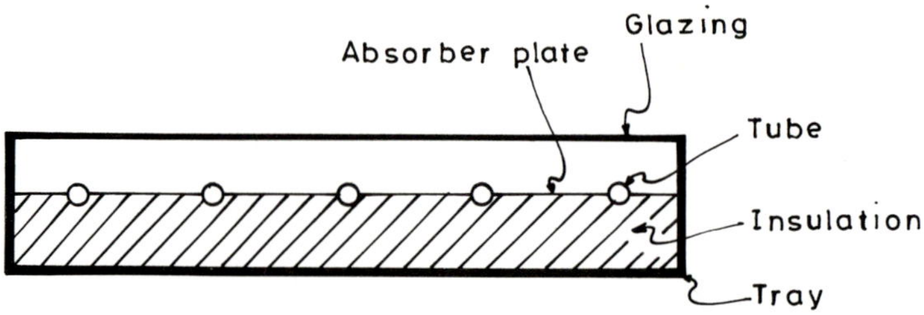

FIG.1 - TYPICAL FLAT PLATE COLLECTOR

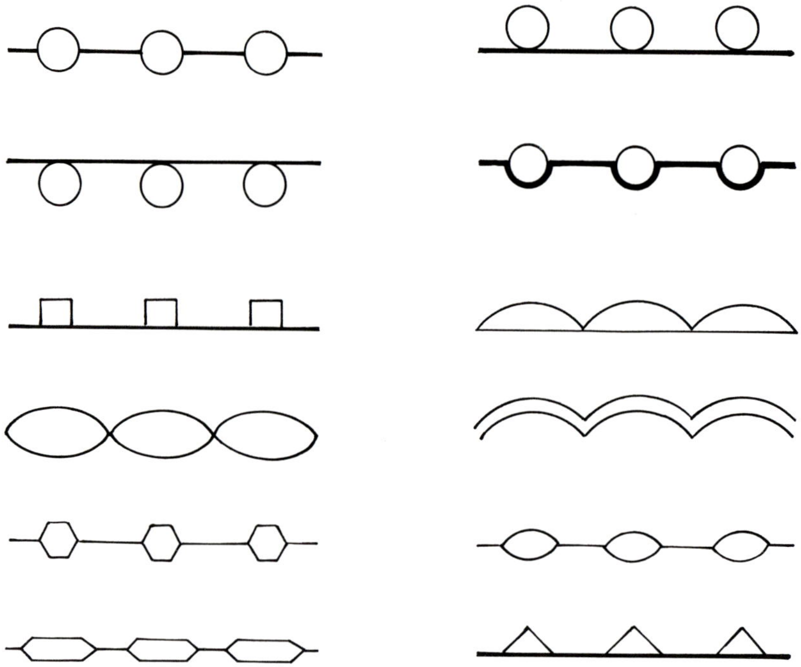

FIG.2 - ABSORBER PLATE CONFIGURATION

plates are:

1. Pipe and Fin type
2. Full Water Sandwitch type
3. Semi Water Sandwitch type

Each of these three basic collector type has got some advantages and disadvantages as discussed in Table 1. It is seen from Table 1 that pipe and fin type of collector using composite materials has got advantages over the other pipe and fin type using same material for pipe and fin and other semi sandwitch and full sandwitch type collectors. There are several mechanisms of bonding the tubes to the absorber plate. A few typical bonding mechanisms are shown in Fig.4. In the conventional

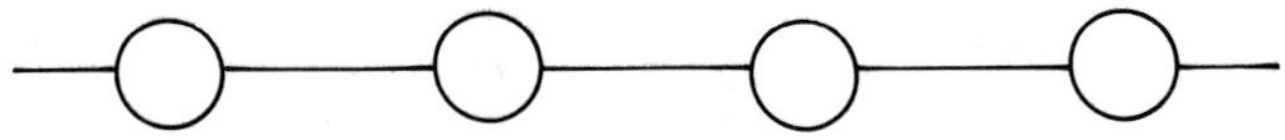

(a) Pipe and fin type absorber plate

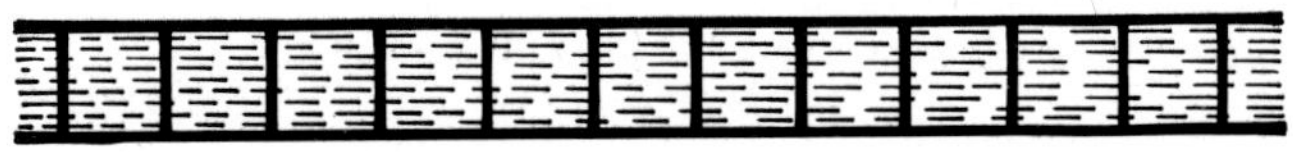

(b) Full sandwich type absorber plate

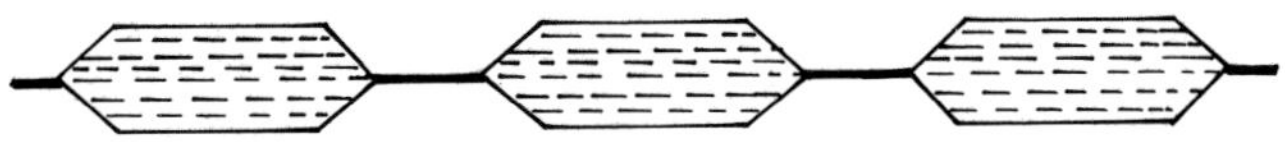

(c) Semi - sandwich type absorber plate

FIG.3 - BASIC COLLECTOR TYPES

TABLE 1 - Advantages and Disadvantages of Absorber Plate Types

Absorber type	Advantages	Disadvantages
1. Water sandwitch	Cheap, easy to produce, light, high thermal performance	Bulging under pressure, low life, unreliable, high thermal capacity
2. Semi-water sandwitch (roll-bond aluminium)	Light weight, cheap, fairly high thermal performance	Highly susceptible to internal corrosion
3. Pipe and fin (all copper)	Low thermal capacity, good corrosion resistance	Very expensive
4. Pipe and fin (composite materials)	Cheap, light weight, good corrosion resistance, low thermal capacity and flexibility in the choice of materials	Special bonding techniques required

integral construction where the absorber plate forms the tube wall as shown in Fig.4(a), the tube wall thickness remains half the thickness of the absorbing plate. Therefore, the tube does not withstand high pressure and resulting thick absorbing plate thereby increasing the cost with a little gain in efficiency. The flow tubes can be attached with a mechanical joint to the absorbing plate in many ways such as: tying tube to the absorbing plate with the help of binding wire, absorbing plate spring clamped to the tubes, tack welding of tubes to the absorbing plate at certain spaces, tying the tubes to the grooved absorbing plate with

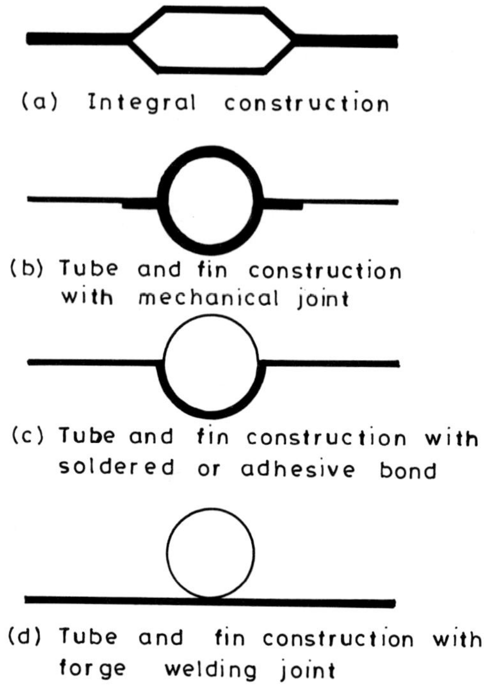

FIG.4 - DIFFERENT BONDING MECHANISMS

hooks, sandwitching the tube between two grooved metal sheets and then revetting the two sheets together. In all these mechanical joining techniques (Fig.4(b)) care is taken so that the tubes remain in good thermal contact with the absorbing plate. No doubt such mechanical joints are simple in construction but provides poor heat transfer between absorbing plate and tube. To overcome this problem the contact area is made large and joint is made uniformly tight. But due to age effect or corrosion, generally a gap is developed resulting in poor bond between the absorbing plate and the tube.

Generally soft-soldering is done to make a good bond between the absorbing plate and tubes (Fig.4(c)). Soft soldering technique is an excellent bonding technique but suffers from disadvantages such as: It is expensive technique due to soldering of large number of tubes on absorbing plate, and due to thermal stresses the bond may crack resulting in poor bond. A similar problem is experienced where an adhesive is used to provide thermal bond between absorbing plate and tube. In such cases due to large thermal cycling (heating and cooling) and ageing the bonding agent can deteriorate resulting in poor heat transfer.

Recently, methods have been developed to join copper and aluminium metallurgically. In this process copper tubes are inserted in channels of roll-bond aluminium plate and are jointed together by pressure-temperature techniques. In this configuration also tube and sheet of sufficient thicknesses are required making the absorbing plate costly.

With the invention of forge welding or high frequency welding any tube material can be welded to any plate material irrespective of tube and plate thicknesses. In this case no third material is used in welding and therefore the heat transfer becomes 100 percent even the width of the joint is only twice the thickness of the absorbing plate. This type of bonding which is recently introduced has high thermal performance and long life compared to all other types of bondings.

As seen earlier, the pipe and fin type of flat plate collectors can be used for a variety of solar energy applications due to their long life, low maintenance and capital cost, good corrosion resistance, and withstanding high pressure. Therefore, here, we are concerned with the design and construction aspects of the pipe and fin, which greatly influence the performance (efficiency) and cost of collector.

3. SYSTEM OPTIMIZATION

The flat plate collector functions as a heat exchanger between a radiant source and the fluid. Garg and Gupta (Ref.1) earlier tried to optimize the system, considering the plate efficiency factor which is a design constant and can be determined by the construction details of the collector, and does not depend greatly upon operating conditions such as temperature, sunshine intensity or wind velocity etc. (Ref.2).

In this direction work has also been done by Mcgregor (Ref.3), Williamson (Ref.4) and Holland and Barjoni (Ref.5) for optimizing the collector configuration. They have stressed on the tube spacing and plate thickness with their conductivity values. But here the author has tried to see the effect of various design parameters, and flow conditions on various duct geometries. Increasing the efficiency is not the main goal of designer, but the cost reduction, which is a great obstacle to the more widespread use of solar collector is also one of the important criteria. The system should be with minimum cost per unit efficiency.

Here in this paper a most general theoretical investigation is carried out for optimizing the collector cost/effectiveness, by considering various tube diameters, its materials and tube spacings, fin materials and its thicknesses for black as well as selective paint.

Bond conductance for various joining techniques is taken into account. Various tube geometries are considered and correspondingly the convective heat transfer coefficients are taken. Efficiency and cost for all possible combinations is computed and by keeping all the other parameters into consideration the optimal system is proposed.

3.1 Factors Influencing Collector Thermal Efficiency

3.1.1 Plate efficiency factor. The rate of useful heat collection is given by the rate of incident solar radiation minus outward heat losses:

$$q_u = q_a - U_L (T_p - T_a) \tag{1}$$

The thermal efficiency factor, F_R, as given by Bliss (Ref.2) is

$$F_R = \frac{\text{Actual heat collection rate}}{\text{Heat collected if collector were at incoming fluid temperature}}$$

AND THE PLATE EFFICIENCY FACTOR, F_P:

$$F_P = \frac{\text{Actual heat collector rate}}{\text{Heat collected if collector were at average fluid temperature}}$$

Using the above definitions, equation (1) can be written in the form:

$$q_u = F_R \left[q_a - U_L (T_1 - T_a) \right] \tag{2}$$

The expression for F_R in terms of F_P, U_L and flow rate per unit collector area was derived by Bliss (Ref.2) and is given by

$$F_R = F_P \left[\frac{GC_p}{F_P U_L} \left\{ 1 - e^{-(F_P U_L / GC_p)} \right\} \right] \tag{3}$$

where

$$F_p = \frac{1}{\dfrac{WU_L}{\pi d h_c} + \dfrac{WU_L m_t}{\pi d K_t} + \dfrac{WU_L}{C_b} + \dfrac{WU_L}{U_L (b + F'(W-b))}} \tag{4}$$

and

$$F' = \frac{\tanh a(w-b)/2}{a(w-b)/2} \tag{5}$$

and

$$a^2 = \frac{U_L}{k_p m_p} \tag{6}$$

Here F' is the fin efficiency of the collector which is of parabolic nature in temperature.

The right hand denominator of equation (4) is the heat transfer resistance from the plate to the fluid, defined as:

(i) $\frac{WU_L}{\pi d h_c}$ - Transfer of heat from inner surface of the tube to the water.

(ii) $\frac{WU_L m_t}{\pi d k_t}$ - Conduction of heat from outside to inside of tube through tube wall.

(iii) $\frac{WU_L}{c_b}$ - Conduction of heat from the fin to the tube through the tube fin bond.

(iv) $\frac{WU_L}{U_L\left[b+F'(w-b)\right]}$ - Conduction of heat along the fin towards the pipe.

Thus from all above description we conclude that the tube spacing, its diameter, its material; fin material and its thickness; heat transfer coefficient; bond conductance; heat loss coefficient are all directly related to the system performance. Therefore the aim of the designer should be the best cost effectiveness, a function of efficiency and cost. The main scope for reducing the cost lies in selecting the optimum combination of pipe spacing and fin thickness for a particular material of pipe and fin. Material cost will be reduced by increasing the spacing between pipes and by making the plate thinner. However, this leads to a reduction in fin efficiency, plate efficiency factor and overall system performance. Therefore the aim should be to determine the combination of pipe spacing and plate thickness, which will minimize the ratio of cost to useful energy collected by the system.

3.1.2 Cost of the collector. The total cost of the system can be expressed as (Ref.5):

$$C = V + R$$

where C = total cost of the system per unit area
V = cost of absorber plate (dependent on W and m_t) per unit area
R = remaining system cost (independent of W and m_t) per unit area
$V = P/W + Qm_t$

where P = cost/length of the pipes and associated fittings and fixtures, together with the cost of assembly and attachment.
Q = cost/volume of the plate material

Thus $$C = P/W + Q\,m_t + R \qquad (7)$$

Therefore the aim here is to minimize the ratio C/F_P, i.e. the cost/efficiency.

3.2 Collector System

The collector system considered here is of the pipe and fin type (Fig.5),

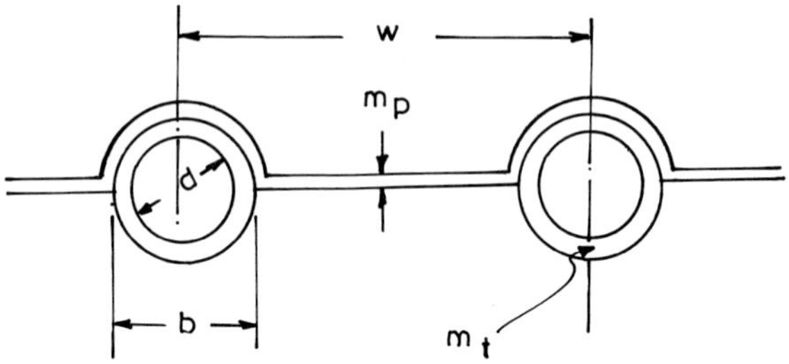

FIG.5 - COLLECTOR CONFIGURATION

which is supposed to be the best choice for domestic as well as industrial water heating requirements. The possible materials of the fin (K_p) may be copper, aluminium, steel or galvanized iron of thicknesses (m_p) 0.091 cm, 0.071 cm, 0.056 cm, 0.046 cm and 0.038 cm. Similarly the pipe may be of copper, aluminium, steel or galvanized iron of inner diameter (d) 1.27 cm, 1.91 cm and 2.54 cm, spaced (W) at 2.5, 5.0, 7.5, 10.0, 12.5, 15.0, 17.5, or 20.0 cm. The bond conductance is taken as 10, 20, 30, 40 ($W/m^{o}c$) as suggested by Whillier (Ref.6) and Garg and Usha (Ref.7). The various clamping arrangements are considered and the corresponding measured values of the bond conductance are used (Table 2). The overall heat-loss coefficient is calculated by following Klein (Ref.8) and as discussed by Garg and Datta (Ref.9) for dull black as well as selective paint with 0, 1 and 2 glass covers under similar operating conditions (Table 3). The analysis given here is based on earlier publications.

3.3 Film Heat Transfer Coefficient

The system is generalized to a great extent. Other than the circular tube which is generally used, triangular, square, rectangular, hexagonal and oval shaped ducts are considered and, by taking corresponding equivalent hydraulic diameter, the film heat transfer coefficient from the duct inner surface to the fluid has been evaluated analytically. For a fully developed laminar flow in a solar collector, Oliver (Ref.12) has given a correlation which takes into account non-uniformity of the surface wall temperature. The non-uniformity occurs since the heat collected by the fin is passed into the tube at the sides where the tube-wall temperature remains higher, while it decreases towards the top and bottom of the tube. Using Oliver's correlation, Baker (Ref.13) has shown that the heat transfer coefficient for a circular tube of 12 mm diameter is 564 $W/m^{2}\,^{o}C$.

The underlying case refers to natural circulation, where the mass flow rate is 40 $Kg/hr.m^2$, the average velocity is 0.0897 m/sec. Because of small heat flux, the flow rate in flat-plate collectors is usually low and is laminar. For laminar flow in tubes or ducts, the following empirical relations given by Sieder and Tate (Ref.14) and modified by

TABLE 2 - Value of Bond-Conductance for Different Types of Bonds Tested

Pipe	Fin	Bond Conductance W/m°C
G.I. Pipe	Wrapped and clamped aluminium fin	11.65
G.I. Pipe	Wrapped and revetted aluminium fin	15.15
G.I. Pipe	Self clamped fin as used in M.I.T.Solar House	11.65
G.I. Pipe	Wrapped and wired aluminium fin	25.60
G.I. Pipe	Wrapped and wired aluminium fin with thin aluminium foil in the interspace	31.46
G.I. Pipe	Wrapped and wired aluminium fin with heat conducting epoxy	39.00
G.I. Pipe	Wrapped and wired aluminium fin with butamin mixed with copper dust in the interspace	35.00

TABLE 3 - Effect of the Number of Glazings Used and of the Emissivity on the Overall Heat Loss Coefficient U_L;

$\alpha = .0.95$, $s = 45^o$, $w = 5.5556$ m/sec, $H = 900$ W/m^2,

$K_r = 0.0379$ W/m^{o}C, $T_p = 70^o$C, $T_a = 40^o$C

Emissivity of absorber plate / Number of glass covers used	Overall Heat Loss Coefficients (W/m^2°C)							
	0.05	0.1	0.2	0.4	0.6	0.8	0.9	0.95
n = 0	28.26	28.71	29.73	32.40	36.52	43.65	49.61	53.72
n = 1	4.463	4.812	5.477	7.757	8.412	8.712	9.153	9.364
n = 2	3.163	3.395	3.798	3.401	4.843	5.178	5.317	5.381

Kreith and Kreider (Ref.15) and by Shah and London (Ref.16) are used to determine h_c;

$$Nu_{DH} = \frac{h_c DH}{K_b} = 1.86 \left[\frac{Re_{DH}.Pr.DH}{L} \right]^{1/3} . \left(\frac{\mu_b}{\mu_w}\right)^{0.14} \tag{8}$$

When $Re.Pr.DH/L > 10$

and

$$Nu_{DH} = \frac{h_c DH}{K_f} = \frac{Re_{DH} Pr.DH}{4L} \ln \left[\frac{1}{1 - \dfrac{2.654}{Pr^{0.167} \left(\dfrac{Re_{DH}.Pr.DH}{L}\right)^{0.5}}} \right]$$

When $Re.Pr.DH/L > 100$ (9)

The working fluid is taken as water. The film heat transfer coefficient, h_c, for all six duct shapes and for 12 mm diameter is calculated by using equations (8) and (9). The results are given in Table 4.

TABLE 4 - Hydraulic Diameter, Reynold Number, Nusselt Number, Film Heat-Transfer Coefficient, Friction Factor and Pressure Drop for Various Duct Shapes.

Duct shape	Shape	Dimensions (m)	Hydraulic diameter (m) (DH)	Reynold number (Re_{DH})	Nusselt number (Nu_{DH})	Film heat transfer coefficient ($W/m^2\,^oC$)	Frictional loss (f)	Pressure drop (Pascals)
Triangular		x=0.012 y=0.0157 z=0.0155	0.0091	1965	6.83	443.57	0.0094	15.93
Square		x=0.01	0.01	1863	7.26	429.61	0.0086	13.21
Circular	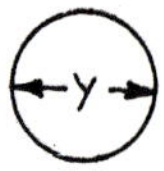	x=0.0127	0.0127	2365	8.51	396.04	0.0065	7.86
Hexagonal	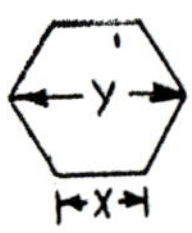	x=0.0054 y=0.0108	0.0107	1993	7.60	419.80	0.0080	11.53
Oval	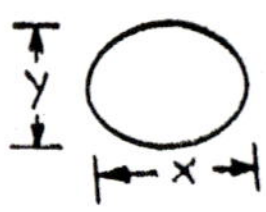	x=0.0064 y=0.005	0.0111	2067	7.78	414.57	0.0077	10.71
Rectangular	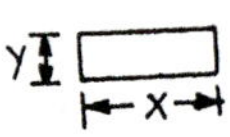	x=0.0071 y=0.0141	0.0097	1807	7.12	434.04	0.0089	14.10

The hydraulic diameters, Reynold number and Nusselt numbers are also given. It is assumed that the calculated values of h_c for 12 mm diameter for a particular duct are the same as for the 19 and 25 mm diameters. The skin-friction factor, f, is given by

$$f = 16/Re_{DH} \tag{10}$$

and pressure drop according to Eckert and Drake (Ref.17) is

$$\Delta P = 4f \cdot \frac{L}{DH} \cdot \frac{\bar{V}^2}{2g_c} \cdot \rho \tag{11}$$

The calculated values of f and ΔP for various duct shapes are shown in Table 4.

4. RESULT AND DISCUSSION

The efficiencies F', F_p, F_R, the cost C and the ratio of cost to efficiency, C/F_p, were computed for different set of parameters. The numerical values of various parameters are as follows:

U_L = 4.8 W/m^2°C, W = 10 cm, m_p = 0.55 mm,
d = 0.0127 m, m_t = 1 mm, G = 40 Kg/hr.m^2, C_b = 20 W/m°C,
h_c = 396 W/m^2°C (circular tube), K_t = 63 W/m°C (G.I. tube),
K_p = 201 W/m°C (aluminium plate), P = 10 Rs/m,
C_p = 4190 J/Kg°C, Q = 81300 Rs/m^3 and R = 250 Rs/m^2.

In Table 5, comparison has been made of the cost of the collector plate per unit area for transferring of the same amount of heat with copper sheet taken as the base. From the Table it is clearly seen that aluminium sheet is the best material to be used for the fin since it minimizes the cost at high performance. The second best material is galvanized iron.

TABLE 5 - Effect of Collector Plate Types

Material	Thermal conductivity (W/m°C)	Equivalent thickness (mm)*	Cost per unit area (Rs/m^2)
Copper	385	0.20	100
Aluminium	201	0.36	24
Steel	150	0.50	250
G.I. Sheet	63	1.20 = .047"	55

*A 0.20 mm copper is used as reference material.

The measured values of bond conductance for different clamping arrangements for the galvanized pipe are given in Table 2. It is clear from this table that, by using a heat-conducting epoxy resin, the value of bond-conductance can be increased. The effect of the number of glazings (glass sheets) and collector plate emissivity on the overall heat-loss coefficient is shown in Table 3. Without glass cover, the collector heat-loss coefficient becomes very high and decreases with the emissivity of the plate. For a normal flat-plate collector with a black board paint (dull black) and a single glass cover, the overall heat-loss coefficient may be taken as 9-10 W/m^2°C. From Table 4, where the values of film heat transfer coefficient and pressure drop for different pipe geometries are given, it appears that triangular pipe transfers the maximum amount of heat, but at the same time, it also produces the maximum pressure drop. Table 6 shows the effect of the duct diameter on the thermal efficiencies F', F_p and F_R. From this Table, it is clear that the tube diameter has only a slight effect on

TABLE 6 - Effect of Duct Hydraulic Diameter on Thermal Efficiencies

d	p	F'	F_p	F_R	C	C/F_p
0.0127	10	0.9745	0.9287	0.9282	394.71	425.04
0.0191	15	0.9781	0.9417	0.9407	444.71	472.26
0.0254	20	0.9814	0,9497	0,9490	494.71	520.94

F', F_p and F_R but the cost-to-efficiency ratio increases as the tube diameter increases. The effect of tube spacing on thermal efficiency factors is shown in Fig.6. It is seen that the efficiency decreases with the increase in spacing. The minimum value of C/F_p is obtained when the spacing is about 15 cm. The conductivity of the tube also plays an important role and its effect is shown in Table 7. It is clearly seen that the galvanized iron tube gives the minimum value of C/F_p.

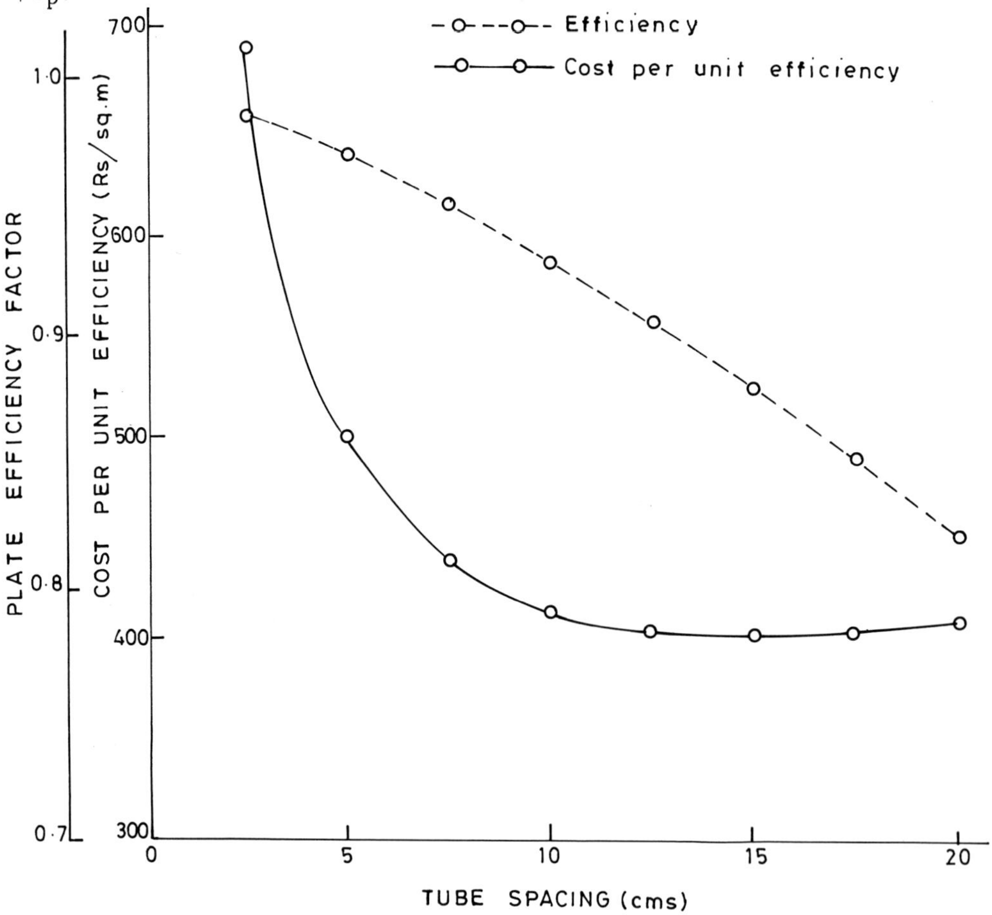

FIG.6 - EFFECT OF TUBE SPACING ON COLLECTOR PERFORMANCE

TABLE 7 - Effect of Duct Material on Thermal Efficiencies

k_t	P	F'	F_p	F_R	C	C/F_p
Copper	35	0.9745	0.92865	0.9282	644.71	694.14
Aluminium	15	0.9745	0.92875	0.9282	444.71	478.82
Steel	30	0.9745	0.92875	0.9282	594.71	640.34
G.I.	10	0.9745	0.92879	0.9282	394.71	425.04

A decrease in absorber thickness appears to decrease F', F_p and F_R by a very small amount, as is shown in Table 8. The plate material

TABLE 8 - Effect of Plate Thickness on Thermal Efficiencies

m_p	F'	F_p	F_R	C	C/F_p
0.0009	0.9842	0.9361	0.9324	405.17	432.81
0.0007	0.0700	0.9328	0.9304	392.91	421.24
0.00055	0.9745	0.9287	0.9282	383.71	413.19
0.00046	0.9696	0.9249	0.9241	378.20	408.88
0.00038	0.9636	0.9203	0.9198	372.29	405.64

plays an important role since it transfers the heat absorbed by the plate to the tube. This effect is shown in Table 9. The most cost

TABLE 9 - Effect of Plate Material on Thermal Efficiencies

K_p	F'	F_p	F_R	C	C/F_p
Copper	0.9865	0.9379	0.9365	839.49	895.11
Aluminium	0.9745	0.9287	0.9282	394.71	425/03
Steel	0.9662	0.9222	0.9199	698.22	757.83
G.I.	0.9237	0.8895	0.8866	384.58	452.35

effective collector plate is seen to be made of aluminium. The film heat-transfer coefficient, which is responsible for the transfer of heat from inner surface of the tube to the fluid, has little effect on the thermal efficiencies, as is shown in Table 10. This result may be because of small tube diameter or small temperature difference. The overall heat-loss coefficient affects the performance of the collector considerably, as is shown in Table 11. The thermal efficiency factors F', F_p and F_R decrease with increase of overall heat-loss coefficient. The thermal efficiency factors also increase with increase of bond conductance, but the effect is not very large (Table 12). Table 13 shows 1-2 percent increase in F_R for a fourfold increase in flow rate.

TABLE 10 - Effect of Film Heat-Transfer Coefficient on Thermal Efficiencies

h_c	F'	F_p	F_R	C	C/F_p
396.0	0.9745	0.9207	0.9282	383.71	413.19
414.6	0.9745	0.9298	0.9282	383.71	413.19
419.8	0.9745	0.9301	0.9282	383.71	412.53
429.6	0.9745	0.9307	0.9282	383.71	412.28
434.0	0.9745	0.9309	0.9282	383.71	412.17
443.6	0.9745	0.9314	0.9282	383.71	411.94

TABLE 11 - Effect of Overall Heat-Loss Coefficient on Thermal Efficiencies

U_L	F'	F_p	F_R	C	C/F_p
3.4	0.9818	0.9484	0.9461	394.71	416.21
4.8	0.9761	0.9331	0.9321	394.41	423.02
5.4	0.9714	0.9205	0.9176	394.71	428.82
9.4	0.9514	0.8695	0.8683	394.71	453.95

TABLE 12 - Effect of Bond Conductance on Thermal Efficiencies

C_p	F'	F_p	F_R	C	C/F_p
10	0.9745	0.9084	0.9074	383.71	422.40
20	0.9745	0.9287	0.9282	383.71	413.19
30	0.9745	0.9356	0.9324	383.71	410.12
40	0.9745	0.9391	0.9365	383.71	408.59

Table 13 shows 1-2 percent increase in F_R for a fourfold increase in flow rate.

The effect of the duct or tube geometry and of the spacing on the plate efficiency factor is shown in Table 14. From this Table, it is seen that, for a particular spacing, the tube geometry has a negligibly small effect on the plate efficiency factor, while, for a particular geometry, the plate-efficiency factor decreases with an increase of spacing.

TABLE 13 - Effect of Flow Factor on Thermal Efficiencies

Flow rate G(Kg/hr)	20	40	60	80
Film heat transfer Coefficient ($W/m^2 {}^oC$)	315.71	396.00	451.61	498.47
F'	0.9745	0.9745	0.9745	0.9745
F_p	0.9220	0.9287	0.9319	0.9341
F_R	0.9199	0.9282	0.9282	0.9323
C	350.01	350.02	350.03	330.04
C/F_p	379.61	376.91	375.61	374.76

5. OPTIMUM COLLECTOR CONFIGURATION

The optimum collector configuration for all the six geometries considered is shown in Table 15. The minimum value of C/F_p, for each geometry, is obtained, for the same set of parameters. It is clearly seen that, for all tube geometries, the value of C/F_p is only slightly different. But, since the pressure drop (which is also of great importance) varies greatly and is minimum for the circular case and maximum for the triangular case for the same mass flow rate, it is likely that the circular pipe configuration is best.

TABLE 15 - OPTIMUM COLLECTOR CONFIGURATION

$U_L = 4.8\ W/m^2 {}^oC$, $d = 0.0125$ m, $k_t = 63\ W/m^oC$,
$C_b = 40\ W/m^oC$, $m_p = 0.00038$ m, $k_p = 201\ W/m^oC$,
$W = 0.1500$ m, $p = 10$ Rs/m, $Q = 81300\ Rs/m^3$

Shape	h ($W/m^2 {}^oC$)	C/F_p	Friction factor (f)	Pressure drop (Pascals)
Circular	396.00	398.98	0.0068	7.86
Oval	414.60	398.27	0.0077	10.72
Hexagonal	419.80	398.08	0.0080	11.53
Square	429.60	397.74	0.0086	13.22
Rectangular	434.00	397.59	0.0089	14.10
Triangular	443.60	397.28	0.0094	15.93

For the optimum case
Plate material (K_p) = Aluminium; Plate thickness (mp) = 28 gauge;
Pipe diameter (d) = 1/2"; Pipe spacing (W) = 15 cm
Pipe material (K_t) = Galavanized iron.

TABLE 14 - Effect of Various Duct Shapes and Duct Spacing on Thermal Proficiencies

Duct spacing W	Duct shape h_c	Circular 396	Oval 414.6	Horizontal 419.8	Square 429.6	Rectangular 434	Triangular 443
	F_p	0.9863	0.9867	0.9868	0.9869	0.9870	0.9871
	C/F_p	704.30	704.07	704.00	703.89	703.84	703.73
0.050	F_p	0.9704	0.9710	0.9713	0.9716	0.9717	0.9720
	C/F_p	509.79	509.45	509.36	509.20	509.13	508.98
0.075	F_p	0.9511	0.9520	0.9520	0.9527	0.9529	0.9533
	C/F_p	450.07	449.63	449.52	449.31	449.22	449.03
0.100	F_p	0.9287	0.9298	0.9301	0.9307	0.9310	0.9315
	C/F_p	425.04	424.50	424.36	424.10	423.99	423.75
0.125	F_p	0.9037	0.9051	0.9055	0.9062	0.9065	0.9071
	C/F_p	414.63	413.99	413.82	413.62	413.39	413.10
0.150	F_p	0.8768	0.8784	0.8788	0.8788	0.8799	0.8806
	C/F_p	412.14	411.40	411.20	410.85	410.70	410.37
0.175	F_p	0.8485	0.8503	0.8507	0.8516	0.8519	0.8527
	C/F_p	414.65	413.82	413.59	413.20	413.02	412.65
0.200	F_p	0.8193	0.8212	0.8217	0.8225	0.8229	0.8237
	C/F_p	420.72	419.78	419.54	419.08	418.89	418.48

NOMENCLATURE

b = outer diameter of the tube (m)
C_p = specific heat of the fluid (J/Kg°C)
C_b = bond conductance (W/m°C)
d = inner diameter of the tube (m)
DH = hydraulic diameter (m)
F' = fin efficiency factor
F_p = plate efficiency factor
F_R = heat removal efficiency factor
f = friction factor
G = mass flow rate (Kg/hr.m^2)
g_c = proportionality constant (= 1 Kg-m/N.sec.^{2})
h_c = film heat transfer coefficient (W/m^2°C)
H = solar intensity (W/m^2)
K_b = fluid thermal conductivity at the bulk temperature (W/m°C)
K_f = fluid thermal conductivity (W/m°C)
K_p = thermal conductivity of absorber plate (W/m°C)
K_t = thermal conductivity of tube (W/m°C)
L = length of the pipe or duct (m)
m_p = thickness of the absorber plate (m)
m_t = thickness of tube (m)
n = number of glass covers
Nu_{DH} = Nusselt number for a particular hydraulic diameter
ΔP = pressure drop (Pascals)
Pr = $\frac{\mu c_p}{K}$ = Prandtl number
Q_u = total useful energy gain from the collector (W/m^2)
Re_{DH} = $\frac{\rho \bar{V} D_H}{\mu}$ = Reynold number
s = tilt of the collector (degrees)
T_1 = inlet water temperature (°C)
T_a = ambient temperature (°C)
T_p = temperature of the absorber plate (°C)
T_w = average temperature of fluid (°C)
U_L = overall-heat loss coefficient (W/m^2°C)
$\bar{V}$ = average velocity of the fluid (m/sec)

W = tube spacing (m)

w = wind velocity (m/sec)

ρ = density of the fluid (Kg/m^3)

μ = viscosity of the fluid (Kg/m-sec)

μ_b = viscosity of the fluid at the bulk temperature half way between the inlet and outlet (Kg/m-sec)

μ_w = viscosity of the fluid at the wall temperature (Kg/m-sec)

α = absorptivity of plate

τ = transmittance of glass cover

REFERENCES

1. H.P. Garg and C.L. Gupta, (1968), 'Design data for direct solar utilization devices - 1: System data', Journal of the Institute of Engineers (India) XLVIII(9), Pt.GE, 3(11).
2. R.W., Bliss, (1959), 'The derivation of several plate efficiency factors, useful in the design of flat plate solar heat collectors', Solar Energy, 3(4), 55.
3. A.N.K. MacGregor, (1978), 'Economic use of materials in the design of solar water heating collector plates of the pipe and fin type', International Solar Energy Congress, New Delhi (India), Jan.16-21.
4. A. Williamson, (1976), 'Solar energy - Thermal resistance of pipe-fin junctions', Final year project report of Higher National Diploma in Engineering, Napier College of Commerce and Technology,Edinburgh, Scotland.
5. T.H. Holland and J.T. BaRGONI, (1975), 'Design, fabrication, testing and delivery of a solar energy collector system for residential heating and cooling', Project Report, Contract No.NASA-31327 by Horneywell Inc., Energy Resources Center, Minneapolis, Minnesota.
6. A Whillier, (1964), 'Thermal resistance of the tube-plate bond in solar heat collectors', Solar Energy, 8(3), 65.
7. H.P. Garg and U. Rani, (1981), 'Calorimetric determination of bond conductances in pipe and fin type flat-plate collectors', Applied Energy, 8(3), 65.
8. S.A. Klein, (1974), 'Calculation of flat-plate collector loss co-efficients', Solar Energy, 17, 79.
9. H.P. Garg and G. Datta, (1984), 'The top loss calculation for flat-plate solar collectors', Solar Energy, 32(1), 141-143.
10. H.P. Garg, Usha Rani and Ram Chandra, (1981), 'Optimization of fin and tube parameters in a flat-plate collector', Energy, 6, 83-92.
11. Usha Rani, (1981), 'Analysis and optimization of solar collectors for heating of fluids', Ph.D. Thesis, Indian Institute of Technology Delhi.
12. D.R. Oliver, (1962), 'The effect of natural convection on viscous flow heat transfer in horizontal tubes', Chem. Engg. Sci., 17, 335.
13. L.H. Baker, (1967), 'Film heat transfer coefficients in solar collector tubes at low Reynold numbers', Solar Energy, 11(2), 78.

14. E.N. Sieder and C.E. Tate, (1936), 'Heat transfer and pressure drop of liquid in tubes', Ind. Eng. Chem., 28, 1424.
15. F. Kreith and J.F. Kreider, (1978), Principles of Solar Engineering, McGraw Hill Book Company, New York.
16. R.K. Shah and A.L. London, (1974), 'Thermal boundary conditions and some solutions for laminar duct flow forced convection', Trans. ASME Journal of Heat Transfer, 96, 159.
17. K.R.C. Eckert and R.M. Drake, Jr., (1972), Analysis of Heat and Mass Transfer, McGraw Hill Book Company, New York.

HOT WATER STORAGE SYSTEMS

S.P. Sukhatme
Department of Mechanical Engineering
Indian Institute of Technology
Bombay
India

1. INTRODUCTION

Sensible heat storage is one of the basic methods for storing thermal energy. In this method, the energy is stored by heating a liquid or a solid which does not change phase. The amount of energy stored (E) can be expressed in the form

$$E = m\, C_p\, (T_2 - T_1) \tag{1}$$

where m is the mass of the material, C_p is the specific heat, and T_1 and T_2 represent the lower and upper temperature levels between which the storage operates. The difference (T_2-T_1) is referred to as the temperature swing.

Water is the most commonly used medium in a sensible heat storage system. Most solar water-heating and space-heating systems use hot water storage tanks located either inside or outside the buildings or underground. The sizes of the tanks used vary from a few hundred litres to a few thousand cubic metres. An approximate thumb rule followed for fixing the size is to use about 75 to 100 litres of storage per square metre of collector area.

Water storage tanks are made from a variety of materials like steel, copper, concrete and fibreglass. The tanks are suitably insulated with glass wool, mineral wool or polyurethane. The thickness of insulation used is large and ranges from 10 to 20 cm. Because of this, the cost of the insulation represents a significant part of the total cost and means to reduce this cost have to be explored.

If the water is at atmospheric pressure, the temperature is limited to $100^{o}C$. It is possible to store water at temperatures above $100^{o}C$ by using pressurized tanks. This has been done in a few instances.

Cost (including the cost of insulation) vary from Rs.2000 to 10,000 per m^3 (US$ 1 = Rs.12). The lower value is applicable for large tanks having volumes greater than 20 m^3, while the higher value would represent a typical cost for a tank having a volume around 1 m^3.

H. P. Garg (ed.), Solar Water Heating Systems, 113–123.

2. ANALYSIS OF A WATER STORAGE TANK (Ref.1)

2.1 Well-Mixed Situation

Consider the insulated liquid-storage tank shown in Fig.1 receiving energy (when available) from an array of collectors and discharging energy (when needed) to a load for use in an application. We assume

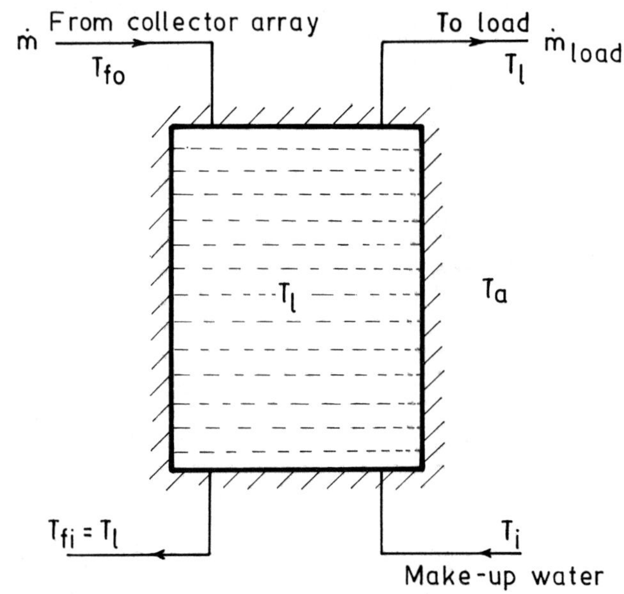

FIG.1 - ANALYSIS OF A WELL-MIXED WATER STORAGE TANK

that the liquid in the tank is always well-mixed and consequently is at a uniform temperature, T_1, which varies only with time. An energy balance on the tank yields the following equation:

$$\left[(\rho VC_p)_1+(\rho VC_p)_t\right]\frac{dT_1}{dt} = q_u - q_{load} - (UA)_t(T_1 - T_a) \tag{2}$$

where $(\rho VC_p)_1$ represents the heat capacity of the liquid in the tank, $(\rho VC_p)_t$, the heat capacity of the tank material, q_u, the rate of useful heat gain received from the collectors, q_{load}, the rate at which energy is being discharged to the load, $(UA)_t$, the product of the overall heat-transfer coefficient and surface area of the tank, and T_a, the ambient temperature around the tank. The heat capacity term $(\rho VC_p)_t$ is likely to be of importance for small-sized tanks only. For large tanks, its value may be negligible in comparison to $(\rho VC_p)_1$. Denoting the sum of the two heat capacities by the symbol $(\rho VC_p)_e$ and integrating differential equation (2) under the assumption that q_u, q_{load} and T_a are constants, we get, subject to the initial condition $t = 0$, $T_1 = T_{1i}$

$$\frac{q_u - q_{load} - (UA)_t (T_1 - T_a)}{q_u - q_{load} - (UA)_t (T_{1i} - T_a)} = \exp\left[-\frac{(UA)_t t}{(\rho V C_p)_e}\right] \tag{3}$$

The assumption that q_u, q_{load} and T_a are constants is valid if the time interval of the integration is kept reasonably small (say one hour or less). Equation (2) can be used to determine the variation of the temperature T_1 with time if the variation of q_u and q_{load} is given. Alternatively the values of q_u and q_{load} may be calculated from the inlet and outlet temperatures and flow rates as

$$q_u = \dot{m}C_p(T_{fo} - T_{fi}) = \dot{m}C_p(T_{fo} - T_1) \tag{4}$$

and

$$q_{load} = \dot{m}_{load} C_p (T_1 - T_i) \tag{5}$$

Equation (2) then becomes

$$(\rho V C_p)_e \frac{dT_1}{dt} = \dot{m}C_p(T_{fo} - T_1) - \dot{m}_{load} C_p (T_1 - T_i) - (UA)_t (T_1 - T_a) \tag{6}$$

It is important to note again that either q_u or q_{load} may be zero at any particular time. For example, when $T_{fo} < T_1$, the flow through the collectors would be stopped and we would have $\dot{m} = 0$ and $q_u = 0$. Similarly if no energy is required on the load side, $\dot{m}_{load} = 0$ and $q_{load} = 0$.

Example 1: A cylindrical hot water storage tank, 1.73 m in diameter and 2.1 m high, is made from a steel plate ($\rho = 7800$ Kgm^{-3}, $C_p = 0.46$ kJ/Kg-K) 6 mm thick. Apart from the mass of steel required for making the surface, an additional 200 Kg of steel is required in the form of angles, etc. for strengthening the tank, which is insulated all round with glass wool insulation 20 cm thick (k = 0.040 W/m-K). The initial temperature of the water in the tank is 50.0°C at 0700 h in the morning on a particular day and the variation of q_u and T_a up to 2000 h is as shown.

Time (hr)	7-8	8-9	9-10	10-11	11-12	12-13	13-14
q_u (kJ/h)	16 660	37 493	54 890	60 070	69 890	70 120	62 205
T_a (°C)	17.8	21.9	25.1	27.4	29.1	30.2	30.9

Time (hr)	14-15	15-16	16-17	17-18	18-19	19-20
q_u (kJ/h)	55 490	41 070	19 195	0	0	0
T_a (°C)	31.2	31.3	30.8	28.9	26.0	24.0

The load requirement is such that energy is continuously withdrawn from the tank at a constant rate of 27000 kJ/h for 18 hours a day starting at 0500 h. Assuming that the water in the tank is always well-mixed, calculate the variation of its temperature.

We shall first determine the values of $(\rho V C_p)_e$ and $(UA)_t$.

$$\text{Internal volume of the tank} = \frac{\pi}{4} \times 1.7^2 \times 2.1 = 4.767\ m^3$$

$$(\rho V C_p)_1 = 971.8 \times 4.767 \times 4.195 = 19432\ kJ/K$$

$$\text{Mass of steel used} = \left[(\pi \times 1.7 \times 2.1 + 2 \times \frac{\pi}{4} \times 1.7^2) \times 0.006 \times 7800\right] + 200 = 937.0\ Kg$$

Therefore,

$$(\rho V C_p)_1 = 937.0 \times 0.46 = 431\ kJ/K$$

Hence

$$(\ V C_p)_e = 19432 + 431 = 19863\ kJ/K$$

In calculating the value of $(UA)_t$, we shall assume that the flow of heat through the cylindrical surface and through the flat top and bottom is one dimensional. We shall also assume that the thermal resistance offered by the insulation dominates and that the other thermal resistances to the flow of heat at the inner and outer surfaces can be neglected by comparison. The overall heat-transfer coefficient U_1 (based on the inner diameter of the tank) for the flow of heat through the cylindrical surface is given by

$$\frac{1}{U_1} = \frac{r_1}{k_i} \ln \frac{r_2}{r_1}$$

where r_1 is the radius of the tank, r_2 is the radius with the insulation added on, and k_i is the thermal conductivity of the insulation. Substituting we get

$$\frac{1}{U_1} = \frac{0.85}{0.04} \ln \frac{1.05}{0.85} = 4.49$$

Therefore, $U_1 = 0.223\ W/m^2\text{-}K$

Similarly the overall heat-transfer coefficient U_2 for the flow of heat through the flat surfaces is given by

$$U_2 = \frac{0.04}{0.2} = 0.2\ W/m^2\text{-}K$$

Hence,

$$(UA)_t = (0.223 \times \pi \times 1.7 \times 2.1) + (0.2 \times 2 \times \frac{\pi}{4} \times 1.7^2) = 3.409\ W/K$$

We now apply equation (3) over a one hour interval of time. Substituting the given values, we have from 0700 to 0800 h,

$$\frac{18660-27000-3.409\text{x}3.6(T_1-17.8)}{18660-27000-3.409\text{x}3.6(50.00-17.8)} = \exp\left[-\frac{3.409\text{x}1\text{x}3600}{19863\text{x}1000}\right]$$

$$(T_1)_{0800\ h} = 49.56^{o}C$$

Continuing in the same way, we obtain the following variation for T_1 up to 2000 h.

Time (h)	8	9	10	11	12	13	14
T_1(°C)	49.56	50.07	51.46	53.11	55.25	57.41	59.16
		15	16	17	18	19	20
		60.58	61.27	60.86	59.48	58.10	56.72

The results are plotted in Fig.2. It is seen that from 0700 to 0800 h, the energy withdrawn exceeds the useful heat gain and that the water

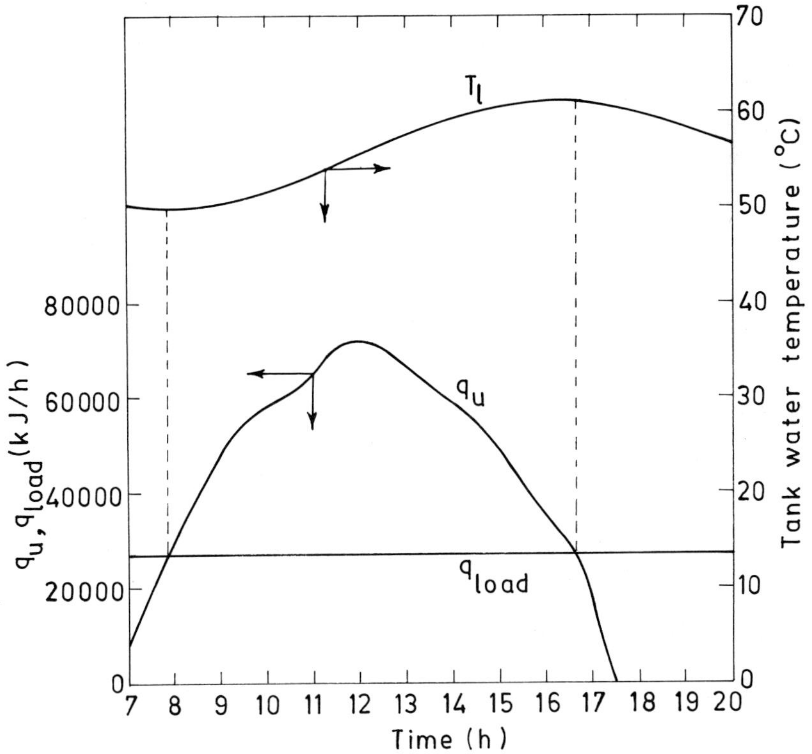

FIG.2.EXAMPLE 1-VARIATION OF TEMPERATURE IN A WELL-MIXED WATER STORAGE TANK

temperature decreases. Thereafter, the useful heat gain (whose variation is typical of a clear sunny day) exceeds the energy withdrawn up to 1600 h. As a result, the tank water temperature starts increasing steadily and touches a peak around 1600 h. After this the useful heat gain decreases to zero and since energy is still being withdrawn, the tank water temperature decreases. From the load demand, it is clear that if the calculations were continued over a 24 h period, the tank water temperature would be around 50°C at 0700 h on the next day. This indicates that there is an approximate matching between the useful heat gain and the energy withdrawn over a day and that the purpose of the storage tank in the present case is only to take care of the short term mismatch between supply and demand of energy over a day.

It should also be noted from Fig.2 that the tank temperature fluctuates by about 12°C. The magnitude of this fluctuation essentially depends upon the volume of water in the tank. If the volume of the water were to be doubled, the fluctuation would be approximately halved.

The preceding analysis has been performed under the assumption that the liquid flowing through the collectors and to the load and that stored in the tank is the same and that one stream mixes with the other. This may not always be the case. For example, in certain cases, it may be necessary from the point of view of corrosion to pass specially treated water mixed with chemical inhibitors through the collector circuit. It may also be necessary to add antifreeze compounds to the water passing through the collectors if temperatures below 0°C are likely to be encountered. In such cases, the fluid streams have to be separated and heat has to be transferred from the collector-circuit liquid to the storage-tank liquid through a heat exchanger. The heat exchanger may be a separate unit external to the storage tank or may be in the form of a coil immersed in the storage tank. We will now analyse the second situation shown schematically in Fig.3.

For an immersed coil, it is easy to show from heat exchange theory that

$$\frac{T_{fo}-T_{fi}}{T_{fo}-T_1} = 1 - \exp\left[-(UA)_e/\dot{m}C_p\right] \qquad (7)$$

where $(UA)_e$ is the product of the overall heat-transfer coefficient and the area for the heat exchanger coil. Thus, the expression for the rate of useful gain becomes

$$\begin{aligned} q_u &= \dot{m}C_p(T_{fo}-T_{fi}) \\ &= \dot{m}C_p(T_{fo}-T_1)\left[1-\exp\left\{-(UA)_e/\dot{m}C_p\right\}\right] \end{aligned} \qquad (8)$$

Substituting equation (8) for q_u and equation (5) for q_{load} into the energy balance equation (2), we get

$$\begin{aligned} (\rho VC_p)_e \frac{dT_1}{dt} &= \dot{m}C_p(T_{fo}-T_1)\left[1-\exp\left\{-(UA)_e/\dot{m}C_p\right\}\right] \\ &\quad - \dot{m}_{load}C_p(T_1-T_i)-(UA)_t(T_1-T_a) \end{aligned} \qquad (9)$$

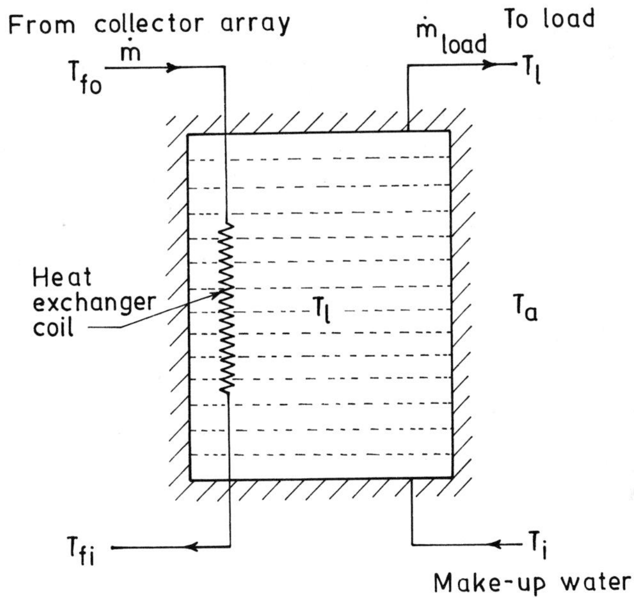

FIG.3 - ANALYSIS OF A WELL-MIXED WATER STORAGE TANK WITH AN IMMERSED HEAT EXCHANGER COIL

In equation (9), the specific heats of the two liquid streams have been assumed to be the same. In case they are different, they would need to be distinguished in an appropriate manner.

It is important to realize that the use of a heat exchanger for a given application not only increases the complexity of the set-up but also raises the temperature level of operation of the collector array by a few degrees. Since collector efficiency decreases with increase in temperature, more collector area is required for meeting the given energy demand.

2.2 Thermal Stratification

In a thermally stratified situation, the temperature of the contained liquid varies from the bottom to the top, being less at the bottom and more at the top. This situation is in contrast to that obtained in a well-mixed tank in which the liquid temperature is uniform throughout.

Thermal stratification is obviously desirable if the temperature difference (T_1-T_i), refer Fig.1, is significant. If the line taking liquid to the load at the required temperature T_1 is located approximately near the top of the storage tank, then only the liquid near the exit port has to be at the temperature T_1. The rest of the liquid in the storage tank can be at a lower temperature at all times. As a result, heat losses from the tank are reduced. A second advantage is that the collectors operate at a lower temperature level and deliver a higher collection efficiency.

In some applications, thermal stratification is obtained naturally as a matter of course. An example is a natural circulation water heating system. In such a system, the flow rates are low and a certain degree of thermal stratification is always obtained. Thermal stratification can also be obtained in a forced circulation water heating system if the fluid inlet and outlet configurations are suitably located and shaped. Ring distributors are used so that the velocities entering and leaving the tank are low and are in such a direction as to reduce mixing. In some cases, a floating inlet, made of wide, thin-walled flexible plastic hose connected to the inlet port, is used. This type of inlet delivers hot liquid from the collectors at a level at which the inlet liquid temperature is equal to that in the storage tank.

The analysis of a thermally stratified tank is complicated by the fact that the temperature profile has to be solved as a function of time. In addition, in most practical situations, the problem is not one-dimensional in nature and involves consideration of the heat conduction through the walls of the storage tank.

A simplified analysis is possible by assuming that the storage tank consists of a number of well-mixed sections at different temperatures and writing down energy balance equations similar to equation(6) for each section. Consider the situation in which the storage tank is assumed to consist of two well-mixed sections (Fig.4) at temperatures

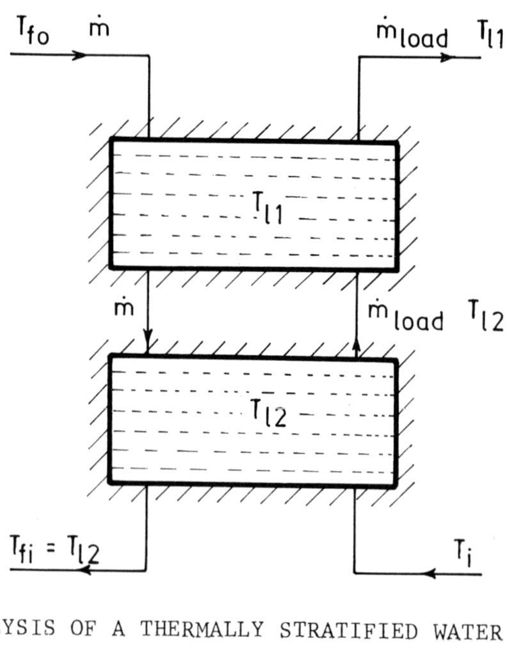

FIG.4 - ANALYSIS OF A THERMALLY STRATIFIED WATER STORAGE TANK

T_{11} and T_{12}. Writing energy balances on each of the sections, we obtain

$$(\rho VC_p)_{e1}\frac{dT_{11}}{dt} = \dot{m}C_p(T_{fo}-T_{11})-\dot{m}_{load}C_p(T_{11}-T_{12})-(UA)_{t1}(T_{11}-T_a) \qquad (10)$$

$$(\rho VC_p)_{e2}\frac{dT_{12}}{dt} = \dot{m}C_p(T_{11}-T_{12})-\dot{m}_{load}C_p(T_{12}-T_1)-(UA)_{t2}(T_{12}-T_a) \quad (11)$$

Differential equations (10) and (11) have to be solved simultaneously for the unknowns T_{11} and T_{12}. One way of solving them is to express them in finite difference form and to choose a suitable time interval. Values of T_{11} and T_{12} are then obtained from one time interval to the next.

Equations (10) and (11) have been written under the assumption that the flow from the collectors can enter the storage tank only at the top of section 1. This occurs when $T_{fo} > T_{11}$. When $T_{fo} < T_{11}$, no flow occurs.

In case the tank is also provided with an inlet between the two sections, three possibilities exist:

(1) Flow $\dot{m}$ enters at the top of section 1 when $T_{fo} > T_{11}$.

(2) Flow $\dot{m}$ enters between sections 1 and 2 when $T_{11} > T_{fo} > T_{12}$.

(3) There is no flow and subsequently no energy to be stored, when $T_{fo} < T_{12}$.

Equations (10) and (11) can be asily modified to account for this situation.

Example 2: The temperatures in a hot water stratified tank are to be determined by assuming that the tank consists of two equal well-mixed sections with inlets at the top and in between the two sections. The following data re given:

Hour	$\dot{m}$(Kg/h)	T_{fo} (^oC)	$\dot{m}_{load}$(Kg/h)	T_a (^oC
1200-1300	2000	80	250	22
1300-1400	2000	80	240	23
1400-1500	2000	76	220	22

It is also given that:

(1) Mass of water in the tank = 5000 Kg

(2) $(UA)_{t1} = (UA)_{t2}$ = 30 kJ/h-^oC

(3) Make up water enters at 20^oC at the same rate as the rate of withdrawl to the load

(4) At 1200 h, T_{11} = 70^oC, T_{12} at 65^oC.

Calculate the values of T_{11} and T_{12} at 1500 h.

Neglecting the values of the heat capacity term $(\rho VC_p)_t$, equations (10) and (11) can be written in finite difference form as follows:

$$(\rho VC_p)_{\ell 1}\left(\frac{T_{11.f}-T_{11.i}}{\Delta t}\right) = \dot{m}C_p\left(T_{fo} - \frac{T_{11.f}+T_{11.i}}{2}\right)$$

$$-\dot{m}_{load}C_p\left(\frac{T_{11.f}+T_{11.i}}{2} - \frac{T_{12.f}+T_{12.i}}{2}\right)$$

$$-(UA)_{t1}\left(\frac{T_{11.f}+T_{11.i}}{2} - T_a \quad (12)\right.$$

$$(\rho VC_p)_{\ell 2}\left(\frac{T_{12.f}-T_{12.i}}{t}\right) = \dot{m}C_p\left(\frac{T_{11.f}+T_{11.i}}{2} - \frac{T_{12.f}+T_{12.i}}{2}\right)$$
$$-\dot{m}_{load}C_p\left(\frac{T_{12.f}+T_{12.i}}{2} - T_i\right)$$
$$-(UA)_{t2}\left(\frac{T_{12.f}+T_{12.i}}{2} - T_a\right) \qquad (13)$$

In these equations, $T_{11,f}$ and $T_{11,i}$ represent the final and initial values of T_{11} over the time interval Δt. A similar meaning is attached to the symbols $T_{12,f}$ and $T_{12,i}$. The symbols T_{fo}, T_a and T_i now represent average values over the time interval Δt.

We take Δt = 1 h and note the presence of the inlet in between the two sections while substituting the given data from one hour to the next.

<u>Time interval 1200-1300 h</u>: $T_{11,i} = 70^\circ C$; $T_{12,i} = 65^\circ C$

Substituting into equation (12), we get

$$2500 \times 4.19 \times (T_{11,f}-70) = 2000 \times 4.19 \left(80 - \frac{T_{11.f}+70}{2}\right)$$
$$- 250 \times 4.19 \left(\frac{T_{11.f}+70}{2} - \frac{T_{12.f}+65}{2}\right)$$
$$- 30 \left(\frac{T_{11.f}+70}{2} - 22\right)$$

$$\text{or} \quad 3628.6\ T_{11,f}-125\ T_{12,f} = 264282 \qquad (14)$$

Similarly, substituting into equation (13), we get

$$- 1000\ T_{11,f}+3628.6\ T_{12,f} = 164300 \qquad (15)$$

Solving equations (14) and (15)

$$T_{11,f} = 75.10^\circ C; \quad T_{12,f} = 65.98^\circ C$$

<u>Time interval 1300-1400 h</u>: Now $T_{11,i} = 75.10^\circ C$; $T_{12,i} = 65.98^\circ C$

Substituting into equations (12) and (13), we get

$$3623.6\ T_{11,f}-120\ T_{12,f} = 271457$$

and

$$-1000\ T_{11,f}+3623.6\ T_{12,f} = 170882$$

Solving $T_{11,f} = 77.18^\circ C$; $T_{12,f} = 68.46^\circ C$

<u>Time interval 1400-1500 h</u>: Now $T_{11,i} = 77.18^\circ C$; $T_{12,i} = 68.46^\circ C$

Since $T_{11,i} > T_{fo} > T_{12,i}$, the flow from the collectors will bypass section 1 and enter in between sections 1 and 2. Thus, the first term of the right hand side of equation (12) will be zero and in the first term of the right hand side of the equation (13), we would put T_{fo} instead of

$$\left(\frac{T_{11,f}+T_{11.i}}{2}\right)$$

Substituting, we get

$$2613.6\ T_{11,f} - 110\ T_{12,f} = 191875$$

and

$$3613.6\ T_{12,f} = 251469$$

Solving $T_{11,f} = 76.34^{\circ}C$; $T_{12,f} = 69.59^{\circ}C$

REFERENCE

1. S.P. Sukhatme, (1984), Solar Energy - Principles of Thermal Collection and Storage, Tata McGraw Hill Publishing Co., New Delhi, India.

THE USE OF FIBREGLASS IN SOLAR WATER HEATING SYSTEMS

A. Sharif
FGP Limited
9 Wallace Street
Bombay 400001
India

1. INTRODUCTION

The fuel crisis and the rising prices of petroleum products have been a boon for the development of products associated with alternate sources of energy. The enlightened policy of the government has quickened the uses of many of these products. Food is being cooked on gobar gas. Villages are being lighted with solar energy. The energy of the sun is also being used for heating water and cooking food.

India is blessed with abundant sunshine. There are about 250 to 300 days of useful sunshine per year in most parts of the country. Thus solar water heating system presents a practical solution for providing hot water wherever required for home or industry.

The solar water heating system consists of the following:

i) Solar collectors - the most common being the flat plate type
ii) Piping system
iii) Hot water storage tank
iv) Temperature, flow and energy indicators, where required
v) Heat exchangers, where required.
vi) Support structure

Here I shall deal mainly with the selection criteria of the proper insulation for the solar hot water system.

In the solar water heating system, insulation is used for the insulation of the solar collector, pipes and the water storage tanks. The solar collector is insulated to prevent heat loss and insulation up to an 'R' value of 10 is generally specified for the back and sides of the flat plate collector. The efficiency of the insulation will depend upon its thermal conductivity. But this is not the only criteria considered whilst selecting an insulation product. Other parameters are its thermal stability. It should be non-hygroscopic. It must be resistant to settling. It should not contain substances which can lead to corrosion of the surface on which it is applied. It should be non-combustible and unaffected in the temperature range of 0^oC to 200^oC generally experienced in flat-plate collectors.

H. P. Garg (ed.), Solar Water Heating Systems, 125–128.

2. THERMAL STABILITY

This is an important characteristic of the insulation material. The insulation should be thermally stable in the temperature range of $0^{o}C$ to $200^{o}C$. This means that expanded foam insulations with temperature ranges below $200^{o}C$ are suspect as they may soften up. Amongst the Bonded mineral fibre insulations it has been observed by DNES that the resin tends to decompose at around these temperatures, gas is released and the transmittance of the glass can be affected. Unbonded fibre-glass has been found to be the best solution.

3. SETTLING

The solar collectors are installed angularly and the solar cookers are subjected to a great amount of handling so the insulation used should not settle down after a period of time thereby impairing the efficiency of the system.

Some of the factors that affect the insulation and are directly or indirectly responsible for causing settling are moisture and shot content.

4. NON-HYGROSCOPIC

Because the solar collectors are usually kept in the open, there may be an ingress of moisture during the collector temperature cycle. The insulation should therefore be non-hygroscopic. This means it should not absorb moisture. The reason is that a hygroscopic material will tend to absorb moisture and get wet which in turn will affect its thermal efficiency adversely. Moisture can also affect the structural stability of the insulation. Mineral fibre insulations which are based on slag contain a high percentage of lime. In presence of moisture the fibres tend to disintegrate, they become brittle, the wool loses its resilience and settle down.

The insulation shall not contain 'shot' or coarse fibres as it will affect the structural stability of the insulation. This is because 'shot' is a weak link in the insulation. 'Shot' are tiny globules of glass caused by premature cooling. The size of the shot becomes larger if there is an increase in the viscosity of the molten slag. It ranges from microscopic globules adhering to the fibres to as coarse as 3 mm. Shot is generally found in slag based products whether in loose form or bonded. The insulation shall not contain any impurities which can cause corrosion to the surface on which they are applied or be the cause of contamination of food stuff as in the case of solar cookers. It has been observed by DNES on tests carried out with various insulation products that mineral wool or any material which has a percentage of slag tends to cause corrosion of the metal surfaces.

Slag based products contain sulphur which can attack the black coating of the absorber plate and may also contaminate food stuffs in solar cookers.

The fibres of loose glass wool are brittle, they tend to crush easily and the insulation settles down. Since the material is loosely laid by hand the thermal efficiency can vary from applicator to applicator.

Taking all these factors into account it appears that FG-Crown insulation may be the best suitable insulation material for solar water heating systems. Fibreglass Crown comes under the generic term mineral wool. It is also called glass wool. But its similarity to glass wool ends here.

Fibreglass Crown Insulation is a product specially designed and developed to suit the needs of the insulation industry. It is made from specially formulated glass by the sophisticated TEL process wherein raw materials like sand, silica, alumina, borax and other chemicals are mixed to a predetermined formula. The quality of glass can be controlled at all stages so that it is possible to make a high grade product, free from impurities. Direct flame heating converts the raw material in molten glass and in this state it flows under carefully controlled conditions from the main furnace to the forehearth.

The TEL process is based on a spinner which is a steel alloy dish in which molten glass flows from forehearth. The spinner rotates very rapidly and the molten glass is thrown out in fibrous form by centrifugal force through hundreds of small holes in its side. These fibres are directed downwards on to a conveyor belt by a jet of air and sprayed with a resin binder. In case of white wool no resin is sprayed. The fibres, now, in the form of a mat are carried on the conveyor belt through the curing oven to the trimmers and cutters before packed for despatch.

This is a fully automatic process in which the fibre diameter, density and thickness are completely controlled. This is a very important factor as changes in fibre diameter, improper density and thickness will affect the thermal performance.

A wide range of products in various densities can be made, one of these is Fibreglass Crown Solatherm.

5. FIBREGLASS CROWN SOLATHERM

5.1 Thermal Efficiency

Fibreglass Crown Solatherm provides light weight product. Its fine fibres criss cross to form stabilized air cells which provide the insulating effect. Its ratio of glass fibres to air is perfectly controlled to provide maximum thermal efficiency.

5.2 Structural Stability

Fibreglass Crown Solatherm is not made from scrap u-like loose glass wool or slag based product. Specially formulated glass makes it chemically inert. It is non-hygroscopic nor chemically affected by moisture in any way. It has NIL shot content. Perfect temperature and chemical control of the liquid glass, the double seeving effect of the spinner and attenuation of the fibres by the hot gases ensure fine fibres free from'shot' and coarse fibres. It will not settle down. The arrangement of the fibres during the spinning process ensures good resilence of the product. The product will not settle down when subjected to vibration and jolting.

It does not contain any impurity. Special formulation of the glass takes care of this property. It will not corrode the surface on which it is applied. Moreover, this special glass composition ensures that the product can be used within the temperature range from $0^{o}C$ to $200^{o}C$. In fact it can be used from cryogenic temperatures up to $532^{o}C$.

Being glass it is non-combustible. Thus when all these facts are taken into account, Fibreglass Crown Solatherm emerges as the ideal choice.

6. FRP PRODUCTS

Just as selecting the proper insulation can extend the life of the Solar water heating system, selecting fibreglass reinforced plastics for collector housing and storage tanks can extend the life of the system by making it corrosion resistant and maintenance free.

FRP housings require little maintenance, they can be self-coloured by adding pigment to the resin and they will not corrode. They are light weight. Their use will not lead to corrosion, as for instance if an aluminium collector were to be attached to a steel support "the difference in the electrochemical potential between the two will cause the aluminium to leak within a few years". Both the inner storage tank and the outer casing of the storage tank can be made from Fibreglass reinforced plastic. Again there will be no corrosion and maintenance problems.

Although glass is the most popular glazing used for the collector system, FRP sheets have been used abroad and may one day be used in India.

The ideal long life solar collector and storage systems will consist of low iron glass glazing, fibreglass insulation, FRP casings and storage tanks.

INSULATION MATERIALS FOR SOLAR HEATING SYSTEMS AND THEIR APPLICATION

T. Udayakumar
Lloyd Insulations (India)
M-13, Connaught Place
New Delhi 110 001
India

1. INTRODUCTION

Solar heating systems mainly consist of the collectors, tanks and pipe lines. In order to minimise the heat losses, from the back and sides of the absorber in the collector, from the surface of the tank in which the heated water is stored and from the pipe lines, thermal insulation is provided. There are certain specific requirements of insulation materials considered for the above applications. This paper, apart from detailing the above requirements, lists out various indigenously available insulation material, and analyses their suitability for such applications.

2. INSULATION FOR SOLAR APPLICATIONS

The object of using thermal insulation in solar applications is to minimise heat losses from the system and thus increase the efficiency.

Thermal insulation in solar applications are mainly used:

(a) in the collectors to minimise heat losses from the back and sides of the absorber,

(b) on the tank surfaces to minimise the heat losses from the stored hot water to the ambient,

(c) on the pipe lines carrying the hot water, again, to minimise the heat losses to the ambient.

3. REQUIREMENTS TO BE MET BY THE INSULANTS FOR SOLAR APPLICATIONS

The insulation material chosen for solar applications has to meet certain specific requirements in addition to the general requirements of any insulation material. The general requirements are:

- Low thermal conductivity
- Uniformity
- Good compression strength
- Low water vapour permeance

H. P. Garg (ed.), Solar Water Heating Systems, 129–132.

- Low water absorption
- Good fire resistant characteristics
- Resistance to corrosion
- Resistance to microbes and pests
- Prolonged life
- Ease of application
- Economical

In addition to the above basic requirements, the following are to be met with, when the insulation material is considered for application in the solar collectors:

- The material should be stable at least up to 200°C.
- The material should not give rise to degassing when subjected to 200°C.
- The material should have no tendency to settle in course of time even when applied on inclined surfaces.

4. INSULATION MATERIALS

Thermal insulation materials may be broadly classified into two types viz.:

(i) Mass type insulation
(ii) Reflective insulation

Mass type of insulation material is based on interposing a mass of material with built-in capacity to retard heat flow. The reflective insulation is based on providing a series of reflective surfaces with the intervening space either evacuated or filled with a thin layer of mass type insulant. The reflective type is used for cryogenic applications. For solar applications mass type of insulation is used.

Indigenously available mass type of materials can further be classified into:

- Fibrous material (like Mineralwool, Glasswool, Asbestos fibers, Ceramic fibers, Woodwool etc.),
- Organic foams (like polyisocyanurate foam, polyurethane foam, phenolic foam, expanded polystyrene etc.),
- Cellular material (like foam concrete),
- Powders (like Saw dust, Rice husk etc.),
- Granular material (like Calcium Silicate, 85% Magnesia, etc.).

Out of the above, material falling under the type 'powders', Asbestos fibers, Ceramic fibers, Woodwool and 85% Magnesia cannot be used for solar applications for obvious reasons. Thus the materials for selection could be short-listed as:

Mineralwool, Glasswool, Polyisocyanurate and Polyurethane foams, Expanded polystyrene and Calcium Silicate.

Now, keeping in mind the requirements already enumerated, the insulation material selection is to be made.

4.1 FOR COLLECTOR APPLICATION

Since expanded polystyrene is suitable for continuous service temperature of only up to a maximum of 80°C, its use in collectors is totally ruled out.

While all other materials have low thermal conductivity values in the desired range for solar applications, Calcium Silicate has very high thermal conductivity values in that range requiring large sized collector compartments and due to its higher density, the total weight of the collector will become unviable. Thus, use of Calcium Silicate is also ruled out in the case of solar collectors.

The cellular foam products cannot be stable at 200°C and hence their use in collectors is not advisable.

Considering Glasswool, it is available in loose as well as bonded forms. In case of the bonded glasswool products due to higher binder content, binder migration and hence degassing occurs at temperatures far below 200°C, leaving a deposit on the underside of the collector plate. Though the material is self supporting, crumbling of brittle and coarse fibres will be there in course of time. Further due to low density, the material does not possess enough compression resistance. Thus, the range of bonded glasswool products cannot be used for collector applications.

In case of unbonded Crown glasswool, though problems of degassing will not be there, the material being unbonded, uniformity cannot be obtained. In addition it settles in course of time and thus cannot be used in collectors.

Other unbonded glasswool products available in this country are of totally inferior quality and cannot be used for solar applications. There is a tendency on the part of nodal agencies, at times, to mention in their purchase specifications the insulation material to be 'glasswool'. This will lead to the supply of unbonded glasswool which is not the product for such applications.

Considering the mineralwool products, these again, are available both in unbonded as well as bonded forms.

The unbonded mineralwool also has the same drawbacks as that of unbonded glasswool and thus is not suitable for solar applications.

The bonded mineralwool being manufactured in this country under the trade name of SPINTEX is a rockwool product. This is made of superfine (3 to 4 microns) rock fibers in the optimum density range. The resin bonded rockwool felts are baked at around 300°C to form self supporting, uniform slabs, with minimum required resin content.

Due to the above, SPINTEX Rockwool slabs possess all the important characteristics to meet the requirement of solar applications.

SPINTEX Rockwool slabs

- are stable well above 200°C, their service temperature being 750°C;
- does not outgas/give rise to degassing when subjected to a temperature of 200°C, since it is baked at a temperature of about 300°C;
- does not settle even after prolonged periods of operation in inclined applications, since they are self supporting and of high density.

In addition SPINTEX Rockwool slabs have low thermal conductivity, uniformity, good compression strength, low water vapour permeance and water absorption, good resistance to fire, microbes and pests, and prolonged life. All the above factors have been proved beyond doubts by the inspection carried out by DNES on the condition of the solar collectors installed at the National Physical Laboratory, Delhi, more than seven years after installation and satisfactory performance. Thus the proof-of-the-pudding test has categorically established that SPINTEX Rockwool is an excellent insulation material for solar collectors.

4.2 Insulation of Storage Tanks

While SPINTEX Rockwool slabs could well be used for the insulation of tanks also, it will be economical to use lightly bonded machine laid machine stitched rockwool mattresses on tanks.

4.3 Insulation of Pipes

Rockloyd preformed rockwool pipe sections, due to their higher efficiency, ease of application and very low maintenance costs are being extensively used for the insulation of the piping falling under solar applications.

5. CONCLUSION

There are a wide range of insulation materials available indigenously. However, the proof-of-the-pudding tests have established that SPINTEX Rockwool slabs meet all the requirements of solar applications. Lightly bonded machine stitched rockwool mattresses for tanks and preformed rockwool pipe sections for pipe lines are being extensively used.

THERMAL INSULATION IN SOLAR THERMAL DEVICES

B.C. Raychaudhuri
Centre of Energy Studies
Indian Institute of Technology
Hauz Khas, New Delhi 110 016
India

1. INTRODUCTION

Thermal Insulation is a device or a practice which is used in a system for minimising heat losses caused due to transfer of heat from hotter to colder regions. It is one of the cheapest methods of energy conservation, where applicable.

There are three modes of heat transfer: (a) Conduction, (b) Convection, and (c) Radiation.

Conduction heat transfer takes place through a material by molecular interchange of kinetic energy by electric impacts. For metals, free-electron carriers are much more predominant. Conduction heat transfer is directly proportional to the thermal conductivity of the material (K).

Convection heat transfer is caused by mass movement of fluid parcels from hotter to colder regions, acting as heat carriers. Convection are of two kinds, (a) natural or free convection caused by buyoancy forces, and (b) forced convection, caused by forced movement of the fluid using mechanical means (fans, pumps, stirrers, etc.). Convective heat transfer is directly proportional to the surface heat transfer coefficient (h).

Radiation heat transfer is caused by the electromagnetic wave propagation with the speed of light (3×10^{10} cm/s). It follows the optical laws. Every material emits heat radiation, depending on its temperature above absolute zero. It works as an oscillator generating electromagnetic waves.

2. MATERIAL PROPERTIES

All the properties such as absorptivity, reflectivity and transmissivity are dependent on the temperature of the radiation source, whereas the emissivity or the emissive power is dependent on the temperature of the emitting surface and hence on the wavelength, as shown in Figs.1 and 2. Because of the wavelength dependency, the Green-house effect is observed. It could be seen that 90 per cent of the emissive power of solar radiation (6000 K) falls in the region 0.1 and 3 micron. While a body

H. P. Garg (ed.), Solar Water Heating Systems, 133–151.

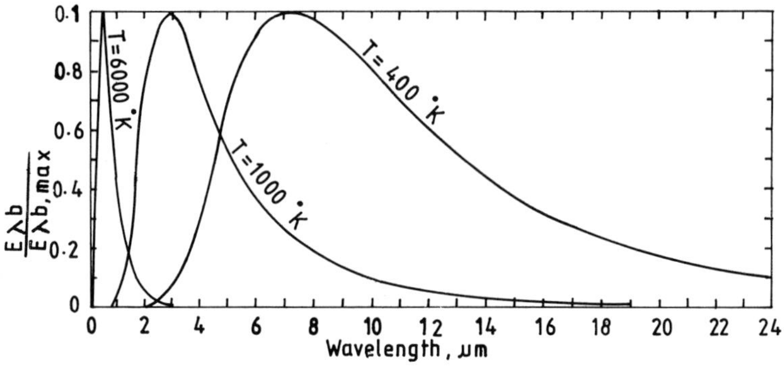

FIG.1 - SPECTRAL DISTRIBUTION OF BLACKBODY RADIATION

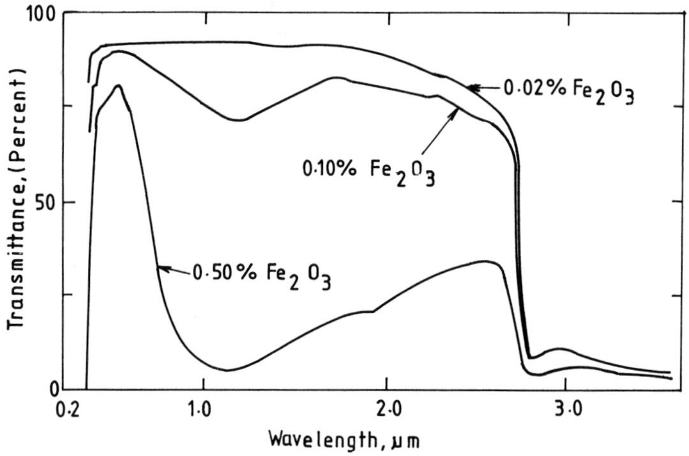

FIG.2 - SPECTRAL TRANSMITTANCE OF GLASS (6mm) WITH VARIOUS IRON OXIDE CONTENTS

at 1000 K emits most of the radiation in the range 1 and 20 micron. It is also seen that glass transmits the solar radiation but it is opaque to the longer wave radiation emitted by the surface inside the green houses which explain the green house effect.

For given geometric configuration factor, the radiation heat transfer is directly proportional to the effective emissivity of the system surfaces.

Some values of thermal conductivity, K($W/m^{o}C$), surface heat transfer coefficient, h($W/m^{2o}C$) and emissivity, are given in Table 1 showing the order of magnitudes.

For a given temperature differential, the reduction of heat loss by conduction would mean use of low thermal conductivity insulation material or increase in its thickness.

Reduction of heat loss by convection would mean reduction of surface heat transfer coefficient which would mean decrease of fluid movement or absence of fluid heat carrier (vacuum). Reduction of forced convection coefficient would mean reduction of fluid flow velocity to the extent possible by controlling the equipment causing the flow. The reduction of free convection coefficient would be done by suitable location of the heated surface, or by reducing the thickness of air layer over the surface to the order of mean free path.

Reduction of heat transfer by radiation would mean reduction of surface emissivity, introducing radiation shields etc.

TABLE 1 - Thermal Properties

A. Thermal Conductivity (W/m^oC) at 100^oC mean temperature

Air	0.034
Rock wool	0.051
Cork	0.045
Wood	0.055
Glass	0.78
Brick	0.69
Aluminium	206
Copper	407
Water	0.60

B. Surface Heat Transfer Coefficient ($W/m^{2o}C$)

Air (free convection)	5-25
Air (forced convection)	10-500
Water (forced convection)	250-10,000
Oil (forced convection)	50-1500

C. Absorptivity (α)/Emissivity (ϵ)

	Short Wave Length (α)	Long Wave Length(ϵ)
Aluminium foil (bright)	0.05	0.05
Aluminium foil (oxidised)	0.25	0.12
Aluminium paint	0.50	0.50
G.I. bright	0.25	0.25
White wash, new	0.12	0.90
White oil paint	0.20	0.90
Black paint (ordinary)	0.85	0.90
Black paint (selective)	0.85	0.10

3. THERMAL INSULATION MATERIALS

These can be broadly classified into two types, (a) mass insulation, and (b) reflective insulation. Mass insulation and reflective insulation are schematically shown in Figs.3 and 4 (Ref.3). Mass insulation are porous solid bodies (cellular, granular or fibrous), filled normally with air and the insulation property is primarily due to the low conductivity of

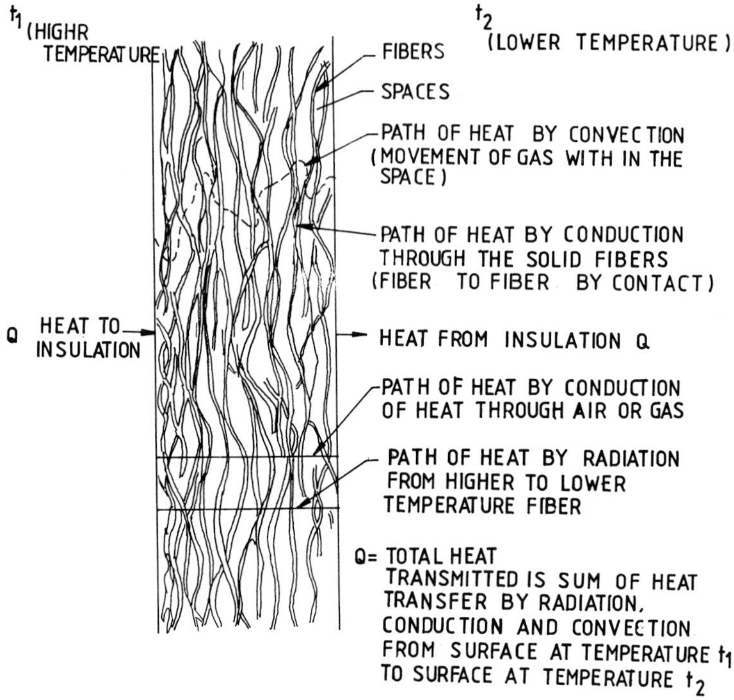

FIG.3 - HEAT FLOW (SCHEMATIC) THROUGH MASS INSULATION

entrapped air (Ref.1,2,3). In addition to conduction, due to porous structure, convection and radiation plays an important role. The density of the material therefore plays a significant part as this indirectly determines the proportionate contribution of the three modes of heat transfer, and there are optimum densities for different materials giving the minimum thermal conductivity. These are illustrated in Figs.5 and 6. Good fibrous insulation materials use 3 to 5 micro fibre diameter.

A reflective insulation consists of thin metallic foils having low emissivity (aluminium foils), which are located and separated by poorly conducting solid frames in such a way so as to minimise the radiation transfer, and reducing convection transfer by keeping the gap dimensions of the order of mean-free path of air at the operating temperature.

The radiation heat transfer with radiation shields gets reduced by $1/(1+n)$ from that without the shields, n being the number of shields.

4. SOLAR THERMAL DEVICES

The most important component is the solar collector. Let us consider a flat plate collector, Fig.7.

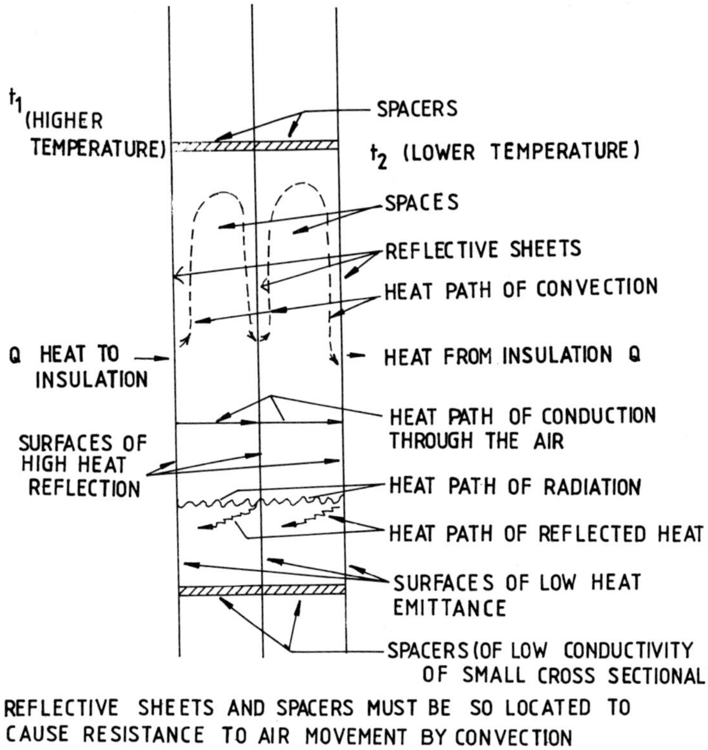

FIG.4 - HEAT FLOW (SCHEMATIC) THROUGH REFLECTIVE INSULATION

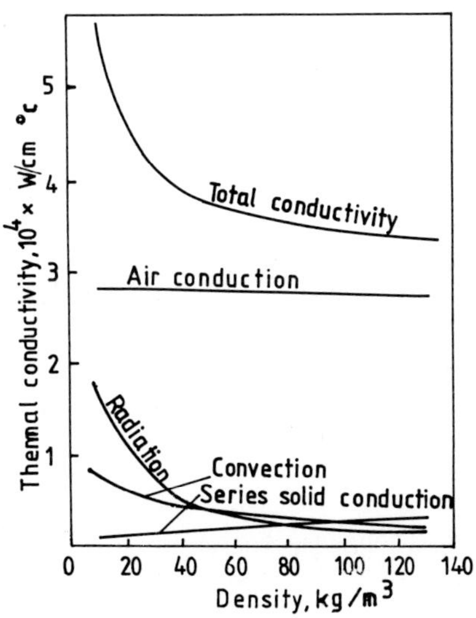

FIG.5 - HEAT TRANSFER THROUGH A-TYPE INSULATION
BY VARIOUS MECHANISMS

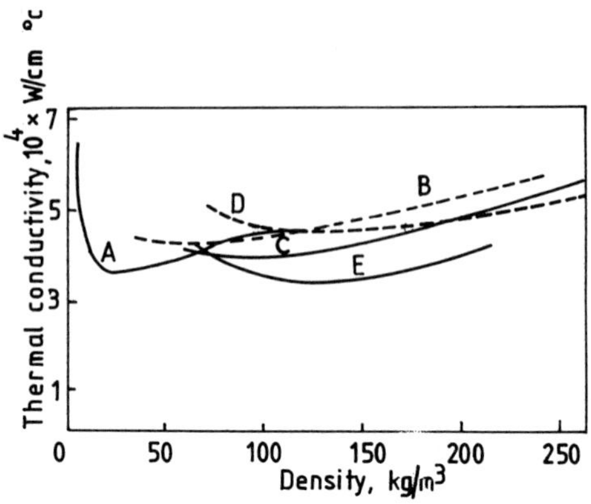

FIG.6 - VARIATION OF THE THERMAL CONDUCTIVITY WITH DENSITY. A, KAPCOCK; B, BAGASSE; C, CORKBOARD, D, SLAG WOOL; E, MINERAL WOOL.

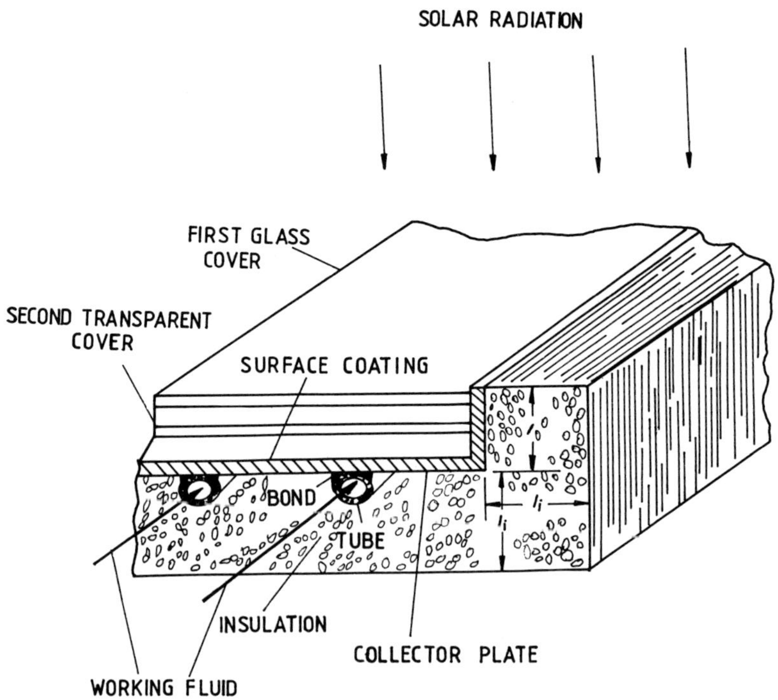

FIG.7 - FLAT-PLATE SOLAR ENERGY COLLECTOR

It may be stated that, in general,

(a) Use of glazing material is done to have the green house effect and to reduce convection losses by using multi-layer glazing.

(b) Use of selective coating on the absorber plate and use of anti-reflection coatings on the glazing increases heat gain to the absorber and reduces heat loss to the ambient.

(c) Use of proper spacings between the glazing and the absorber and also between different layers of glazing material, and evacuation of the annular space in the concentric annular tube collectors reduces the convection heat loss to the ambient.

(d) Use of transparent honeycomb structure between the glazing and the absorber plate suppresses the convection and reduces heat loss to the ambient.

(e) Putting mass insulation material at the bottom and edges of the collector reduces conduction heat loss to the ambient.

In brief all the above methods are used to minimise the overall heat transfer coefficient (U) of the collector and hence may broadly be treated as a part of the thermal insulation system. These are illustrated in Figs.8 and 9.

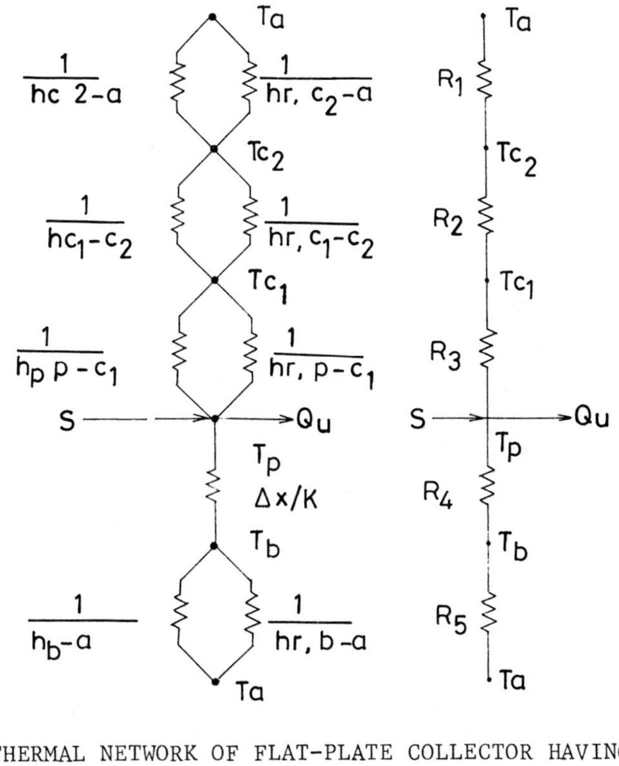

FIG.8 - THERMAL NETWORK OF FLAT-PLATE COLLECTOR HAVING TWO GLAZING

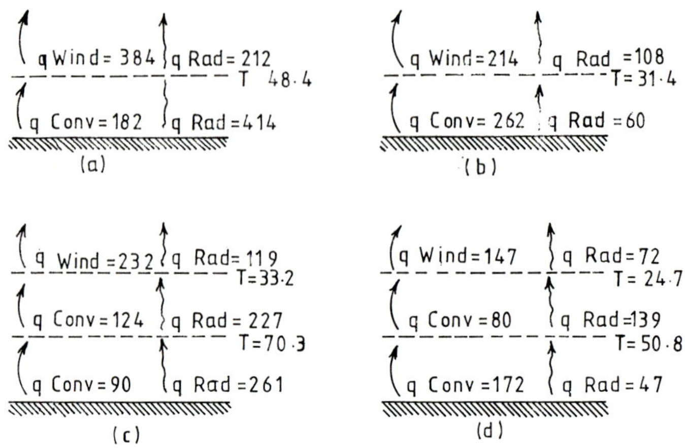

FIG.9 - EFFECT OF GLAZING AND COATING ON THE UPWARD HEAT LOSS FROM FLAT-PLATE COLLECTOR (HEAT LOSS $W/m^2 {}^oC$ AND TEMPERATURE ^oC)

The overall heat transfer coefficient (U) is constituted of top loss coefficient (U_t) and bottom (including edge) loss coefficient (U_b). The bottom loss coefficient is concerned with the mass insulation, which only will be treated in this paper.

Roughly U_b is of the order of 5 to 10% of U and the cost of mass insulation to provide the above U_b is of the order of 10 to 15% of the collector cost. It may be mentioned that dust and shading reduces the collector efficiency of the order of 2%.

5. SOLAR THERMAL SYSTEM

In a solar thermal system, the solar collector is connected to storage tank and to the end use system such as space heating and cooling appliances. These connecting piping and ductings, storage tanks are required to be properly insulated to reduce the heat loss and increase the total system efficiency. These are illustrated in Figs.10 and 11. The temperature profiles of the air as it goes through the inlet duct collector and outlet duct are illustrated in Fig.12.

If hot water coils are used in a Plenum Chamber to heat the air, and the air is forced through ducts by blowers to large rooms like cinema halls, the air ducts are preferably insulated inside the metallic ducts to reduce the noise level. The air velocity effects the U-value of the duct as shown in Fig.13. It has been found by TIMA that at higher velocities, increasing the density of the linear material provided generally better thermal performance than simply increasing the thickness. Higher densities tend to reduce the convection-caused heat losses because of smaller and tighter inter-cellular structure.

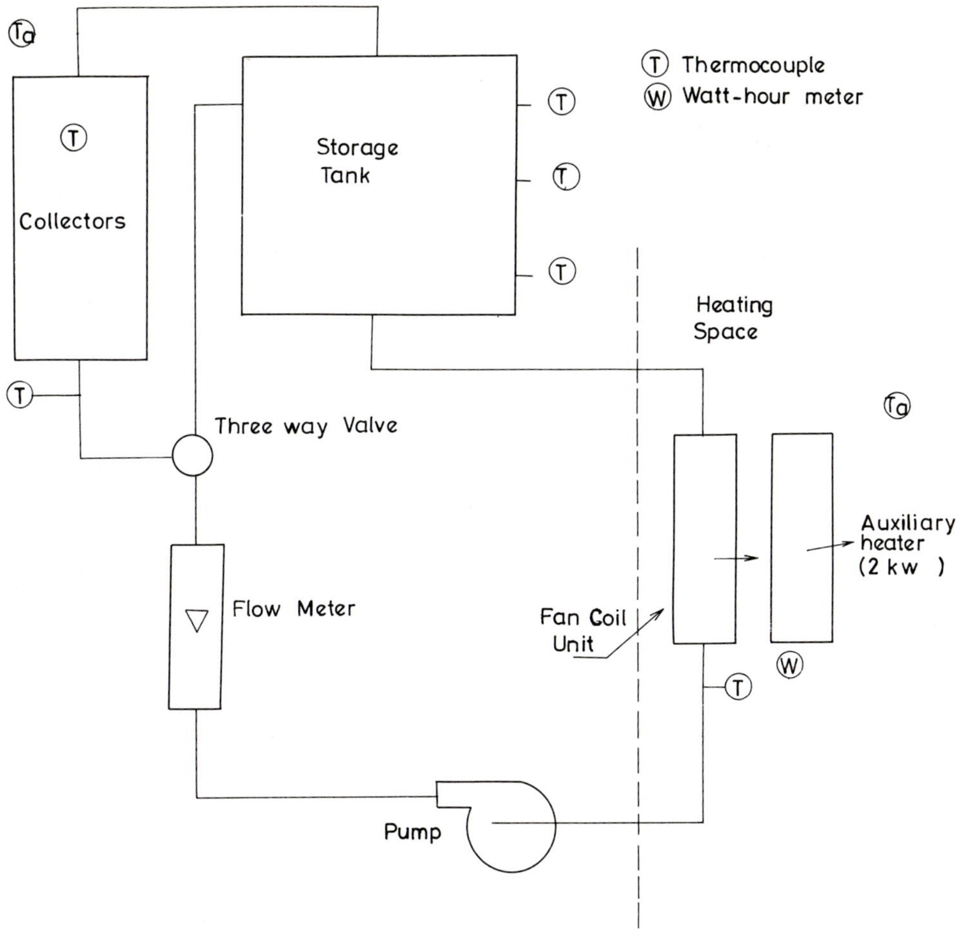

FIG.10 - SCHEMATIC DIAGRAM OF SOLAR SPACE HEATING SYSTEM USING WATER

6. ECONOMICS OF THERMAL INSULATION

The insulation thickness is normally determined for specific purposes such as to achieve specific temperatures and properties of process fluids, to achieve specified heat loss, to achieve specified surface temperatures, to prevent burns and condensation, etc. When there are no rigid specifications as above, the thickness is usually determined from economic considerations.

Thermal insulation reduces heat loss, but the cost of insulation gets added up. With the increase in insulation thickness, the amount of heat loss gets progressively reduced and the cost of insulation gets increased. Beyond a critical thickness, addition of insulation does not become cost effective as illustrated in Figs.14 and 15 (Ref.3).

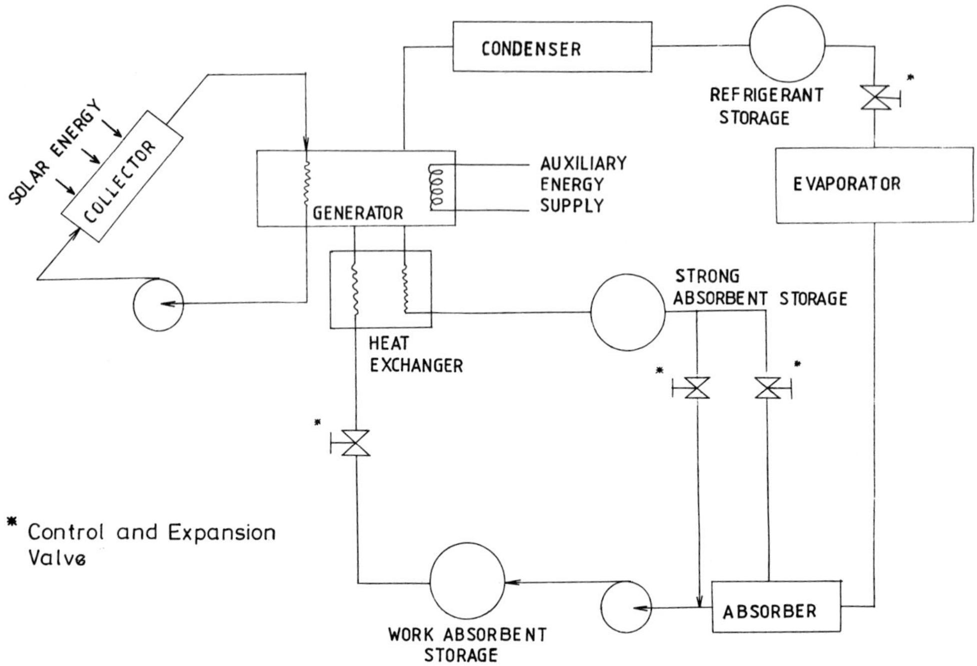

FIG.11 - SOLAR POWERED ILES ABSORPTION CYCLE HEAT PUMP SYSTEM

Cost factors involved in economic thickness calculations are:

Heat - Fuel cost, capital investment cost, interest, depreciation, maintenance, duration of operation, etc.

Insulation - Capital investment cost, interest, depreciation, maintenance etc.

These may be expressed graphically and minimum total cost worked out or may be obtained by use of computers. Marginal pay-back analysis determines the insulation thickness beyond which any additional thickness will not pay for itself for additional heat savings over a predetermined pay-back period.

Sometimes the economic thickness provides an insulation exposed surface temperature, which might not be acceptable from safety standards point of view. Then the insulation thickness has to be increased to take care of the above. However, the surface temperature is very much dependent on the surface resistance. For normal wind conditions, if surface emissivity is decreased, surface temperature increases without much difference in heat transfer as shown in Figs.16 and 17. Instead of steady state heat transfer calculations, if unsteady calculations are done, which are more appropriate and practical, the cost of heat loss and hence the cost of total heat loss show a lower value as illustrated in Fig.18.

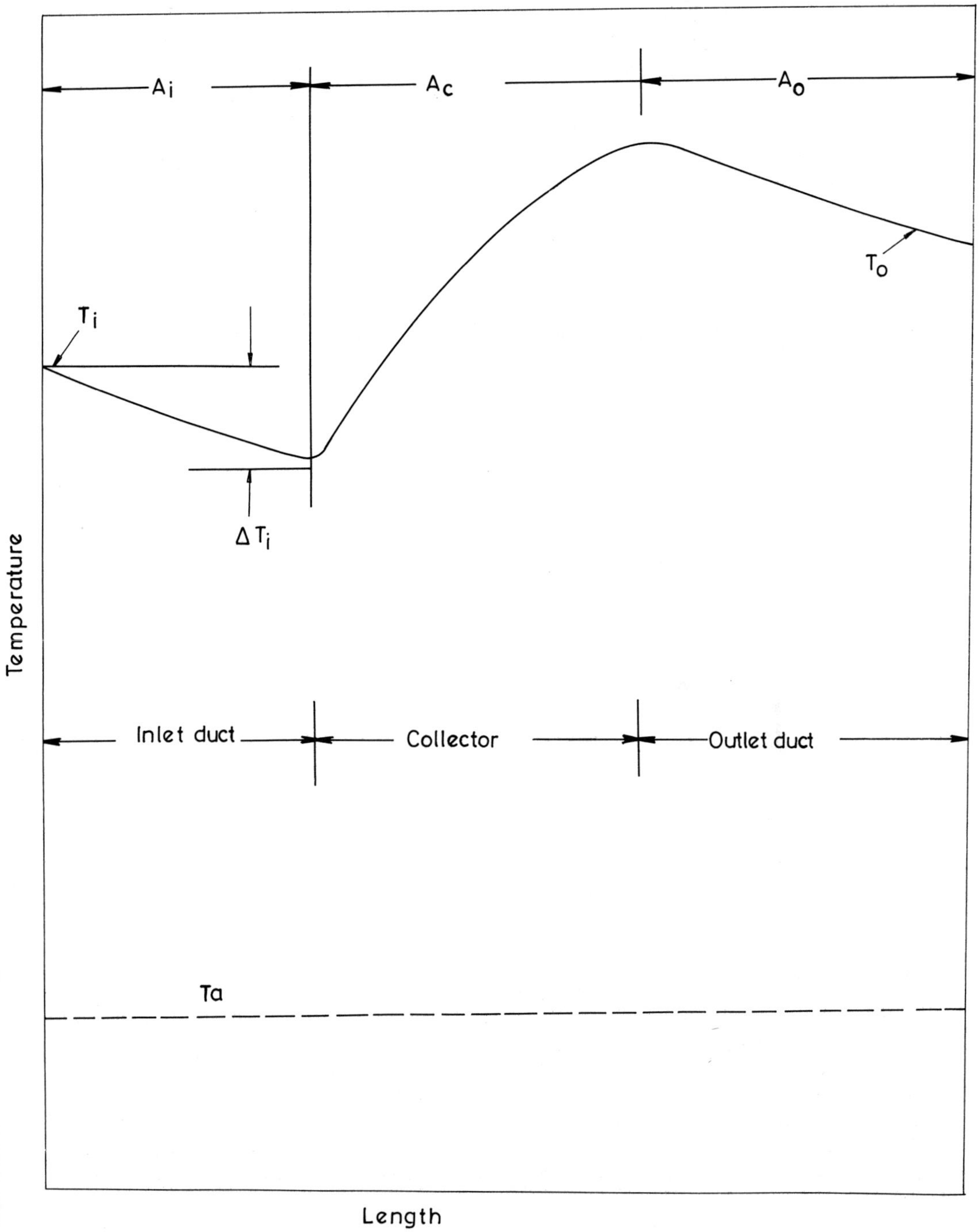

FIG.12 - TEMPERATURE PROFILE IN A DUCT

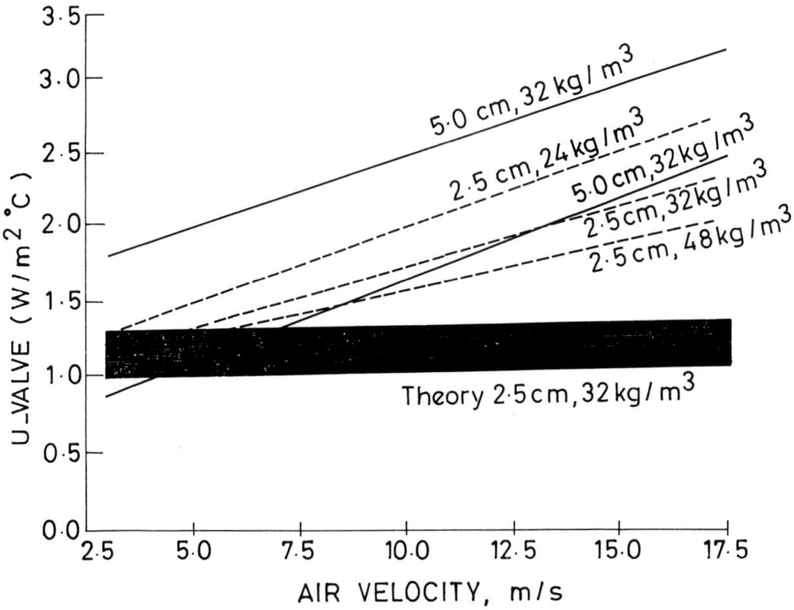

FIG.13 - EFFECT OF AIR VELOCITY ON THE OVERALL HEAT LOSS COEFFICIENT

7. COST EQUATIONS

$$\text{Total cost} = \text{Insulation cost} + \text{cost of heat}$$

$$= W \times \left[B.t + C + C_{uh}.n.H.q \right]$$

$$W = \text{Worth factor} = \frac{(1+i)^N - 1}{i(1+i)^N}$$

where

i = rate of interest (per cent)

N = plant life (years)

B = insulation cost, labour cost, efficiency of labour, transportation, depreciation and maintenance

t = insulation thickness

C = plant capital overhead

cost of heat = cost of useful heat x evaluation periods x heat loss

$$\text{cost of useful heat } (C_{uh}) = \frac{\text{cost of fuel/Kg}}{\text{conversion efficiency x calorific value/Kg}}$$

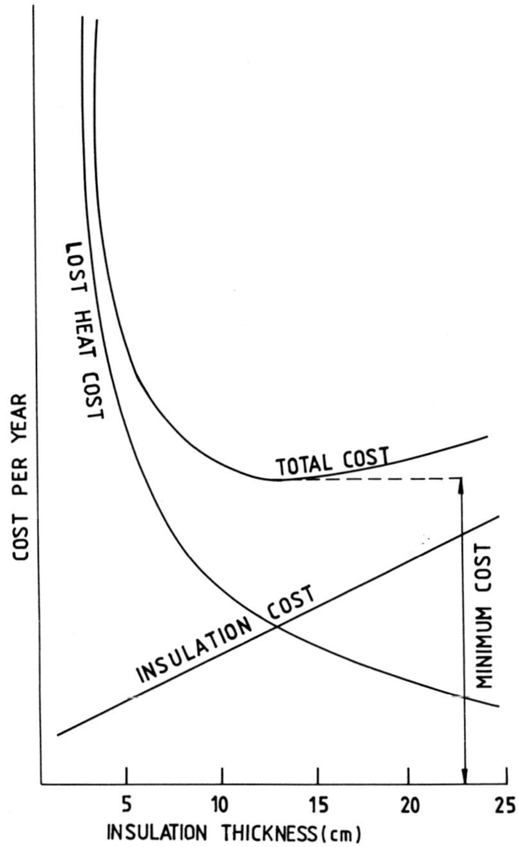

FIG.14 - COST OPTIMIZATION

$$\text{repayment period } (n) = \frac{1}{\text{return of capital} + \dfrac{1}{\text{plant life (years)}}}$$

evaluation period = n x H, where H is hours of operation per year

heat loss = q

Repayment period may be considered also as pay-back period.

8. HEAT LOSS EQUATIONS (Ref.2)

$$\frac{q}{A_o} = \frac{T_i - T_o}{\dfrac{r_3}{r_1 h_1} + \dfrac{r_3 \ln(r_2/r_1)}{K_1} + \dfrac{r_3 \ln(r_3/r_2)}{K_2} + \dfrac{1}{h_o}}$$

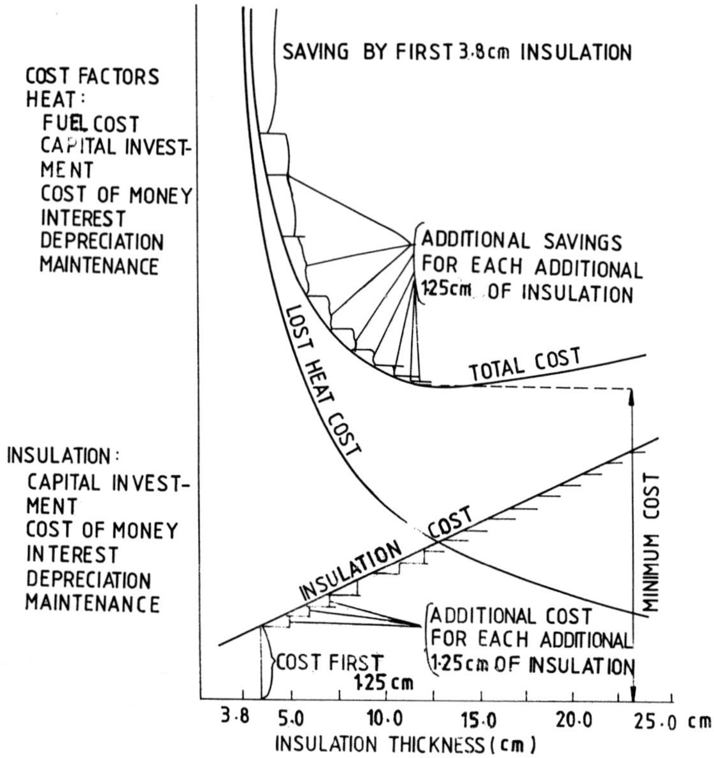

FIG.15 - ECONOMIC THICKNESS OF INSULATION

where,

r_1 = pipe radius, (outer) (m)

r_2 = radius of the inner insulation layer (m)

r_3 = radius of the outer insulation layer (m)

T_i = flowing fluid temperature (^oK)

T_1, T_2, T_3 = temperatures of the pipe surface, outersurface of the inner layer, outer surface of the outer layer respectively (^oK)

T_o = ambient air temperature (^oK)

A_o = outer area of the insulated pipe per unit length (m^2)

K_1 & K_2 = thermal conductivity of the inner and outer insulated layer at the mean temperatures of each insulation layer (W/m) (pipe metal conductivity is neglected)

h_1 & h_o = surface heat transfer coefficients of the inner fluid-pipe and the outer insulation-ambient air temperature interface (W/m^2K)

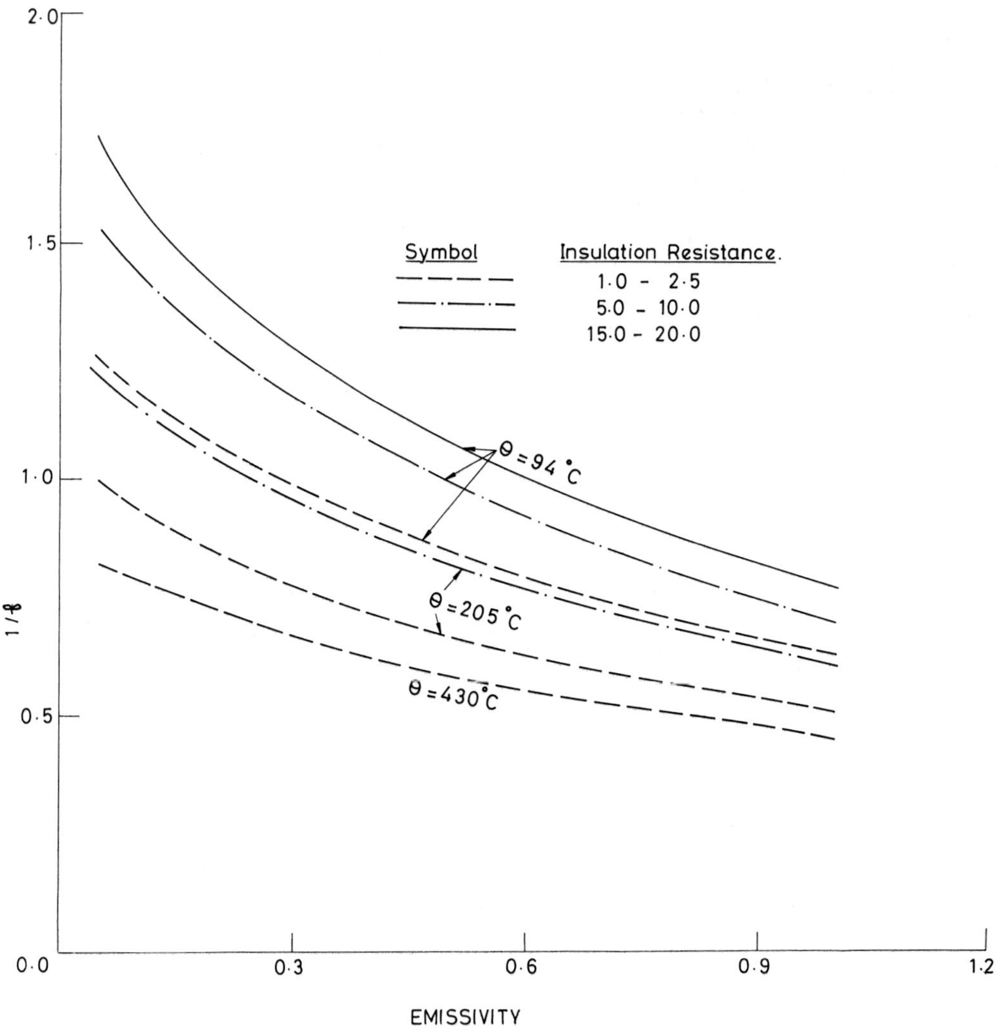

FIG.16 - CHART FOR OUTSIDE AIR RESISTANCE (1/f) AND OPERATING TEMPERATURE (θ) AT DIFFERENT INSULATION RESISTANCE AND EMISSIVITY

Assuming h_i = infinity, $T_i = T_1$, steady state conduction heat loss across the insulation is given by

$$\frac{q}{A_o} = \frac{T_1 - T_3}{\dfrac{r_3 \ln(r_2/r_1)}{K_1} + \dfrac{r_3 \ln(r_3/r_2)}{K_2}}$$

This can be equated to heat transfer from the insulated pipe surface

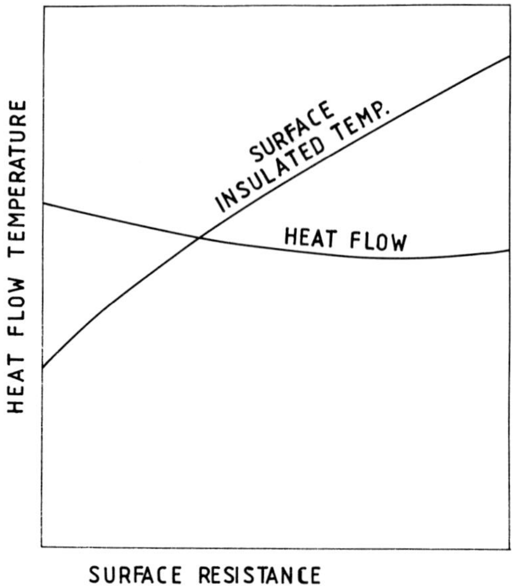

FIG.17 - EFFECT OF SURFACE RESISTANCE ON THE HEAT FLOW

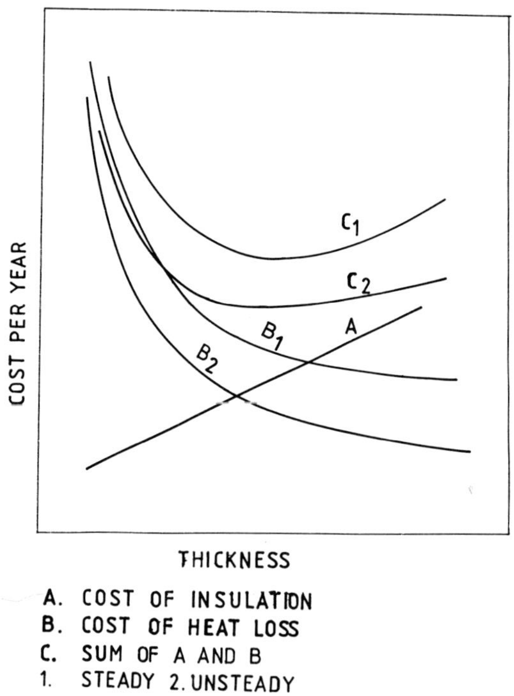

FIG.18 - EFFECT OF INSULATION THICKNESS ON THE COST UNDER STEADY AND UNSTEADY HEAT FLOW CONDITIONS

to the environment by convection and radiation using modified Langmuirs equation and Steffan Boltzmen equation

$$\frac{q}{A_o} = 3.152591 \left[C\left(\frac{0.0354}{2r_3}\right)^{0.2} \times \left(\frac{0.55}{T_{av}}\right)^{0.181} \times (1.84\ \ T_m)^{1.266} \right]$$

$$\times \sqrt{\frac{196.85V+68.9}{68.9}} + 5.669\epsilon \left[\left(\frac{T_3}{100}\right)^4 - \left(\frac{T_o}{100}\right)^4 \right], \ W/m^2$$

where, V = air velocity (m/s) on the external surface

ϵ = emissivity of the exposed surface to the environment

C = 1.016 (Horizontal cylinder) = 1.235 (long vertical cylinders)

$$T_{av} = \frac{T_3+T_o}{2}$$

$$T_m = T_3-T_o$$

For specific cases, the above two equations could be equated. T_3 being unknown could be obtained by using iteration technique and heat loss could be calculated. After the computation of heat loss, the cost equations could be used and economic thickness could be worked out. Computer programmes have been made to carry out the entire computation. The results may be presented as Nomograms for ready use by design engineers as illustrated in Fig.19 (Ref.1).

It may be pointed out that the cost factors could be converted into equivalent coal/fuel oil consumption which will indicate the saving of coal/fuel oil, which is the ultimate goal.

Fortunately, the use of solar thermal devices have not started being used recently, and hence upgrading the insulation systems for energy conservation is not necessary as are required in power houses, steel plants, fertilizers and petrochemical industries and other industries which were designed and installed before the oil crisis was focussed and made an important factor. Hence the optimization could be done properly at the design stage. However, the cost calculations should also take care of the rate of inflation so that timely representative life-cycle costing would be available for determining the most economic insulation thickness.

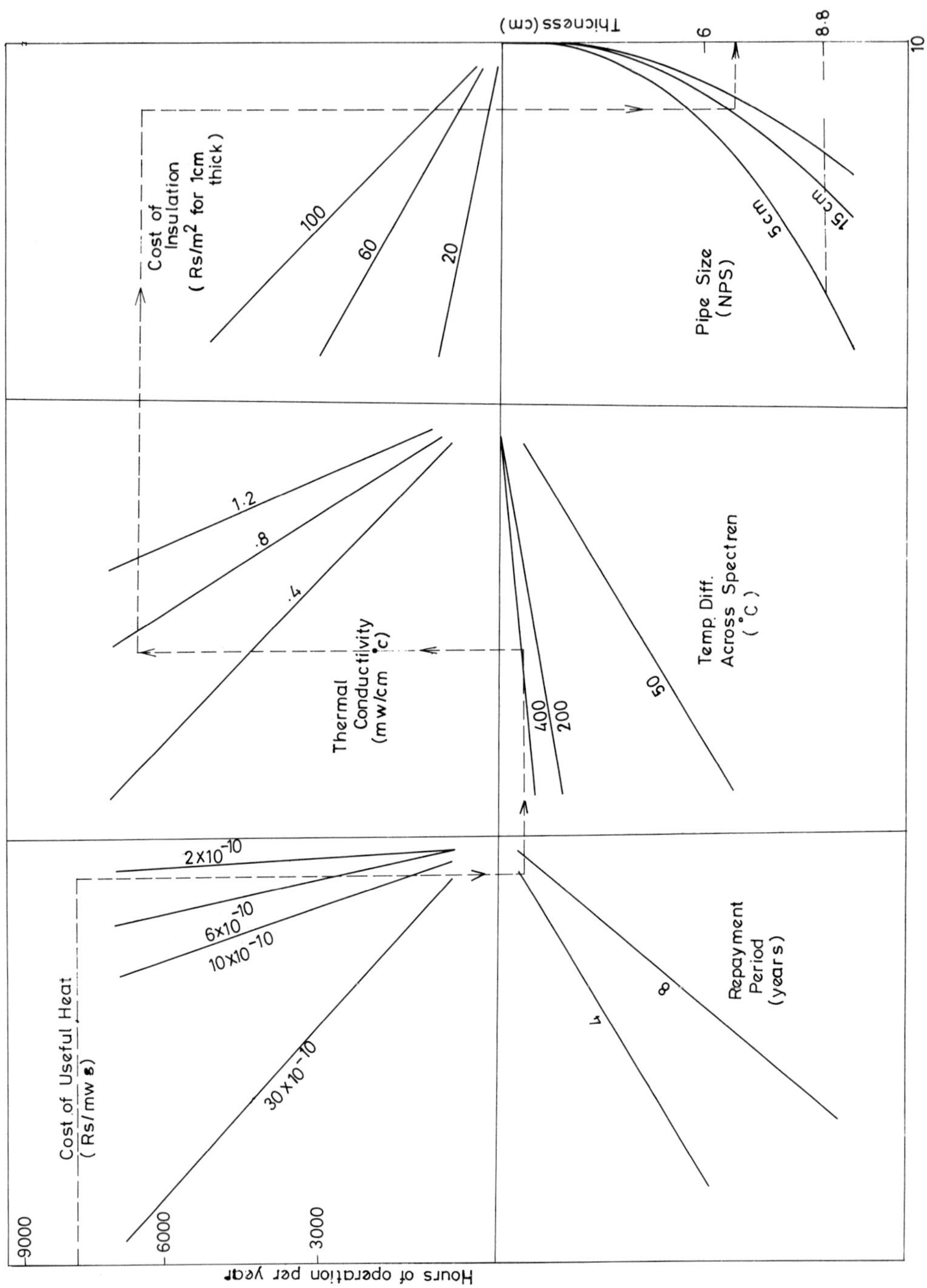

FIG. 19 - CHART FOR DETERMINING THE ECONOMIC THICKNESS

REFERENCES

1. D.K. Sen and B.C. Raychaudhari, (1976), 'Evaluation of economic thermal insulation thickness for hot pipe line', Mechanical Engineering Bulletin (India), 7(4), 106.
2. J.F. Malloy, (1969), Thermal insulation, Van Nostrand Reinhold Publishing Company, London.
3. W.C. Turner and J.F. Malloy, (1981), Thermal Insulation Handbook, McGraw Hill Book Co.

PAINTS AND PAINTING PROCEDURES FOR SOLAR ENERGY COLLECTORS

S.M. Singh
Central Building Research Institute
Roorkee
India

1. INTRODUCTION

A solar collector is a special kind of heat exchanger that transforms solar radiant energy into heat. A large variety of these solar collectors have been designed for trapping the solar energy. These collectors are usually made of copper, aluminium, steel or galvanized iron. A solar collector must have very high absorption coefficient for incident solar radiation and very low emissivity in the long wave length region. Means of enhancing the absorptance values of commonly used solar collectors, black paints and selective absorbers are used. Solar collectors designed for high temperature are given selective coatings, whereas normal paints are applied that are designed for 65 to 70°C. Selective absorbers are usually polished metal surfaces coated with a thin deposit of black oxides of nickel and copper and nickel and zinc sulphides. The most recent commercial selective surface which is being widely adopted in U.S.A. is black chrome. Black paints can economically and efficiently be used up to 65°C and are preferred for solar water heaters. The efficiency of these paints is good as they show a very high absorption coefficient for solar radiation (Table 1). However, they require frequent renewal also.

TABLE 1 - Solar Absorptance of a Few Surfaces

Metal	Normal Solar Absorptance
Aluminium	0.09 - 0.10
Aluminium anodised	0.12 - 0.16
Chromium	0.42
Copper polished	0.35
Iron	0.44
Nickel	0.34 - 0.43
Carbon black in acylic binder	0.94
Lamp black in epoxy	0.96
White zinc oxide	0.12 - 0.18

H. P. Garg (ed.), Solar Water Heating Systems, 153–158.

The application of plastic coatings to metal by utilizing electrostatic spray methods in conjunction with high density powder materials is opening new horizons for product designers and metal working manufacturers. Major types of powder materials are vinyls epoxies and nylons. Other types of materials available for electroplastic coatings include polyester, cellulose and polytetra fluoro ethylene.

As stated above, the solar collectors are generally made of nonferrous metals such as copper, aluminium and galvanized iron. These are more resistant to corrosion compared to iron and steel. Galvanic corrosion of these metals is prevented by taking precautions in the design of solar collector itself. The contact of copper with aluminium and of aluminium with steel is avoided. Where dissimilar metals are used in conjunction with each other, isolating jointing compounds such as bituminous paste plastic sheet and inhibiting primers become essential.

2. PAINTS

Paint is defined as a pigmented liquid composition which is converted to an opaque solid film after application as thin layer. It is made by mixing the pigment and vehicle (binder dispersed in solvent/thinner) grinding the mixture and tinting to correct colour. Drier, plasticizer and other additives are used depending on the type of vehicle and paint. The ratio of pigment to vehicle varies widely depending on specific properties of pigment and vehicle and the requirement of the coating. The ratio is usually expressed as PVC, meaning pigment volume concentration in the dry coating. The properties of pigmented coating that are important for a solar collector are colour and its stability, opacity, degree of flatness or glossiness, smoothness and resistant to light. These characteristics determine whether the light will be reflected, transmitted, refracted, absorbed or will undergo a selected combination of these effects.

3. PIGMENTS

Black pigments are carbon black, lampblack, graphite, bone black, vegetable black, mineral black, iron oxide black and antimoney sulphide. Of these, lamp black is most suitable for pigmentation of black paint for solar collectors. High grade lamp black for coating consists of about 99% carbon. It does not float and is easy to disperse. All lampblacks are readily dispersed in oil, varnish, and lacquers. The colour or mass tone of lamp black is in direct ratio of its particle size; the finer the particle, the greater the intensity of colour. The particle size of lamp black lies between 100 to 200 millicrons. The intensity of colour of the highest grade lamp black does not approach that of carbon, bone or mineral black. However, the soft dull tone of lamp black makes it distinctive from all other black pigments. Lamp black gives a soft semi dull finish. Because of the simplicity of the carbon particle of lamp black, there is very little bodying effect or increase of viscosity of the oil vehicle it is mixed with.

4. VEHICLES

The vehicle is a fluid material consisting of a solution of mixture of binder with a thinner (solvent). The binder is the primary constituent because it cements pigment particles together and fixes the whole mass to the substrate being coated.

4.1 Film Formation Solvent Evaporation

This class of binder consists of stable resinous materials which are thermoplastic, soluble in some solvents and chemically stable during film formation. The resin binder is deposited from solution and the film formation is completed when thinner/solvent has evaporated. It includes shellac, resin, etc. Unfortunately these binders yield films which have inferior weathering properties.

4.2 Film Formation by Chemical Reaction

In this class of binder, evaporation of thinner/solvent alone is inadequate to cause formation of hard film. The binder undergoes chemical reactions which convert it to form hard film. It includes drying oil alkyds and other thermosetting resins such as vinyls, phenolics and epoxies. Oils, oil varnishes and alkyds wet pigments fairly readily by contrast with high polymers. Considerable shearing stress is required to force them to wet or disperse the pigments.

4.2 Proportion of Pigment-Vehicle (Binder)

As discussed earlier a perfect black and non-reflecting coating is a prerequisite for a solar collector. That is a dead matt black coating is required. The film must be homogenious but should have a particulate structure. This is obtained by increasing the pigment/binder ratio. In general such a paint must contain a volume concentration of pigment that is more than its critical volume which depends on the nature of the binder and on the nature of and shape of pigment. However, such a coating shows poor weathering properties on outdoor exposure. The film cracks, and flakes off soon. As the proportion of binder is increased the film becomes more flexible and in general improves protective function but develops a little gloss. A compromise must therefore be made between durability and flatness of the coating.

Another way to increase the durability of the coating is to take resins other than shellac and oil such as alkyds, vinyls and epoxies. Magnesium silicate or any other flating agents may be incorporated to reduce the gloss of the coating. The durability and intensity of blackness of the coating is seriously affected by the degree of dispersion of the pigment in the vehicle. Good dispersion requires considerable mechanical energy to break down the pigment agglomerates into the ultimate particles. Also it is desirable to have a binder with wetting characteristics for the pigment surface. Addition of a wetting agent is sometimes found most desirable. Further it is important to control the particle size of the pigment so that the emissivity of the surface in long wavelength region remains low.

5. PREPARATION OF SURFACE

The chemical and physical nature of a metallic surface has a marked effect on the performance of coatings applied to it. The preparation of the surface may involve many steps such as removal of rust and mill-scale grease, oil, weld deposits etc. A wide variety of surface treatments for metals are available which are given below.

1. Solvent wipe off,
2. Vapour degreasing,
3. Alkali cleaning,
4. Emulsion cleaning,
5. Chemical clearning,
6. Mechanical cleaning,
7. Flame cleaning, and
8. Stripping of old paint.

Rust, if any, is removed by abrasion with emery paper and white spirit. Corrosion products of aluminium is removed by stiff bristled brush. Weld deposits should be ground smooth and fluxing materials used in soldering should be removed by solvents such as trichloroethylene by washing with hot water. It is important as weld flux slag is strongly alkaline and affects adhesion of paint particularly of oil paints to the metal.

5.1 Removal of Oils and Grease

Degreasing is particularly necessary for non-ferrous metals because they readily retain the grease used in their manufacture. New surfaces, particularly if polished, should be roughened or etched on site without weathering. Methods of surface preparation for zinc and galvanized iron, aluminium, and copper are given separately.

(a) Zinc and galvanized iron. Failure of paint on these surfaces is mostly attributed to the smooth and slightly greasy nature of the surface. Greasy surface offers little key to the paint film. As zinc is sensitive to strongly acidic and alkaline conditions, preliminary cleaning of the metal is done by neutral emulsion type cleaners. If time permits, the surface may be left exposed to weather for about six months or less depending on the atmospheric conditions. The surface should loose its luster and become dull in appearance as a result of the conversion of zinc into zinc oxide, a white corrosion product. The surface must then be cleaned free of dust and loose adhering materials.

Etch primer or wash primer is usually used to roughen the surface which is then followed by a normal primary paint. As etch primer consists of polyvinyl butyral resin in a suitable solvent to which phosphoric acid and zinc chromate are added.

Sheradised and electro-galvanized iron have a rougher surface than hot dip galvanized iron. Such surfaces should be smoothened before the application of any pretreatment.

(b) Aluminium and its alloys. In the rolling process and in the fabrication of plates and structural shapes and objects, a thin film of oil or oil emulsion is invariably left on aluminium surface and its presence may have a detrimental effect on the adhesion of paint. Weathering metal for a month or two will probably be enough to remove most of this oil film but degreasing is necessary. Aluminium is not an easy metal to clean because it is sensitive to both acids and alkalis and it is readily corroded by them. It is best to remove oil, grease or wax by solvents such as petrol or a mixture of white spirit and solvent naphtha,or better still, cellulose thinner. For degreasing aluminium it is best to use tetrachloroethylene. Degreasing with petrol and turpentine is usually carried out by rubbing the surface down with rags soaked in the solvent. The solvent should be frequently changed as it gets contaminated by grease and oil. Aluminium can also be degreased with sodium phosphate solution in water.

The second problem with aluminium is in achieving good adhesion of the coating. Slight roughening of the surface is desirable to provide mechanical anchorage to the paint film. On small areas this can be done necessarily by means of wire brushing or rubbing with emery paper. Treatment of etch primer provides the best possible adhesion. Exposure to damp conditions immediately after application of etch primer should be avoided.

(c) Copper and its alloys. The surface is degreased with white spirit or petrol. To improve the adhesion of paint film it is roughened with fine abrasive paper preferably wet or with white spirit. If degreased with sodium phosphate the pH of the solution must be btween 10 and 11.

6. APPLICATION OF PAINT

Painting system consists of priming coat, under coat and finishing coat. The application of the first coat of paint on a surface is known as priming. The main aim of priming is to set up good adhesion between the metal and the surface of the paint coating. Priming is carried out after the surface has been suitably prepared and with the shortest possible delay after this preparation has been finished. In selecting primers, it is necessary to take into account the nature of the metal to be painted. For aluminium, zinc chromate primer is used. Galvanized surfaces are preferably first phosphated and then primed with zinc chromate primer based on alkyds. For copper, primer paint containing inert pigments such as red iron oxide based on oil- or varnish-primers are suitable. Copper can also be painted without using primers.

Primer is usually applied by brush, spray or dipping. It must be applied in a thin uniform layer without any defect. The film thickness must not exceed 15 to 20 microns.

After the priming paint is dried, the surface is sanded with emery paper No.00. Sanding is used to improve adhesion between the successive layers of a coating.

The undercoat for solar collectors may be of the same composition as of the top coat. It would be good if it is diluted with the paint thinner and if necessary two coats may be applied to obtain the desired thickness. The surface is again sanded and wiped with a piece of cloth soaked with white spirit or petrol. It is next painted with finishing coat. Care is taken at each step to obtain a smooth surface, adequate mechanical strength and a good external appearance.

7. PRECAUTIONS DURING PAINTING

Paint should never be applied on damp or wet surface. Dipping and roller coating should be avoided for the application of priming coat. Spraying is quicker and gives a uniform coating. On rough surfaces, brush application gives good results. The thickness of the coating should be adequate to cover the peaks or irregularities in the surface. Most paints tend to recede from sharp edges and from the crown of corrugation and angles. In such places an extra strip coat of the paint should be given before the first full coat is applied over the whole surface.

8. MAINTENANCE

Repainting solar collector is normally necessitated due to dirt deposition and colour fadation as a result of weathering. Repainting should be undertaken as soon as the film starts showing defects such as excessive chalking, greying of paint film, and mild checking. If the oil film is sound, no elaborate surface preparation is necessary. The surface may be merely rubbed down to remove checked film and one coat of paint is applied over it. However, if the film has failed beyond repair, the surface is brought into good condition by removing the old coating. The bare metal is then wiped with a solvent and the repainting is then carried out as for a new surface.

LOW COST, HIGH PERFORMANCE SOLAR SELECTIVE PAINTS

O.P. Agnihotri
Materials Research Laboratory
Physics Department
Indian Institute of Technology Delhi
Hauz Khas, New Delhi 110 016
India

1. INTRODUCTION

Terrestrial solar radiation is a low-intensity, variable energy source arriving at about 1000 wm^{-2}. The economic feasibility of solar energy utilization depends upon efficient collection, conversion and storage. The efficient utilization of solar energy for heating, cooling, and process applications requires the use of flat-plate or focussing collector systems which first capture as much as possible of incoming radiation and deliver a high fraction of the captured energy to the working fluid. The conversion efficiency of a collector system is limited by the thermal losses from the heated absorber due to conduction, convection and radiation. The losses become increasingly significant at higher temperatures. The economical and efficient utilization of thermal energy derived from solar radiation using solar collectors requires an efficient and low cost 'solar selective coating' or 'selective surface'. An efficient solar selective surface is defined as having a high absorptance over the solar spectrum (0.30-2.0 microns) and, in addition also having a low emittance to reduce the thermal radiative heat losses. The achievement of such a surface with wavelength selective properties has been possible due to the fact that the solar spectrum and the thermal infrared spectrum of heated bodies do not overlap to any appreciable extent (for temperatures below 500°C, 98% of the thermal infrared radiation occurs at wavelengths greater than 2 microns). A parameter that has been used to characterise a solar selective surface is the ratio of solar absorptance to thermal emittance. However, the trade-off between absorptance and emittance for various collector systems has been discussed by various authors and it has been shown that an increase of absorptance is more effective in improving the operating efficiency than a corresponding decrease in emittance.

The two common types of coatings generally used are the plated and the paint coatings. Black nickel and black chrome are two important types of plated coatings which have been used for low temperature applications. The economic viability of such coatings is, however, questionable because of the relatively high material and applications cost. The objective of this article is to report on the low-cost

H. P. Garg (ed.), Solar Water Heating Systems, 159–168.

selective paints, whose commercial viability has been established (Ref.1-4). A new and the innovative approach on large area selective surfaces using reflective metal particles will also be discussed.

2. THE ROLE OF SELECTIVE SURFACE IN PHOTOTHERMAL CONVERSION EFFICIENCY OF A FLAT PLATE SOLAR COLLECTOR

The photothermal conversion efficiency, η , of a flat plate solar collector can be defined as:

$$\eta = \frac{\text{solar power absorbed - thermal power emitted}}{\text{solar power incident}}$$

$$= \frac{C\alpha_s E_s - \epsilon_{th} E_{BB}(T)}{CE_s} \tag{1}$$

where C = sun concentration ratio,

α_s = solar absorptivity of absorbing surface,

E_s = total solar power incident per unit area,

ϵ_{th} = thermal emissivity of absorbing surface,

$E_{BB}(T)$ = black body radiation power emitted per unit area at T^oK

and T = absolute temperature of absorbing surface.

Equation (1) can be rewritten as

$$\eta = \alpha_s \left[1 - \frac{1}{C\phi_s} \left(\frac{E_{BB}(T)}{E_s} \right) \right] \tag{2}$$

where ϕ_s = solar selectivity = α_s / ϵ_{th}. The minimum value of ϕ_s is unity for a black body. The maximum value of photothermal conversion efficiency η is α_s, the solar absorptivity of absorbing surface. The efficiency will be maximum if $C\phi_s >> E_{BB}(T)/E_s$. The value of ϕ_s can be maximised by a proper choice of absorber material and C can be maximised by the use of parabolic mirrors, heliostats and solar tracking servos, etc. Hence for maximum efficiency α_s should be as large as possible with a low emittance.

3. OPTICAL PROPERTIES AND COLLECTOR EFFICIENCY

Ramsey et al. (Ref.5) performed calculations using a transient thermal analysis which brings out the effect of the selective optical properties on collector efficiency. It is reported that for solar heating case, 50 per cent collector efficiency can be obtained with a good non-selective coatings (emittance = 0.95), while about 60 per cent efficiency can be obtained with a good selective coating (absorptance = 0.9, emittance = 0.1). For the solar cooling case, the nonselective coating would give a collector efficiency of about 36 per cent whereas the selective coating would give a collector efficiency of about 50 per cent. The calculations assumed a two-cover (glass) collector with efficiency averaged over a one day cycle. A selective coating is more effective in solar cooling applications.

4. SOLAR PAINT COATINGS

A solar paint coating consists of a very thin paint coating which allows a show-through of low emitting (high-reflective) substrates. In this, the optical properties are sensitive to thickness of the coating. The coating consists of a very thin (0.05 mil) paint coating applied to 1.0 mil thick aluminium foil. A pressure-sensitive adhesive is applied to a thickness of 1.0 mil. The coating is applied after removing the liner. Highly reflective metal particles when incorporated in solar paints produce a thickness-intensitive coating.

The solar selective properties of the paints rely on the intrinsic selective properties of the semiconductors. An ideal semiconducting material should have an energy gap between 0.5 eV (2.5 micron) and 1.24 eV (1 micron). Besides relying on the intrinsic solar selective properties, proper control of particle size and void fraction can be used to enhance the solar absorptance of the paints while maintaining a low value for the emittance (Ref.6,7). A polymer with a good transmittance in the infrared is used as a binder material. The pigment is absorbing in the solar spectrum and transparent in the infrared. The absorption of the pigment in the solar spectrum results in high solar absorptance and the low absorption of both pigment and binder in the infrared results in high infrared reflectance or low emittance.

Black inorganic pigments composed of copper, iron, manganese and cobalt oxides which are calcined at various temperatures are good candidates as binder materials. Organic pigments have low absorptance and due to instability problems are not used. In addition to the high solar absorptance and IR transparency, low refractive index, the pigment should also have good thermal stability, be weather resistant, and inexpensive. A good binder material should be transparent in 0.3 to 30 micron range, be adherent to substrates weather resistant and inexpensive. Silicones and aliphatic urethanes are good binder materials.

5. COATING THICKNESS AND PVC

The coating thickness and pigment volume concentration are the key parameters for achieving better optical efficiency. In order to get low emittance thicknesses of 0.1 mil or less are required. A PVC of 30 is reported to give high absorptance and low emittance (Ref.3). The coatings are applied either by means of a spray gun or dip coated using a special dipping machine which provides controlled insertion and withdrawal of the test panels into and out of the coating solution. Spray coatings are reported to be showing (Ref.3) higher reflectance values than coatings applied by brush.

A solar selective paint consisting of polyurethanealkyd lacquer that is pigmented with soot has been reported (Ref.8). The proportions are: 1 volume part of soot to 5 parts of binder and 60 parts of solvents. The solvent consists of a fifty-fifty mixture of xylen and 2-ethoxy-ethylacetate. The soot particles are about 20 nm in diameter and must be separated from each other for optimum performance of the paint. The optimum thickness for solar energy applications is 0.7 micron and is applied by spraying.

6. METAL DUST PIGMENTED PAINTS

In this investigation all the black coatings were deposited on zinc metal dust (size 325 mesh). The zinc powder was first cleaned to remove grease and any zinc oxide film, if present. A dilute solution of sodium carbonate was used to remove grease. A solution of 5 per cent HCl was used to remove zinc oxide layers.

For CuO deposition, a solution was prepared dissolving 10-30 g of $CuSO_4.5H_2O$ in 100 cc of boiling distilled water and adding 0.05 N sodium hydroxide. Upon addition of sodium hydroxide a precipitate forms at first and then dissolves. 100 g zinc powder (325 mesh) was added in a boiled solution. In a few seconds, the zinc metal powder became coated with an adherent black coating. The solution was decanted from the coated particles and then cleaned with distilled water several times using a sintered funnel. The coated particles were then dried in air. For the deposition of copper oxide on aluminium two more types of solution were prepared. One consisted of copper nitrate, concentrated nitric acid and potassium permagnate in 25,3,15 g/l concentrations in water. The deposition of aluminium metal powder (325 mesh) was carried out at 85-90°C for 15 min. In the other type the bath comprised $NaClO_2$ and NaOH solutions in water. The deposition was carried out for 10 min at 100°C. After deposition of the copper oxide on aluminium metal powder, it was cleaned with distilled or deionized water and dried in air.

For CuS deposition, separate solutions of copper tartrate, tartaric acid and sodium hydroxide were prepared in 15, 20 and 35 g/l concentrations and mixed at room temperature. The cleaned zinc metal dust was added at room temperature. Rapid stirring was carried out during the deposition. The coated particles were decanted from the solution and cleaned several times with distilled and deionized water and then dried in air. The coated powder so obtained was treated with a dilute solution of sulphur in CS_2 to convert the coating into sulphide form. For CuS+PbS deposition, the separate solutions of sodium thiosulphate, $5H_2O$, lead acetate, potassium hydrogen tartrate and $CuSO_4.5H_2O$ were prepared in 250, 25, 30 and 20 g/l concentrations respectively and were mixed together. The deposition was carried out by mixing 50 g zinc dust with 100 ml of solution at 50°C. An adherent coating was deposited in seconds onto zinc metal dust.

The coated metal dust was mixed in a resin and applied onto Al, galvanized iron and copper sheet by spraying or painting by brush. The coatings have excellent resistance to temperature, moisture and weathering extremes. They can withstand 16 h exposure at 250°C to 50 per cent hydrochloric, nitric and sulphuric acid and to 20 per cent sodium hydrogen solutions. The pigment to binder material ratio was 3:1. The curing was carried out at room temperature for 12 h. The selective paint was applied to a clean, dry surface at room temperature. The substrate should be clean to prevent poor adhesion before the paint application. A flow chart for the preparation of zinc dust pigmented solar selective paint is given in Table 1.

Figure 1 shows the transmission spectra of the silicon resin used as the binder material in the semiconductor, pigmented solar selective paint. The resin has two absorption peaks, one at 2-3 micron (0.5 micron broad) and the other at 8-9 micron (1.0 micron broad). The overall transmission was averaged to be of the order of 90 per cent. This suggests that the silicon resins are good selective absorbers except in the 8-9 micron and 2-3 micron ranges.

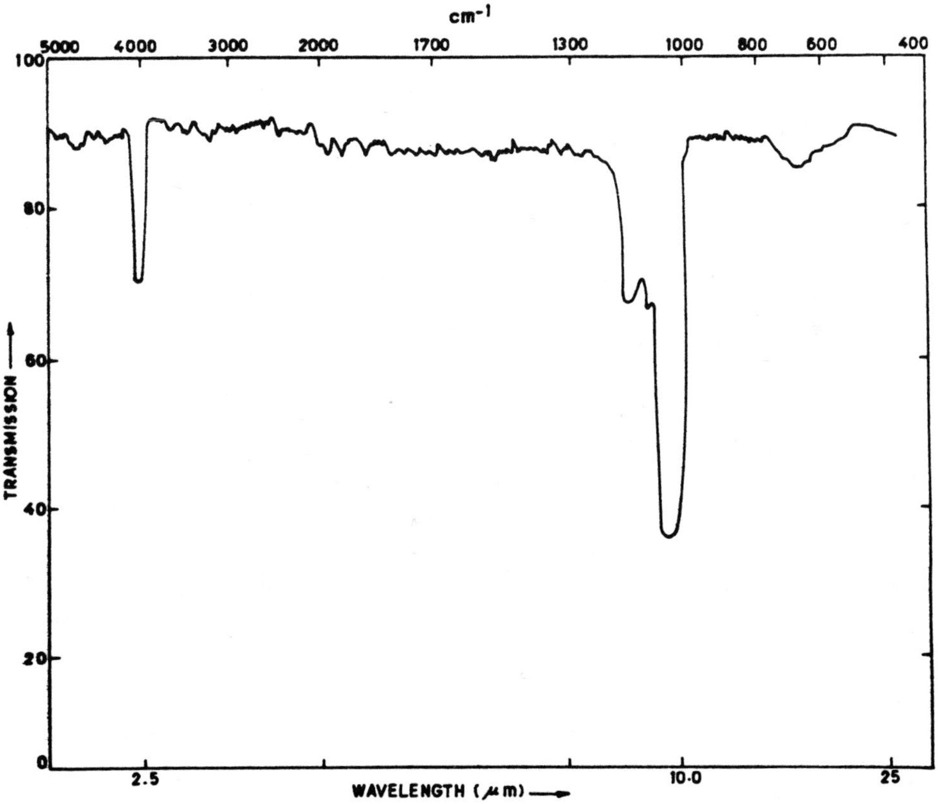

FIG.1 - TRANSMISSION SPECTRA OF SILICON RESIN IN INFRARED REGION

The emittance of the two thicknesses of silicon resin applied to aluminium and copper sheets has been measured. The emittance of the 8.0 micron thick silicon binder coating on copper varies from 0.36 at room temperature to 0.30 at 200°C while the 50.0 micron thick silicon

Table I. Flow chart for zinc dust pigmented solar selective paint

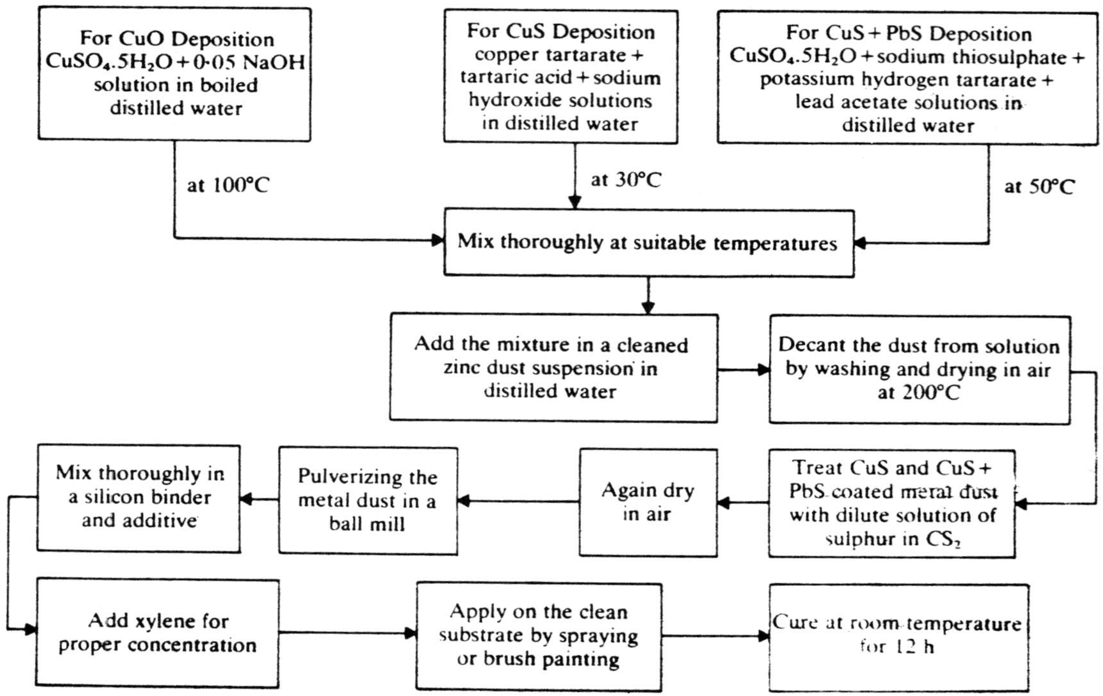

binder coating on copper varies from 0.80 at room temperature to 0.70 at 200°C. The emittance of the copper substrate is found to be almost constant over this temperature range. The decrease in emittance with increase in temperature is attributed to an intrinsic property of the silicon resin. It is obvious that the silicon binder itself has high emittance for thicknesses over 10 micron. Almost similar behaviour has been observed in the aluminium substrate.

The solar absorptance α_s values for the paint coatings were calculated by averaging spectral reflectance data in the visible and near infrared. The data were obtained from an SP8-100 Pye-Unicam and Beckman UV and visible ratio recording type spectrophotometers along with the diffuse reflectance accessory. The integrated sphere gives 8° reflections. The emissivities have been determined using a calorimetric technique. The results of solar absorptivity and thermal emissivity measurements on various samples are summarized in Table 2.

TABLE 2 - Measured Values of Solar Absorptance and Thermal Emittance of Some Selective Paints.

S. No.	Materials (pigment + silicon binder)	Substrate	Preparation technique	α_s	ϵ at 100°C
1*	Cupric oxide coatings on Zn powder by electrodeposition	Aluminium sheet	By brush paint	0.95	0.42
2*	CuS coating on Zn powder by electrodeposition	Aluminium sheet	By brush paint	0.94	0.51
3*	CuS+PbS coating on Zn powder by electrodeposition	Aluminium sheet	By brush	0.96	0.49
4	PbS (Ref.9)	Aluminium sheet	Spraying	0.96	<0.70
5	Ge powder (Ref.9)	Aluminium sheet	Spraying	0.91	<0.70
6	Si powder (Ref.9)	Aluminium sheet	Spraying	0.83	<0.70

* present investigation

Cupric oxide coated zinc dust pigmented solar selective paint gives $\alpha = 0.95$ and $\epsilon = 0.42$. The cupric sulphide and cupric sulphide + lead sulphide coating give $\alpha = 0.94$ and $\epsilon = 0.51$. The thickness of the coatings was of the order of 20 to 30 micron. The emissivity decreases as the thickness increases. The value of solar absorptivity does not change appreciably with the thickness of the coatings. Figure 2 shows a typical reflectance curve obtained for a cupric oxide coated metal dust pigmented solar selective coating applied to aluminium sheet.

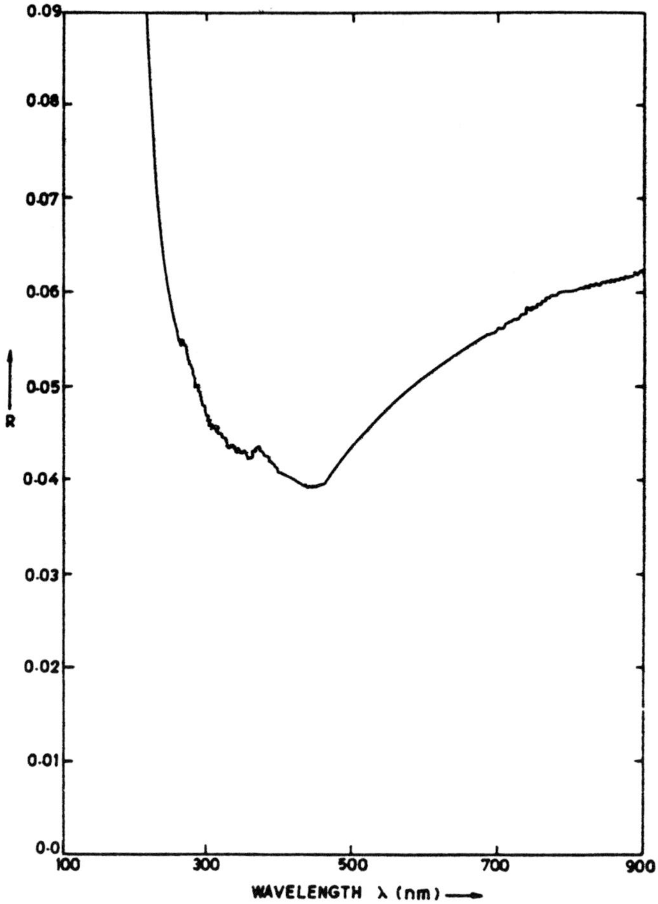

FIG.2 - REFLECTANCE SPECTRA FOR CUPRIC OXIDE COATED METAL DUST PIGMENTED SOLAR SELECTIVE COATING ON ALUMINIUM SHEET

The stagnation temperature of the solar selective surface was measured by enclosing a 30 cm x 30 cm piece in a wooden enclosure insulated on the rear side with glass wool. The measurements were carried out in a maximum solar flux of 80 mW/cm^2 between 20 and 30 January 1979 between 8 a.m. and 5 p.m. A stagnation temperature of 114°C was obtained. The results are shown in Fig.3. The improvement in the collector efficiency, which is the ratio of the temperature, increase above the temperature of the standard panel to the temperature increase of the standard panel above the ambient temperature, is estimated to be around 11 per cent.

Summarizing the results of the present investigations, Table 3 lists the advantages and disadvantages of metal dust pigmented solar selective paints with respect to their use in photothermal solar converters. It seems that these types of coating show considerable promise for large scale commercial production at low cost. The high emissivity values are attributed to an uncontrollable amount of black

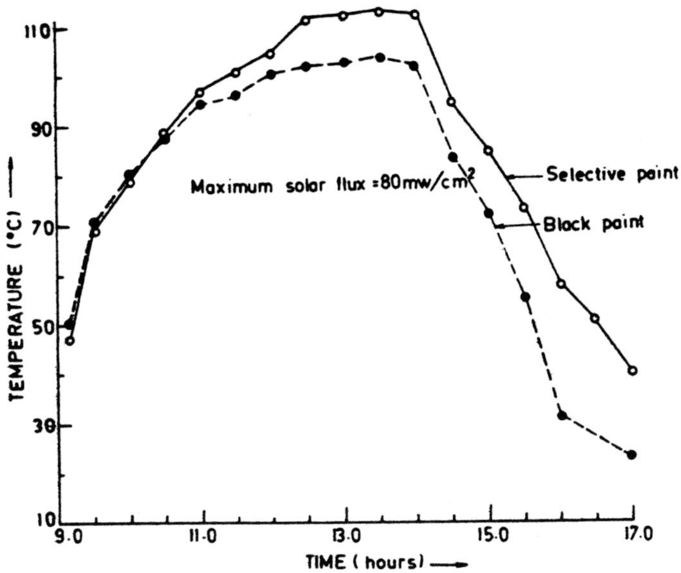

FIG.3 - STAGNATION TEMPERATURE CURVES FOR SELECTIVE AND BLACK PAINT UNDER MAXIMUM SOLAR FLUX OF $80mW/cm^2$ BETWEEN 20 and 30 JAN 1979

TABLE 3 - Advantages and Disadvantages of Selective Paints

S.No.	Advantages	Disadvantages
1.	Nice appearance	1. Poor selectivity
2.	Good corrosion resistance	
3.	Good adhesion to the substrate	
4.	Thermal infrared reflectance of the substrate is not important	
5.	Dry at room temperature	
6.	Large scale production technology exists	
7.	Good thermal stability	
8.	Good heat conductivity	

coating deposition onto fine metal particles. The thickness effect of black coating onto metal particles on the emittance of various selective paints is currently under investigation.

REFERENCES

1. R.B. Petit and R.R. Sowell, (1976), J. Vac. Sci. Technol., **13**, 596.
2. R.E. Peterson and J.W. Ramsey, (1975), J. Vac. Sci. Technol., **12**,174.
3. R.J.H. Lin and P.B. Zimmer, (1977), 'Optimisation of coatings for flat plate solar collectors', Final Report for ERDA under Contract No.EY-76-C-02-2930,000, July 1977.
4. H.Y.B. Mar, R.J.H. Lin, P.B. Zimmer, R.E. Peterson and T.S. Gross, (1975), 'Optical coatings for flat plate solar collectors', Final Report for ERDA under Contract No.NSF-C-957 (AER-74-09104),Sept.1975.
5. J.W. Ramsey, J.T. Borzoni and T.H. Holland, (1975), Development of flat plate solar collectors for the heating and cooling of buildings, NASA-CR-134804, June 1975.
6. D.M. Mattox and R.R. Sowell, (1974), J. Vac. Sci. Technol., **11**, 1973.
7. H. Tabor, (1976), Low Temperature Engineering Applications to Solar Energy, edited by R.C. Jordon (Am. Soc. Heating Refrigeration and Air Conditioning Eng., New York 1969), IV Chap.
8. S. Lofving, (1981), Sol. Energy Mat., **5**, 105.
9. O.P. Agnihotri and B.K. Gupta, (1981), Solar Selective Surfaces, John Wiley & Sons, New York.

SELECTIVE COATINGS FOR PHOTOTHERMAL CONVERSION

L.K. Malhotra and K.L. Chopra
Department of Physics
Indian Institute of Technology
Hauz Khas, New Delhi 110 016
India

1. INTRODUCTION

One of the simplest and most direct methods of harnessing solar energy is to convert the incident solar radiation into heat (called photothermal conversion). A variety of flat plate and concentrating type of collectors for this purpose have been designed and studied, and are being manufactured worldwide. One of the key components in photothermal conversion devices is the absorbing surface and its optical and thermal characteristics. With our present day understanding of the optical properties of surfaces, it is possible to tailor a surface by depositing a coating/film on it so that it yields literally any prescribed optical reflectance, transmittance, absorptance, or emittance spectra. Such surfaces/coatings having a selective response to solar spectrum are called solar selective coatings. Such coatings offer an important and cost effective way to increase the efficiency of photothermal collectors by providing a high solar absorptance (α) in the visible and near i.r. part (0.3-2.5 μm) of the spectrum and a low emittance (ε) to prevent radiation loss in the infrared (beyond 2.5 μm) in which region a heated black surface is expected to radiate.

The subject of selective coatings has been discussed in detail in several books and review articles (Ref.1-7). The Thin Film Laboratory of the Institute has done pioneering work in the development processes for obtaining selective coatings and in understanding their optical and thermal performance. This brief article describes salient features of such coatings with some emphasis on our work.

2. WHEN DO WE NEED A SELECTIVE SURFACE?

For a flat plate collector, significant gains in the collector efficiency are realized by using selective surfaces over the non-selective ones provided the absorber temperature is $\geq 100^{\circ}$C. However, for moderate heating applications, say 60-80°C, there is not much improvement in the efficiency by using a selective absorber. Thus, cheaper non-selective black paints would be preferred for low temperature heating. For

H. P. Garg (ed.), Solar Water Heating Systems, 169–178.

concentrating type collectors, a significant gain in efficiency is expected for selectively coated absorbers for moderate concentrations of about 10 to 50. For very high concentrations (~ 1000), there would again be not much gain in using a selective coating. The efficiency gain for small concentration values would be more marked if evacuated collectors are employed.

3. TRADE-OFF BETWEEN ABSORPTANCE AND EMITTANCE

An acceptable selective surface must have a minimum solar absorptance of 0.9 and a maximum thermal emittance at $100^{o}C$ of 0.2. Efficiency of photothermal conversion can be improved either by increasing absorptance or by decreasing emittance. In most of the practical solar collectors, increasing the value of absorptance is considerably more advantageous than reducing emittance value by the same amount. Jordan and Liu (Ref.8) and Tabor (Ref.9) have discussed the trade off between absorptance and emittance values which would be most effective for a particular collector system and operating temperature. For a single glass cover flat plate collector, if the solar flux is 800 W/m^2 and the mean plate temperature is $328^{o}K$, then a coating which increases absorptance by, say x per cent, is justified even if, at the same time, the emittance increases by 3.88x%. Extensive computations of the trade off conditions for various configurations of flat collector have been carried out by De Winter (Ref.10).

4. MODES OF OBTAINING SPECTRAL SELECTIVITY

The substrate material for a good solar selective absorber is usually metallic since only surfaces with a high conductivity can have a low thermal emittance. Thus, a plastic or a glass absorber would have to be metallized first (tin oxide and other similar coatings are special cases). On such a metallic or metallized substrate, a selective coating with the desired optical properties can be obtained by one, or often more than one, of the following effects:

(i) Intrinsic selectivity
(ii) Textural effect
(iii) Particulate effect
(iv) Interference phenomena
(v) Absorber-reflector tandem
(vi) Quantum size effect.

The above effects can be achieved by one of the following three techniques:

(a) Application of a thin, light absorbing and infrared transmitting coating on the substrate surface;

(b) chemical conversion of a metallic surface into a compound (sometimes followed by an additional deposition step) having properties like those in (a);

(c) modification of the topological features of the surface to provide varying optical properties for the shorter (visible and near i.r. and longer (far i.r.) wavelengths.

A summary of various selective surfaces prepared with their method of preparation and the absorptance and emittance values are listed in Table 1. Individual references can be found in our Review (Ref.1).

TABLE 1 - Summary of Selective Surfaces

Material	Fabrication technique	Absorptance	Emittance (T^oC)
1	2	3	4
INTRINSIC			
HfC	Sputtering	0.65	0.1(100)
Eu_2O_3			
ReO_3			
V_2O_5			
LaB_6			
YB_6	Plasma spray		
ErB_{12}	Plasma spray		
METAL/SEMICONDUCTOR			
a-Si	CVD	0.77	0.1
a-Si	Sputtering and chemical etching	0.94	0.5
A-Ge	Sputtering and chemical etching	0.94	0.55
$Si+Si_3N_4$	CVD + Vac. evap.	0.76	0.07(500)
Si+ITD	Spray pyrolysis	0.80	0.05(100)
CuO_x	Chem. conv.	0.98	0.15(100)
	Chemical etching	0.93	0.11
	Anodization	0.95	0.2
	Spray pyrolysis	0.93	0.11(80)
Cu_2O	Thermal oxidation	0.85	
CoO_x	Anodization	0.93	0.24(260)
	Thermal oxidation	0.87	0.07(60)
	Electroplating	0.95	0.2
	Spray pyrolysis	0.88	0.14
$CoO+Fe_2O_3$	Thermal oxidation	0.90	0.30(140)
ZnO	Chem. conv.	0.93	0.08
	Anodization	0.95	0.08

TABLE 1 (cont'd)

1	2	3	2
WO_3	Sputtering	0.83	0.07
CoS	Electroplating	0.97	0.20
Cu_2S	Chem. conv.	0.79	0.2(200)
	Spray pyrolysis	0.89	0.25
PbS	Vac. evap.	0.98	0.2(24)
	Spray pyrolysis	0.92	0.21
FeC_x	Sputtering	0.80	0.02(150)
Black Nickel	Electroplating	0.95	0.18
	Chem. conv.	0.94	0.15
Black Molybdenum	CVD	0.82	0.08
	Electroplating	0.85	0.11
	Chem. conv.	0.91	0.09
Black Enamel/SnO_2	Spray pyrolysis	0.92	0.15
COMPOSITES			
Al_2O_3-Ni	Vac. evap.	0.94	0.40(150)
	Anodization + Electroplating	0.93-0.96	
Al_2O_3-Pt	Vac. evap.	0.94	0.07(150)
Al_2O_3-Au	Sputtering	0.95	0.025
Al_2O_3-Cu	Sputtering	0.90	0.045
Al_2O_3-Cr	Gas evap.		
Al_2O_3-Co	Vac. evap.	0.94	0.31
Cr_2O_3-Cr	Sputtering	0.92	0.08
	Electroplating	0.95	0.07
	Plasma spray	0.90	0.5(80)
	Ebonizing	0.90	0.50
SiO_2-Fe	Sputtering	0.90	0.03
MgO-Au	Sputtering	0.93	0.09
Cr_2O_3-MoO_2	Electroplating	0.96	
CuO-Cu	Sputtering	0.94	0.04(120)
CaF_2-Si	Sputtering	0.64	0.06

TABLE 1 (cont'd)

1	2	3	4
CaF_2-Ge	Sputtering	0.72	0.10
Cr,Fe,Mo,SS,Ta,Ti, W-Silicides	Sputtering	0.85	0.02
Cr,Fe,Mo,Ni,Ta, W-carbides	Sputtering	0.90	0.06
Ni-Nic	Sputtering	0.90	0.12
INTERFERENCE			
$Al_2O_3/Mo/Al_2O_3$	Vac. evap.	0.95	0.34(100)
	Sputtering	0.95	0.12
Al/Ge/SiO	Vac. evap.	0.79	0.012(100)
Ni/Ge/SiO	Vac. evap.	0.88	0.035(100)
Cr/Ge/SiO	Vac. evap.	0.93	0.11(100)
Al/PbS/SiO	Vac. evap.	0.89	0.02(100)
Ni/Pbs/SiO	Vac. evap.	0.93	0.43(100)
PbS/CdS,PbSe/CdSe, $(PbS)_{1-x}(CdS)_x$	Soln. growth	0.88-0.93	0.10-0.16(100)
PbS/SnO_2,PbS/ZnO, $PbS/CdS/SnO_2$	Soln. growth + Spray pyrolysis	0.88-0.93	0.10-0.16(100)
SiO/Cr/SiO	Vac. evap.	0.88	0.1
$MgF_2/Mo/CeO_2/Mo$	Vac. evap.	0.95	0.07
$MgF_2/Mo/MgF_2/Mc$	Vac. evap.	0.89	0.07
PbS/ZnS	Vac. evap.	0.75	0.027(100)
PbS/Al_2O_3	Vac. evap.	0.71	0.032(100)
WO_2/Al_2O_3	Sputtering	0.93	0.09
TEXTURAL EFFECTS			
Tungsten dendrite	CVD	0.99	0.26
Steel dendrite	CVD	0.89	0.47(260)
Nickel dendrite	CVD	0.95	0.6(100)
	Vac. evap.	0.70	0.35
Rhenium	CVD		
Gold smoke	Gas evap.	0.99	0.1(100)

For large scale applications, a selective absorber should meet simultaneously the following requirements:

(i) a high solar absorptance with a low thermal emittance;
(ii) long term stability at the operating temperature;
(iii) stability under stagnation conditions when no energy is being extracted from the collector;
(iv) stability against atmospheric corrosion;
(v) adaptability for large scale fabrication;
(vi) low cost of production.

As a consequence, even though combinations have been prepared by a wide variety of techniques, on the laboratory scale, the feasibility of a large scale production has been met by only a few coatings. Prominent among them are the electroplated black chrome, nickel pigmented anodized aluminium and black Ni, and black copper prepared by chemical conversion. The main reason for the under-utilization of many of the selective coatings are their higher cost or difficulty in large scale production, or both. A commercially viable technique for fabricating solar selective coatings should have simplicity, low fabrication cost, large area coverage per batch and reproducibility. The technique should also be applicable to various collector shapes (e.g. tubular, parabolic, etc.). Solution growth techniques - extensively studied in our laboratory for over a decade now (Ref.11) meet these requirements in general. Some of the advantages offered by them are: (i) no need of a complicated fabrication system, (ii) fewer maintenance problems, (iii) lower capital and production costs and (iv) easy scale up to larger areas. We have in the last couple of years developed Black Nickel, Black Molybdenum and Black Copper by utilizing a simple immersion technique. Such electroless, simple immersion techniques are expected to play a leading role in the large scale fabrication of absorber panels in the coming years. In the following, we describe salient features of Black Copper, Black Molybdenum and Black Nickel developed by simple electroless processes in our Thin Film Laboratory (Ref.11-14).

5. BLACK COPPER

Selective coatings of black copper on copper have been obtained by dipping clean copper foils or sheets or tubes in an alkaline persulfate solution of potassium peroxodisulfate whose temperature is controlled within $\pm 2^{o}C$. The specular and total reflectance of an optimized coating are shown in Fig.1. The solar absorptance and hemispherical thermal emittance at $100^{o}C$ (ε_{100}) for this coating are 0.98 and 0.15, respectively. The coating has a highly textured surface and gives a velvety look on visual inspection. The SEM study reveals the spacing between two successive dendrites to be about 1.5 μm which is sufficient to allow multiple reflections of solar wavelengths but simultaneously is short enough to make the surface appear smooth to the infrared wavelengths.

The Auger and xPs study shows that the black copper film is composed of a top layer of about 200Å thickness of CuO followed by a region of about 0.7 μm thickness containing predominantly the Cu_2O phase with small amounts of CuO also in this region. This is followed by a graded region

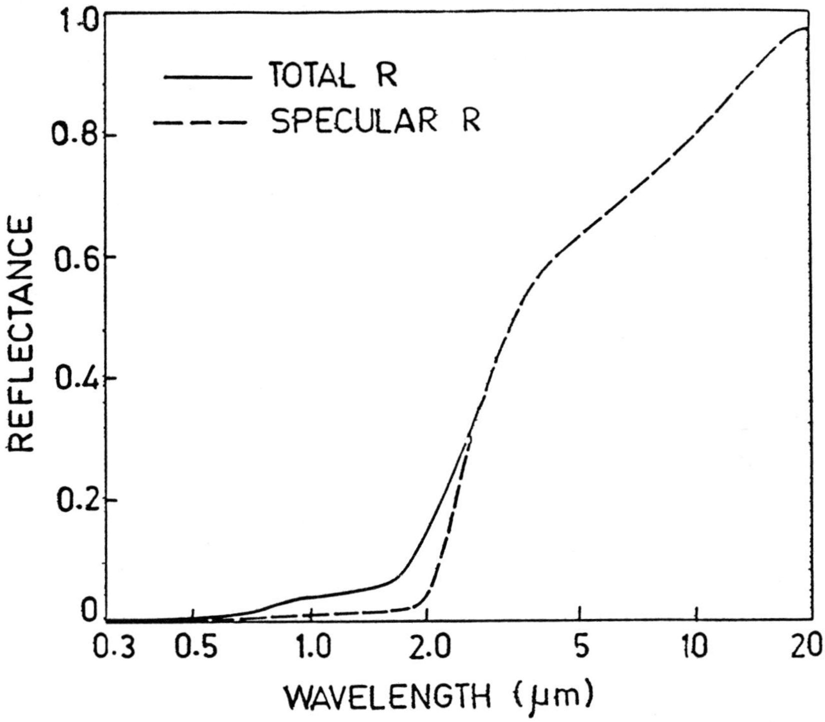

FIG.1 - SPECULAR AND TOTAL REFLECTANCE OF A BLACK COATING ON COPPER

of Cu_2O and Cu down to the substrate with increasing amount of Cu as the substrate is approached. The thin CuO region on the top surface acts effectively as an antireflection layer. Thus, a high solar absorptance is obtained by a combination of antireflective, textural and resonant scattering effects in the film. The coatings are stable in air for temperatures of up to 300°C. The results of various stability tests are shown in Table 2.

6. ELECTROLESS NICKEL BLACK

Nickel black coatings on galvanized iron substrates have been obtained by dipping these substrates in a solution containing nickel sulphate, sodium hypophosphite, ammonium hydroxide sodium acetate and a brightening agent. The coatings have an absorptance $\sim$ 0.90-0.94 with corresponding thermal emittance at 100°C, of the order of 0.07-0.24. The coatings are stable for temperatures less than 250°C. On continued exposure to humidity, white corrosion products start forming. Hence the coating is better suited for evacuated collectors or collectors where sealing of glasspanes is very good. As revealed by AEs, xPs and electron diffraction data, the coatings contain Zn, Ni and Nio in relatively large amounts with small traces of Ni_2O_3 and Ni_7P_3.

TABLE 2 - Values of α and ε_{100} after various Stability Tests on Black Copper.

Test condition				100
Air deposited			0.98	0.15
Air annealed,	$100^{o}C$,	100 hr	0.97	0.14
Air annealed,	$200^{o}C$,	100 hr	0.97	0.15
Air annealed,	$300^{o}C$,	100 hr	0.96	0.15
Air annealed,	$350^{o}C$,	100 hr	0.93	0.14
Air annealed,	$400^{o}C$,	100 hr	coating peels off	
Vacuum annealed,	$200^{o}C$,	10 hr	0.88	0.13
UV exposed,		100 hr	0.97	0.16
Corrosion test*,		100 hr	0.95**	0.23
Steam exposed,		100 hr	0.96	0.17
Outdoor exposed,		4 months	0.97	0.14

*Corrosion test involved boiling the sample in a 5% (wt) NaCl, solu.

**After this test, there were a few white spots on the sample, presumably due to the salt getting trapped at these sites.

Nickel black coatings can be obtained on aluminium also. The deposition process requires an active Zn surface for deposition. Therefore, the aluminium substrates were zincated before coating them with nickel black. The optical properties and stability characteristics are similar to those of the coatings on galvanized iron; however the emittance of coatings on zincated Al is much higher, possibly due to the formation of a thicker film on zincated surface.

7. ELECTROLESS MOLYBDENUM BLACK

Molybdenum black coatings are obtained by dipping clean galvanized iron or zincated aluminium substrates in a solution containing ammonium paramolybdate, sodium acetate sodium hypophosphite and coumarin. Optimized deposition conditions yield films with absorptance $\sim$ 0.90-0.91 on galvanized iron and aluminium substrates. As for electroless nickel black, the emittance of GI (0.09 at $100^{o}C$) is much lower than that on zincated aluminium (0.24 at $100^{o}C$). The films are stable below $300^{o}C$. A graded composition profile of elemental Mo and Zn embedded in a dielectric matrix of MoO_3 seems to be responsible for the optical selectivity.

8. OPTICAL MODELLING

Most of the initial research on selective coatings was product oriented. Selective absorbers were produced by adopting existing processes. The processes were empirically optimized to yield good absorptance and low emittance. Not much effort was directed towards the understanding of the basic absorption and emission mechanisms. The trend changed in late seventies and emphasis was laid on understanding the basic physics with a view to tailor make good selective absorbers. A number of standard methods are now available to design theoretically a good selective surface. For a comprehensive study, the reader is advised to refer to our Review (Ref.1).

8. CONCLUSIONS

(i) Application, material economics, energy inputs and desired system life would determine the choice of an effective coating-substrate-deposition technology package.

(ii) Significant increase in the efficiency of thermal conversion with a marginal increase in the cost of selective surfaces over black paints makes their use absolutely essential and desirable for medium and high temperature applications.

(iii) Chemical conversion of substrate materials to obtain their oxides which are selective is one of the most economical and convenient methods. Black copper, black nickel and black molybdenum prepared this way have given excellent performance.

(iv) Coatings of any desired R, T and ε characteristics on a number of substrate materials can be designed, modelled and tailor made today.

(v) Practical and efficient selective coatings are based on a mixed tandem, gradient index and composite structures which are sensitively dependent on deposition and structural parameters. This makes the selective coating technology a high level technology.

REFERENCES

1. K.L. Chopra, D.K. Pandya and L.K. Malhotra, (1984),'Solar Selective Coatings',Chapter 12, Reviews of Renewable Energy Resources, Vol.2 (Edited M.S. Sodha, S.S. Mathur and M.A.S. Malik), Chapter 3, Wiley Eastern Limited, New Delhi.
2. C.G. Granqvist,(1981), Appl. Opt., **20**, 2606.
3. M.M. Koltun, (1981), Selective Optical Surfaces for Solar Energy Converters, Allerton Press Inc., New York.
4. O.P. Agnihotri and B.K. Gupta, (1981), Solar Selective Surfaces, John Wiley and Sons, New York.
5. C.M. Lampert, (1979), Solar Energy Materials 1(1979)319; 2(1979)1.
6. B.O. Serphain in Topics in Applied Physics, Vol. 31 (Ed. B.O.Seraphin), Springer Verlag, Berlin (1979); in Physics of Thin Films, Vol.10 (Ed. Hass and Francombe), Academic Press, New York (1979).

7. A.B. Meinel and M.P. Meinel, (1976), Applied Solar Energy, Addison Wesley, Reading M.A.
8. R.C. Jordan and Y.H. Liu, (1977), Applications of Solar Energy for Heating and Cooling of Houses, Chapter VI (Ed. R.C. Jordan and Y.H. Liu), ASHRAE, New York, 1977.
9. H. Tabor, Ibid., Chapter IV.
10. F. de Winter Proc. Conf. Am. Sect. ISES, Winnipeg, Canada, Vol. 6, 216-239, Aug. 1976.
11. K.L. Chopra, R.C. Manthla, D.K. Pandya and A.P. Thakoor, (1983), Physics of Thin Films, Vol.12 (Ed. Hass, Francombe and Hoffman), Acad. Press, New York (1983), p. 168.
12. S.N. Kumar, P. Richaria, L.K. Malhotra and K.L. Chopra (To be published).
13. S.N. Kumar, Ph.D. Thesis, IIT Delhi, 1984 (Unpublished).
14. S.N. Kumar, L.K. Malhotra and K.L. Chopra, Solar Energy Materials, 3, 519 (1980).

INSTRUMENTATION AND CONTROLS FOR SOLAR WATER HEATING SYSTEM

M. Ramakrishna Rao
Instrumentation and Services Unit
Indian Institute of Science
Bangalore 560 012
India

1. INTRODUCTION

Solar water heating systems employing flat plate collectors are increasingly in use for domestic and industrial applications. The capacities of this system range from 100 litres to few thousand litres per day in typical applications. Hot water supply from 55°C to 75°C is generally catered for. Two types of SWH systems are used, viz. (i) Thermosyphon system, (ii) Forced circulation system. The former uses minimum of peripheral instruments/controllers and is therefore easy to maintain and operate. The latter employs motor driven pumps with on-off controllers and call for ac/dc power supplies for satisfactory operation. The configuration of a thermosyphon SWH is shown in Fig.1. The essential requirement for satisfactory application of water system is to use storage tank positioned at a relative height of one foot above the height of flat plate collector. Use of gate/globe valves, non-return valves, relief valves/vents and plumbings of specified sizes is an essential feature of a system. It will be noticed that a Thermosyphon system does not use any instrumentation and is generally a low capacity unit for domestic applications. The all day efficiency of the system under favourable conditions of insulation is around 40-50%.

Forced circulation types of SWHs are shown in Figs. 2 and 3. The system of Fig.2 employs a single tank and a differential thermostat for switching on-off the motor driven water pump. A more elaborate system of the forced circulation type, shown in Fig.3, employs heat exchangers and expansion tanks apart from motor driven pumps and differential thermostats, finds application where the heat transfer fluid is either hand water or antifreeze liquid such as Glycol/water mixture.

From the brief description of SWH system given above, it is seen that instrumentation and controls are the inescapable part of the system make-up, though minimal for the thermosyphon system. The various instruments/controllers employed for a SWH system are thus the following

(i) Single channel/multichannel temperature indicators.
(ii) Differential temperature controllers.
(iii) Water flow meters.
(iv) B.Th.U energy meters.

H. P. Garg (ed.), Solar Water Heating Systems, 179–198.

(v) Programmable system controllers.
(vi) Solenoid operated gate/globe valves.
(vii) Motor driven pumps.
(viii) Solar intensity insolation meters/indicators/controllers.

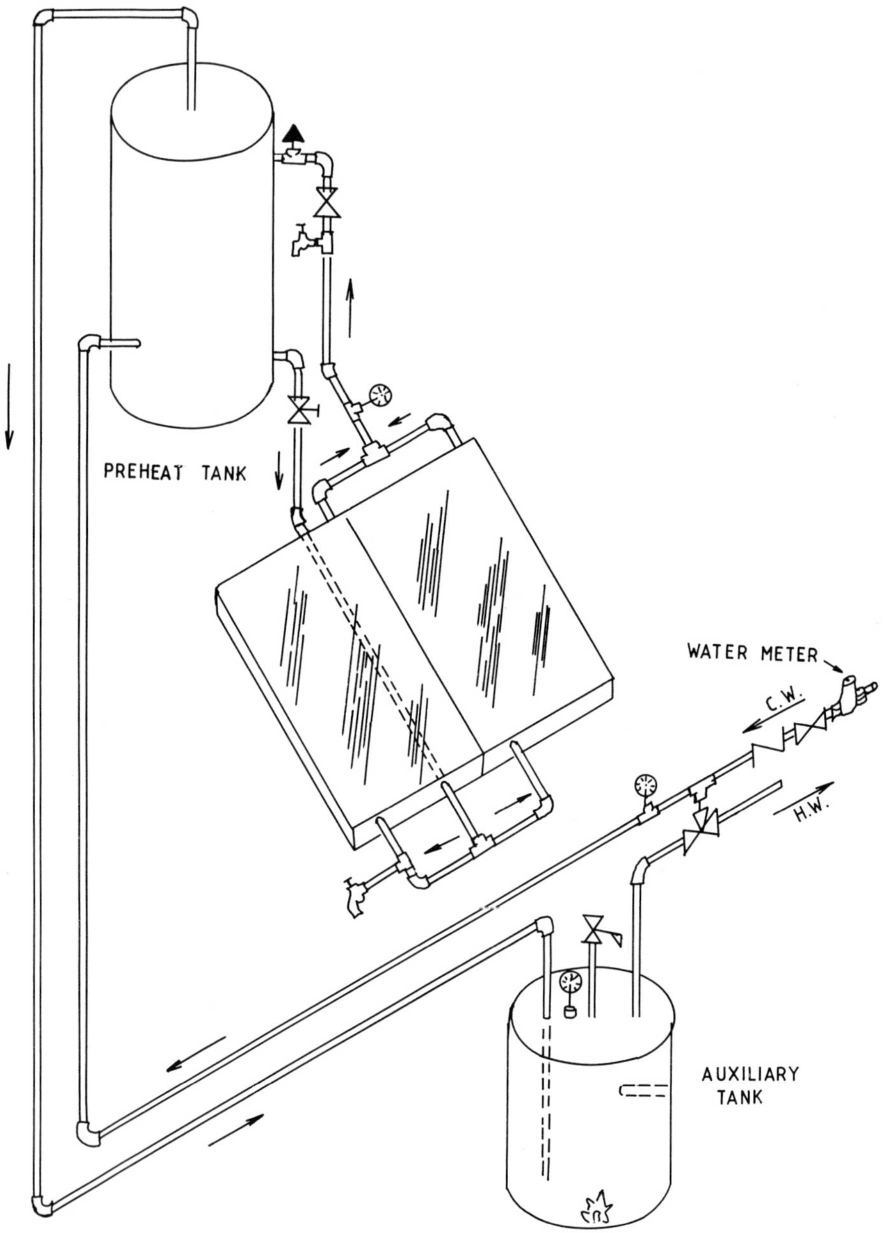

FIG.1 - THERMOSYPHON SOLAR WATER HEATER

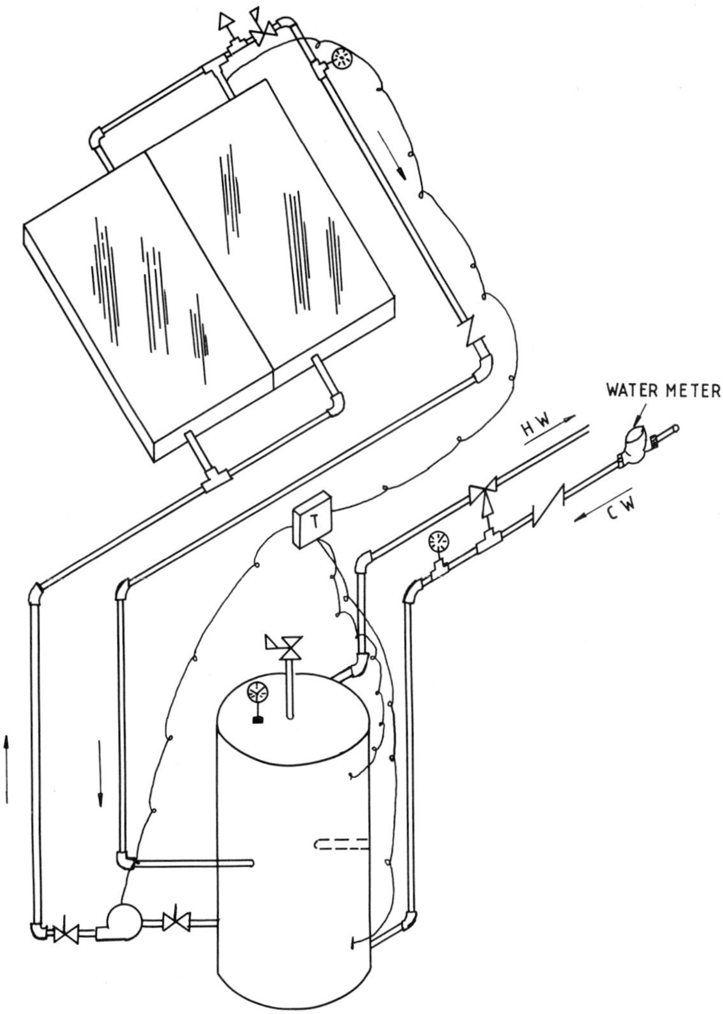

FIG.2 - DIRECT HEATING WITH A SINGLE TANK

In the following paragraphs, we shall now describe the principles of various types of instruments/controllers, which are used with solar water heating systems. The performance characteristics and operation features of each shall be described with a view to help one to select the required type for economical and rugged operation of the SWH system ensuring minimal maintenance with optimal efficiency of operation.

2. TEMPERATURE MEASUREMENTS

Temperature sensors of various types with their ranges of measurements are given in Fig.4. Of these, thermistors and RTDs are resistance types varying nonlinearly/linearly respectively with temperatures. The former

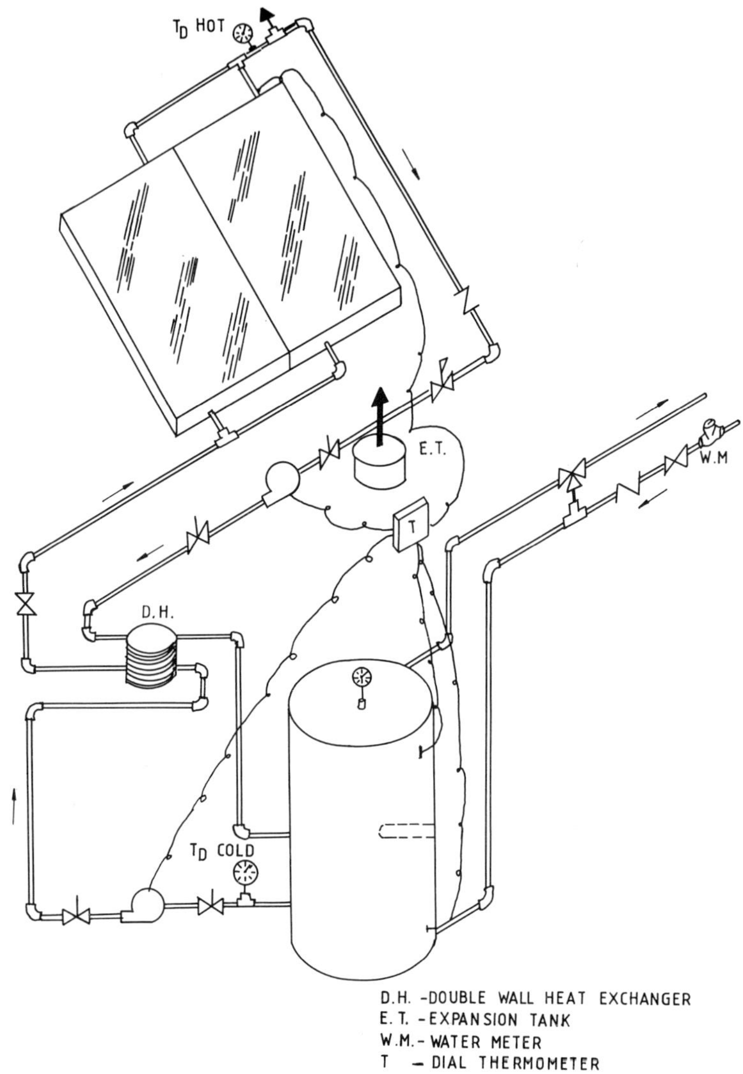

FIG.3 - DIRECT HEATING WITH ANTIFREEZE LOOP AND EXTERNAL HEAT EXCHANGER

has a negative temperature coefficient. However, it is the most sensitive and is well suited for use with SWH systems for temperature measurements in the range of 0°C to 100°C. Its nonlinearity is taken care of by compensation techniques. However, for differential measurements over a small range of temperatures, the nonlinearity does not call for elaborate compensation circuits. Hence, thermistors are invariably employed in the instruments (like temperature indicators and differential thermostats) used with SWH systems. Its resistance temperature characteristic is shown in Fig.5. Thermistors with nominal resistance

SL No.	Temperature Sensors	Temperature Co-efficient	Response	Temperature Range	Sensitivity	Accuracy	Time Constant	Remarks
1	Thermistors	-3% to -4%	Non-linear TVSR	-100 to 300 °C	Good	± 5 %	Fraction of a second to seconds	Ideal for Differential temperature measurement
2	RTD's	·4 % to ·8 %	linear	-200 to 800 °C	—	± 2 %	Seconds	For higher accuracy of measurement
3	Thermo couples	Micro volts per °C (μV/°C)	—	0° to 2000 °C	Good	·5 % to 2%	—	For accuracy & industrial use

FIG.4 - TEMPERATURE SENSORS FOR SOLAR SYSTEM

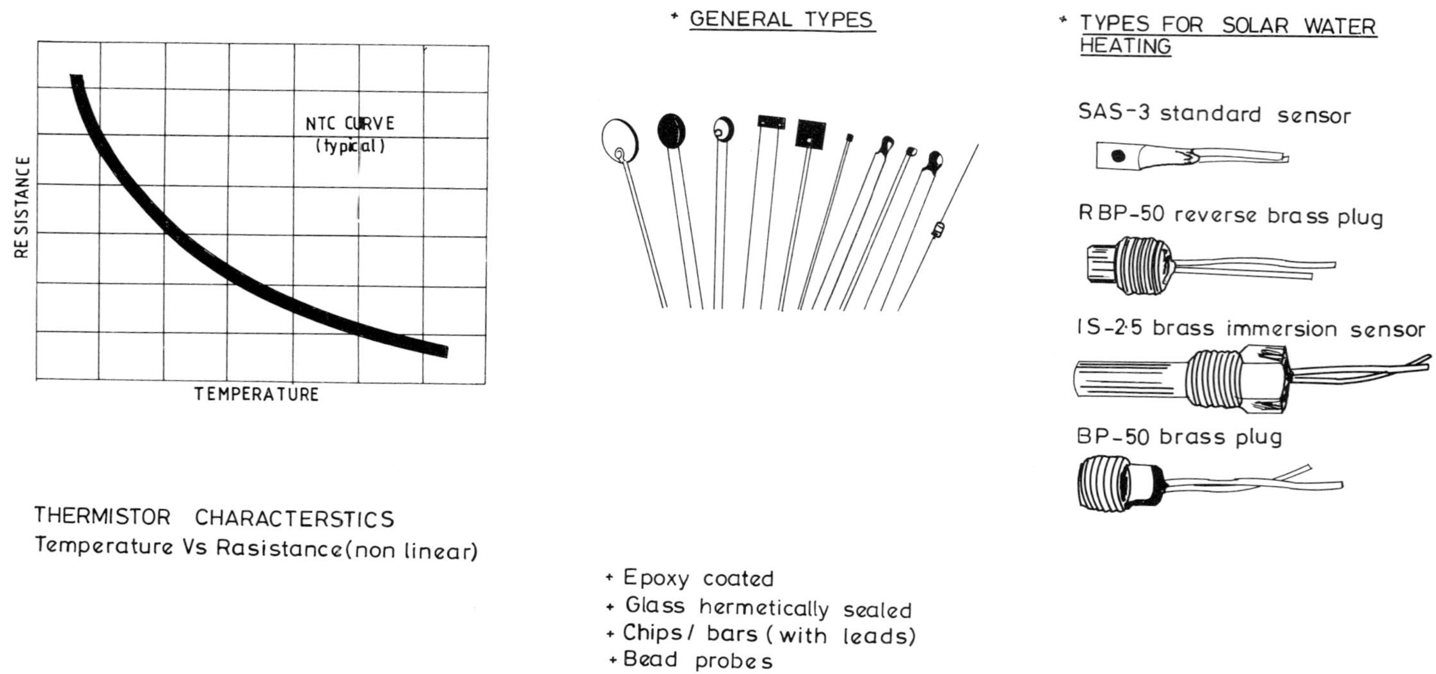

FIG.5 - THERMISTOR TYPE PACKINGS

value of 3 K to 10 K at 25°C (-4 to -6%/°C) are often selected for temperature measurements. Sensor packaging for different functional applications (manner of use) is also shown in Fig.5. Note the tubewell IS 2.5 packaging is employed for use as an immersion sensor.

Platinum wires are generally employed with RTD sensors which are comparatively stable with linear characteristics for wider range of temperature operations. RTDs are used with instruments for precision measurement of temperatures in the laboratory set up of the system. Its packaging is again decided by its manner of use.

Thermocouples are employed at higher temperatures with wider range of measurement and better accuracy (0.5%). Varieties of thermocouples are available, viz. S, B, J, C, E-U types, which are made up of junctions of dissimilar metals generating thermo-emfs in the range of mv/°C to μv/°C. The metals employed are base metals and rare metals. The latter type offers better accuracy at high temperatures. However, thermocouples are rarely used with solar water heating systems. They find use in the industrial applications as well as with high temperature solar energy systems, viz. Heliostats, concentrators etc. The construction and packaging of RTD and Thermocouples are shown in Fig.6.

All of the temperature sensors described above call for detecting and measuring circuit for indication/measurement of temperatures in terms of electrical voltage/currents registered by a meter. Such a circuit is invariably a bridge circuit shown in Fig.7. For detection of a change in resistance (of the sensor) with temperature, a resistance Wheatstone bridge is used either of 2-wire or 3-wire type, the latter for compensating the vitiating effects of connecting wires. A similar type of bridge is used for measurement of thermo-emf's generated by thermocouples. A constant current feed bridge is used for measurement of differential temperatures. It may be noted that the above type of measuring circuits invariably employ an auxilliary power supply of either 6 volts DC or power supply, giving 6 volts/12 volt DC of an AC supply of 230 V. In case of a stand-alone thermosyphon system, it is advisable to use a direct temperature indicator which employs unequal expansion of metals for energising a helical spring with a pointer for indicating the temperature under measurement. Such stand-alone direct acting temperature indicators are also often employed with SWH, under field conditions.

Of course, accuracy of measurement/indication is much lower than that of the types described above. Mercury thermometers which do not call for auxilliary powers are also often used for accurate measurements, but such thermometers are fragile and often malfunction due to blocked capillaries under conditions of usage characterised by fast varying temperatures.

3. FLOW MEASUREMENTS

Water flow rates of the order of 1 to 30 litres per minute are generally required to be measured/registered for monitoring/controlling the operation of solar water heating systems (100 to 1000 litres/day). Varieties of flow meters are available for this purpose. The most common and economical being water meter (2-100 litres/minute), rotameters (few litres

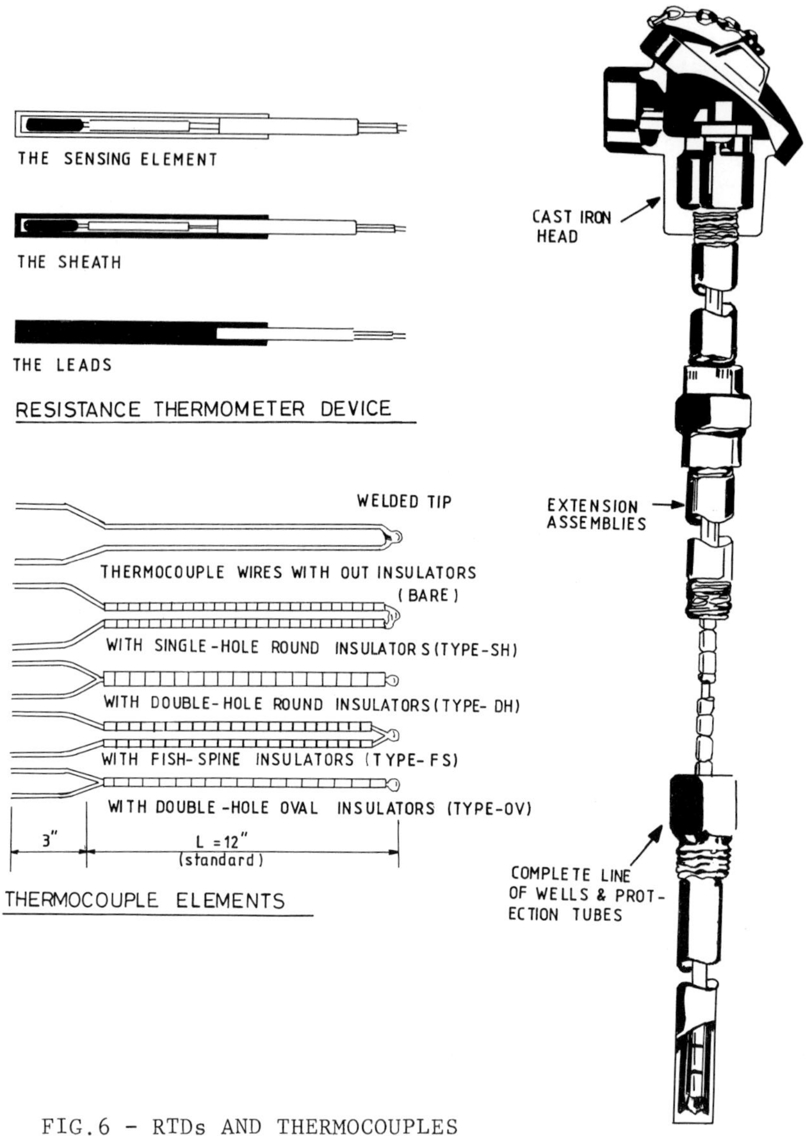

FIG.6 - RTDs AND THERMOCOUPLES

to 400 litres/minutes) and turbine meters few litres to hundreds of litres with a range of 1:10 of flow rates).

All flow rate measurements with SWH systems (except in thermosyphon systems) are under conditions of turbulent flows. As such water meters and turbine meters appear to be the most elegant/economical choice for range of flow meters with solar water heating systems. The quantitative relations governing the measurement of flows with various meters are given in Fig.8. The characteristic performance of water meter is given in Fig.9, while Figs. 10 and 11 give the constructional features of a

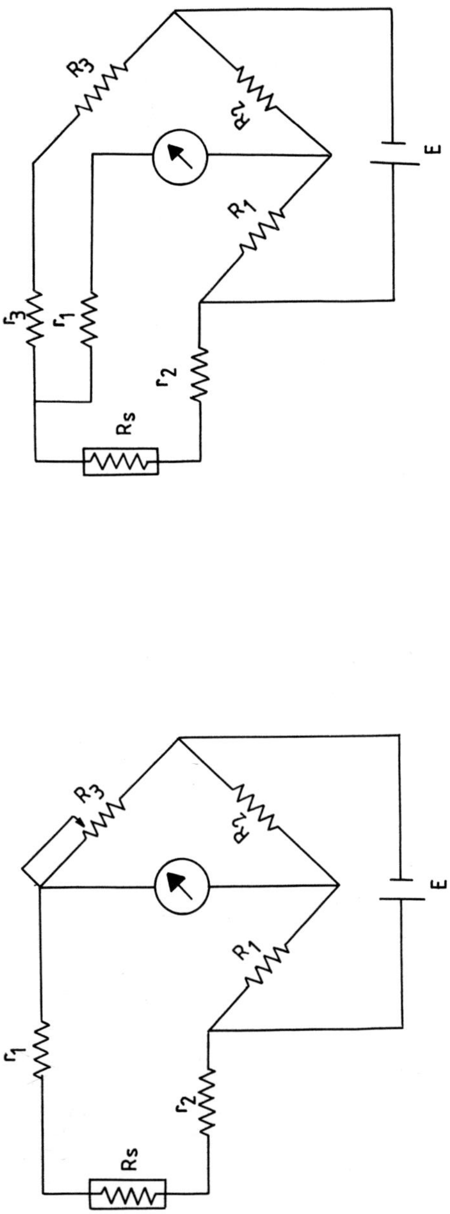

$R = \frac{R_1 R_3}{R_2} - (r_1 + r_2)$

(a) 2 WIRE WHEATSTON BRIDGE

$Rs = R_3 + (r_3 - r_2)$

(b) WIRE WHEATSTON BRIDGE

(COMPENSATION FOR CONNECTING WIRES r_2 & r_3)

FIG.7 - BRIDGE MEASUREMENT

rotameter and a turbine meter. Usage of a flow meter along with a B.Th.U meter for monitoring/registering the total quantity of heat extracted by the solar panel is indicated in Fig.11 for a typical pump driven solar water heating system.

I. INDUSTRIAL LIQUID GAS FLOWS ARE GENERLLY TURBULENT IN NATURE DUE TO HIGH FLUID VELOCITY,

ie - LARGE REYNOLD NUMBER $\gg 2000 = \frac{\rho VS}{\mu}$

WHERE ρ = DENISITY
S = CROSS SEC.
V = VELOCITY
μ = VISCOSITY

II. FLOWS MEASURED : AS,

Qv = VOLUMETRIC FLOW
Qm = MASS FLOW (Qm = ρQv)

III. LAMINAR FLOWS MEASUREMENT:

ROTAMETER

$$Qv = CS\sqrt{2g\,\Delta P}$$

WHERE C = CO-EFFICIENT
S = CROSS SEC. AREA
g = GRAVITY
ΔP = PRESS. DIFFERENCE

IV TRUBINE FLOW RATE :

Qv = P r s p VOLUME UNITS(minute)
WHERE P = SPECIFIC GRAVITY
s = CROSS SEC. OF PIPE
p = PITCH OF TURBINE
r = REVOLUTION PER MINUTE

FIG.8 - FLOW METER

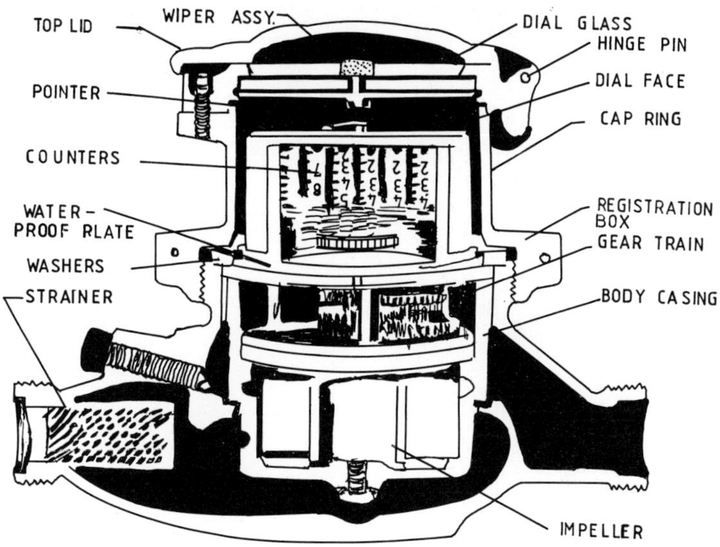

METER SIZE (mm)	15	20	25
NORMAL CAPACITY (l/h)	2,500	3,500	5,500
CONTINUOUS RUNNING CAP. (l/h)	1,500	2,500	3,500
MIN. STARTING FLOW (l/h)	40	60	75
LOWER LIMIT OF FLOW	125	175	275
ACCURACY OF MEASURE-MENT	within ±2	±2	±2

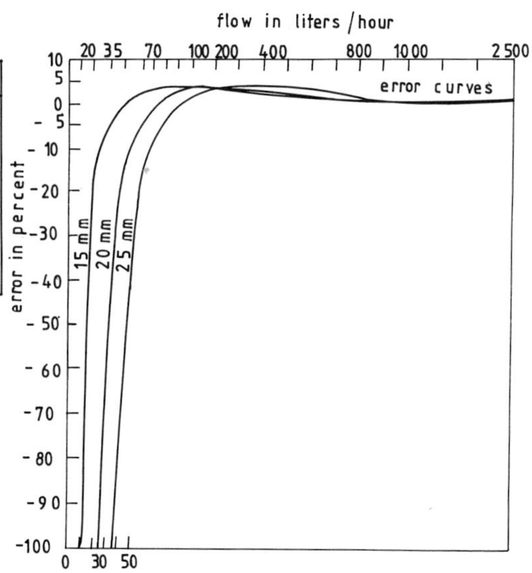

FIG.9 - CHARACTERISTICS

4. ENERGY MEASUREMENTS

Heat energy collected and supplied to the user is generally monitored/ registered during the operation of the SWH systems. The measurement of heat energy calls for the value of operating parameters, viz. water volume-mass, specific heat and differential temperature of hot water outlet and cold water inlet. With these parameter values, B.Th.U meter

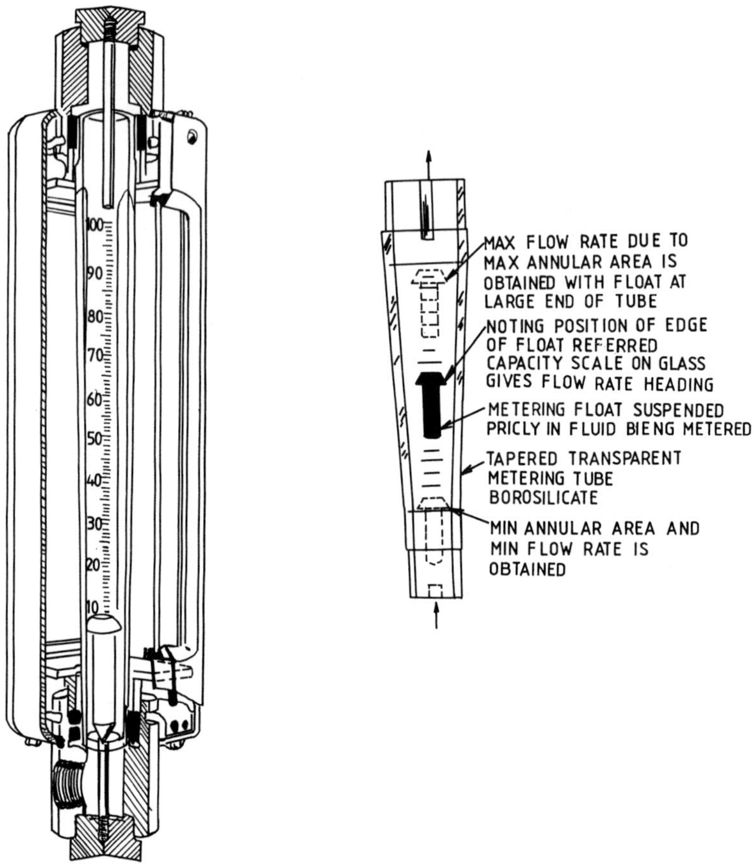

FIG.10 - CONSTRUCTION OF FLOAT TYPE AREA METER

works out the total B.Th.U which is displayed on a panel. The connection of a B.Th.U meter and its working for calculating the B.Th.U units are shown in Figs. 12 and 12A. The relation between B.Th.U and KWH is given by the conversion factor of 0.0003 to be multiplied to B.Th.U to arrive at KWH (1 KWH = 3.333×10^3 B.Th.U).

5. DIFFERENTIAL TEMPERATURE

Temperature differentials between the cold water inlets and hot water outlets (collector output) are employed for a variety of control functions associated with the controlled operation of a SWH, employing driving pump and heat exchangers. The basic circuit for its measurements is the Wheatstone Bridge. The sensors (Thermistors) located in the adjacent arms of the Bridge generate differential resistances as a function of differential temperatures, Fig.13. The resulting differential voltage is employed by the operational amplifier whose output is used to energice a relay for switching on the pump. A delay circuit is incorporated to

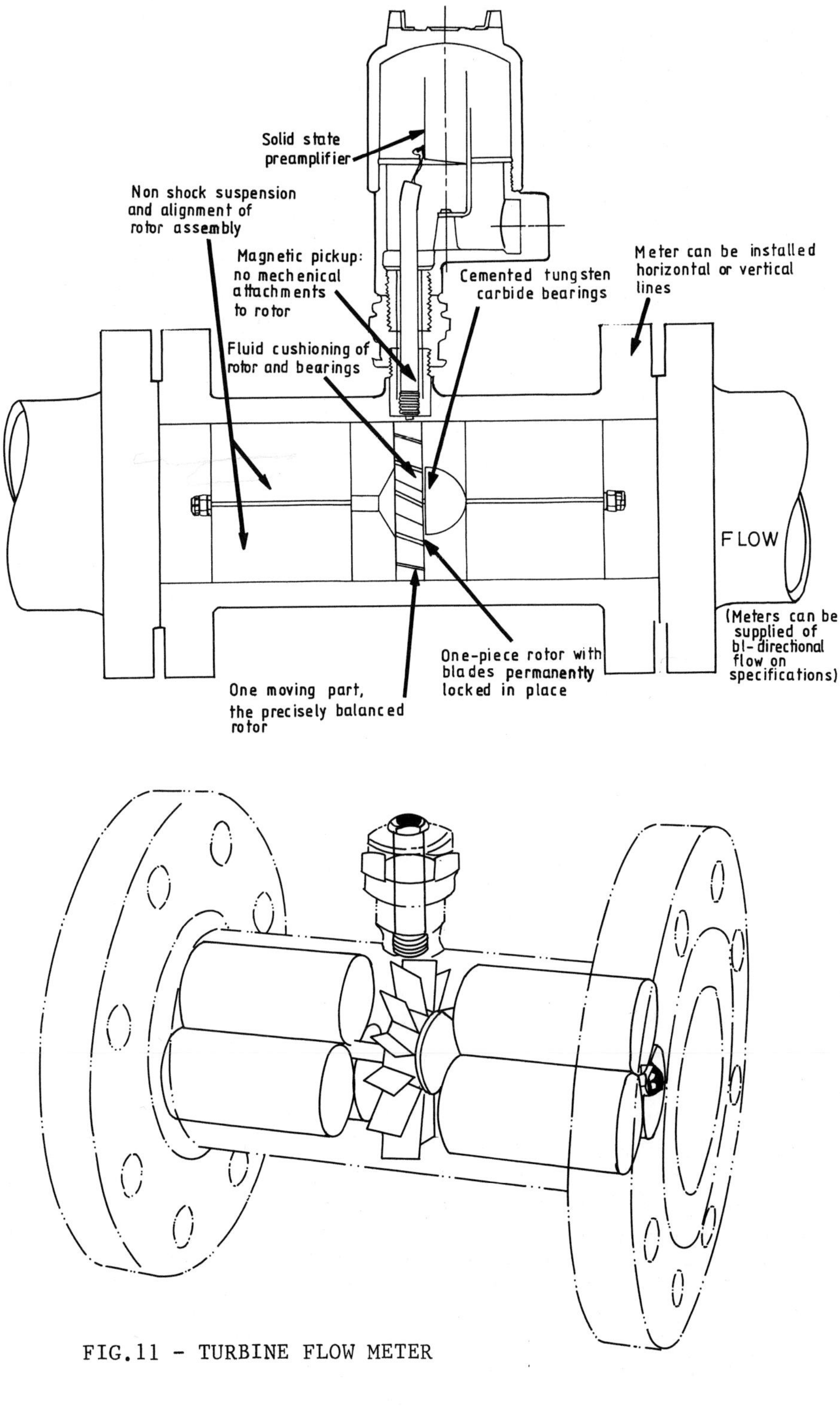

FIG.11 - TURBINE FLOW METER

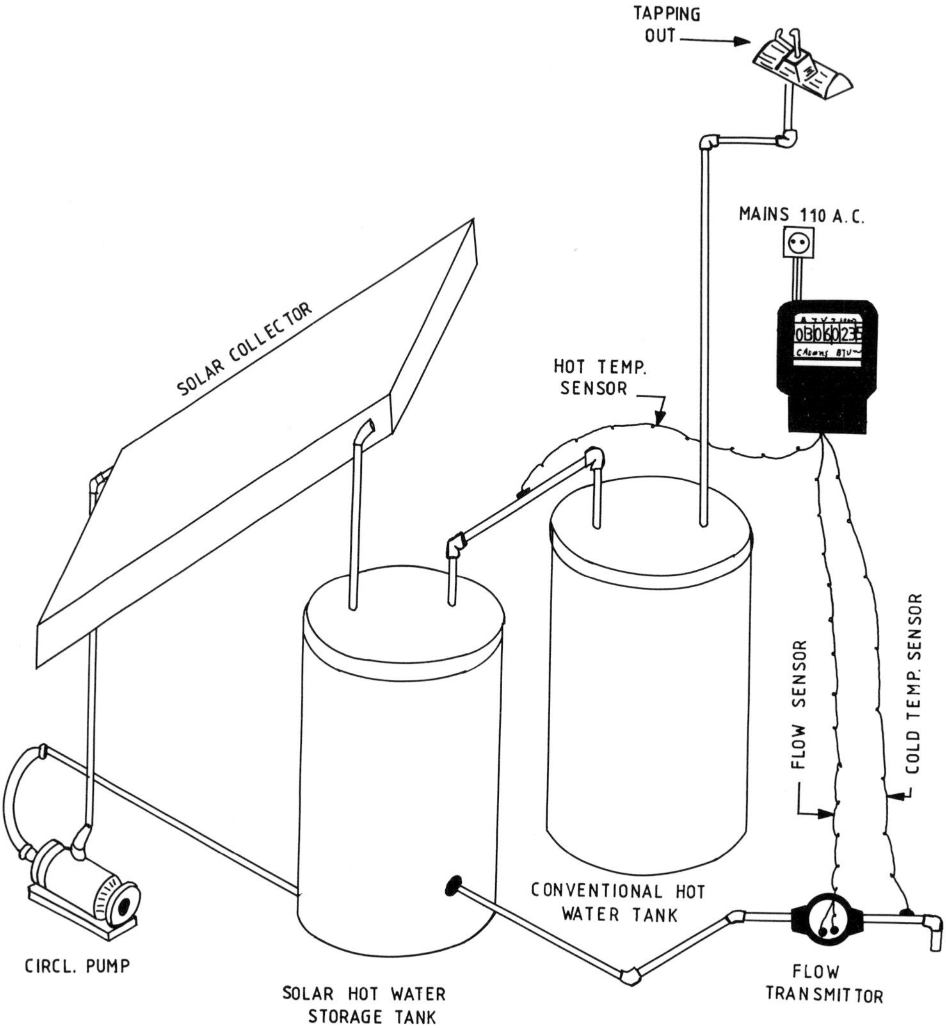

FIG.12 - SWH SYSTEM WITH B.Th. METER

avoid cycling of on-off operation of the pump resulting from drop in differential temperatures due to on rush of cold water causing the collector outlet temperatures to fall from its value employed to initially start the pump. Such differential temperature measurement system employed as a controller, works in the selectable differential temperature range of $2^{o}C$ to $10^{o}C$. In certain applications more than one temperature differentials are required to be measured. Such controllers also implement logics which decide the phased switching on-off of more than one pump used with a more SWH system, viz. one pump for driving water to collector and another for heat exchanger, in such cases differential temperature controllers more than one differential measurement setups along with digital logics for selection and energisation of

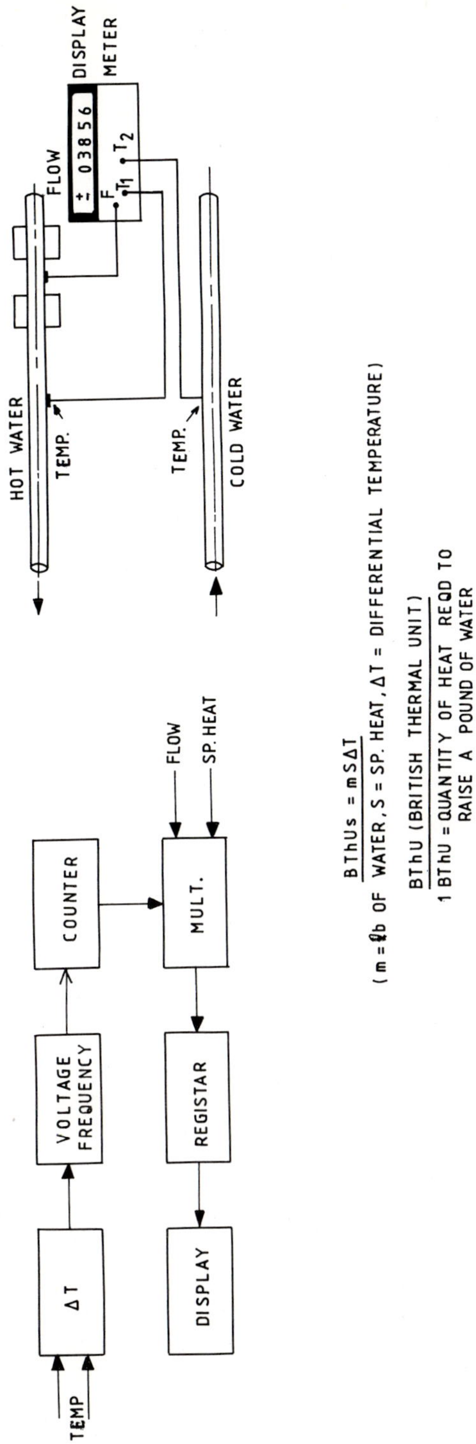

FIG. 12A - BLOCK DIAGRAM OF A B.Th.U METER

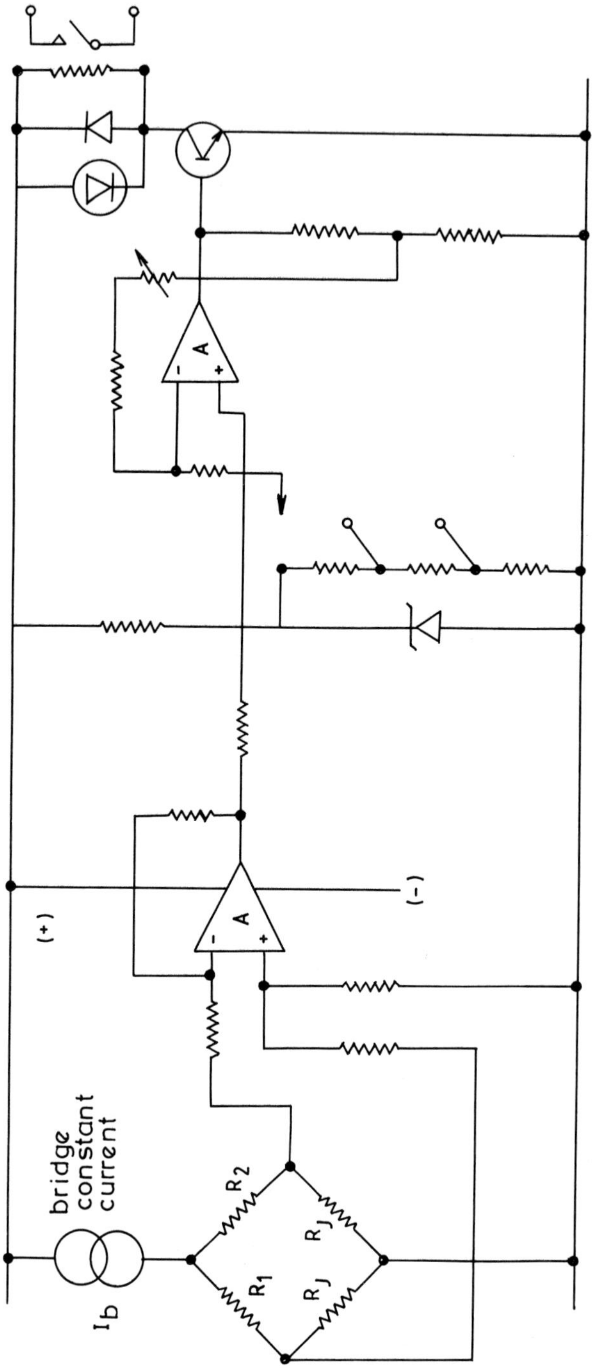

FIG.13 - DIFFERENTIAL TEMPERATURE CONTROLLER

relays controlling the pumping operation. Flow rate control based on emperature differentials or absolute temperatures and solar insolation is the next step towards realisation of highest efficiency of operation of the SWH system. It calls PID (Proportional Integral Derivative Controllers) which are often used with larger industrial process heat systems ensuring constant temperature of hot water supply to the process. Such PID controllers are available and in use with industrial process heat solar energy systems.

6. PROGRAMMABLE CONTROLLERS

Application of the various sensors and measuring devices described in paras 2 to 5 is exemplied in the use of the programmable controller employed for automatic control of a solar water heating system shown in Fig.14.

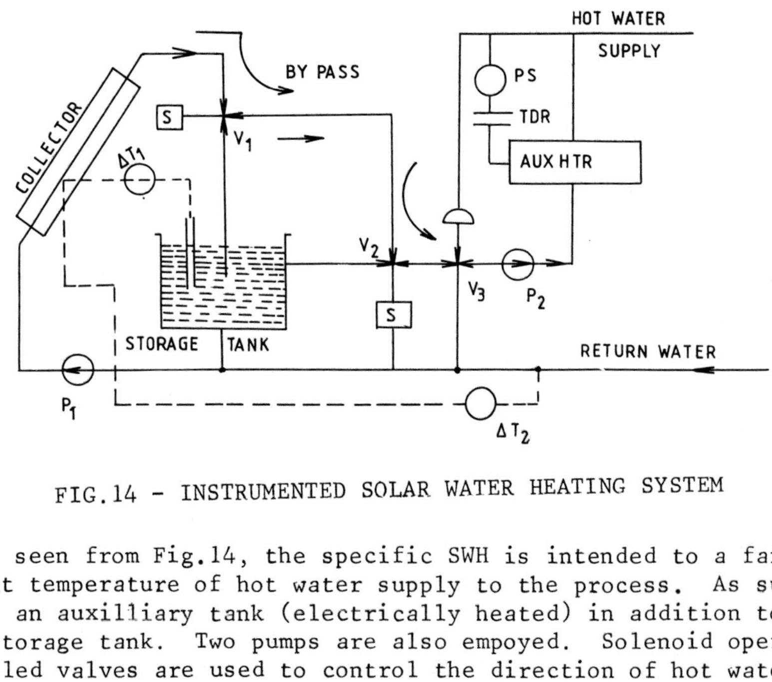

FIG.14 - INSTRUMENTED SOLAR WATER HEATING SYSTEM

As seen from Fig.14, the specific SWH is intended to a fairly constant temperature of hot water supply to the process. As such it employs an auxilliary tank (electrically heated) in addition to the hot water storage tank. Two pumps are also empoyed. Solenoid operated controlled valves are used to control the direction of hot water flow mixed with the cold water for additional heating required under condition of low solar insolation. The system operation is therefore a little more elaborate than a simple thermosyphon system or a single tank forced circulator type SWH. As such it calls for automatic controls in respect of sequenced operations decided by the logic underlying the system operation. A Programmable Controller is, therefore, called for to automatically control this system operation by means of sequencing the steps required to start, run and stop operation of the SWH plant. A ladder diagram for sequenced operation of the plant is shown in Fig.15.

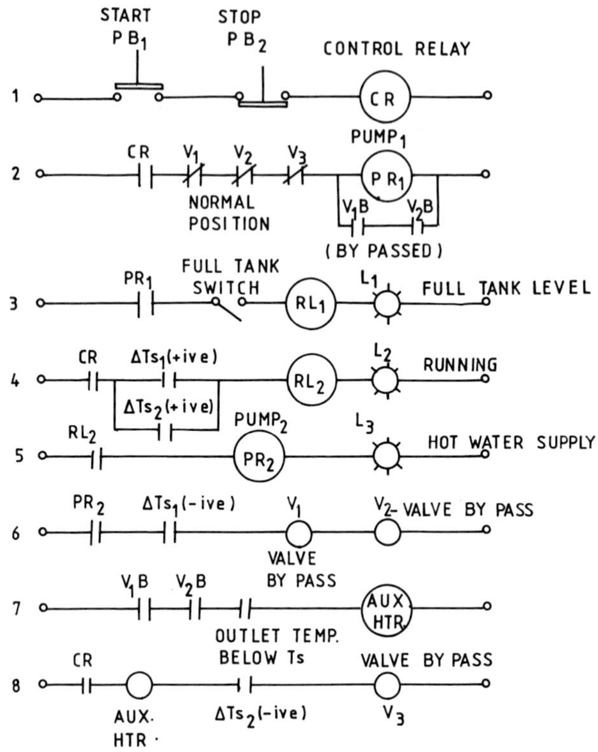

FIG.15 - RELAY LADDER DIAGRAM FOR AUTOMATIC CONTROL OF SOLAR WATER HEATING SYSTEM

The diagram indicates the various steps comprising the total program underlying the system operation. Back step of the programme comprises of several sequenced functions/states/outputs of switches, pumps, relays, valves, and the differential thermostats, pressure switches etc. These steps are next sequenced in the order indicated for automatic control of the various phases (operational) of the SWH.

7. PHOTOVOLTAIC DOMESTIC WATER HEATINC SYSTEM (PV-DHW)

The recently introduced PV-DHW solar water heating system is yet another example wherein a certain degree of automatic control is inbuilt to ensure the closed loop operation (using a heat exchange) at a flow rate which allows for delivery of maximum collected solar energy (70 to 80% at least) to the storage tank in the form of thermal energy. Such an operation calls for control of flow rate as a function of solar insolation, use of efficient heat exchangers, and PV generated power to drive DC pumps (Fig.16). The first function, viz. control of flow rate vs. solar insolation is implemented with the help of an inbuilts programme controlling the flow rate in proportional mode with the sun's insolation monitored by the PV's (Ref.2).

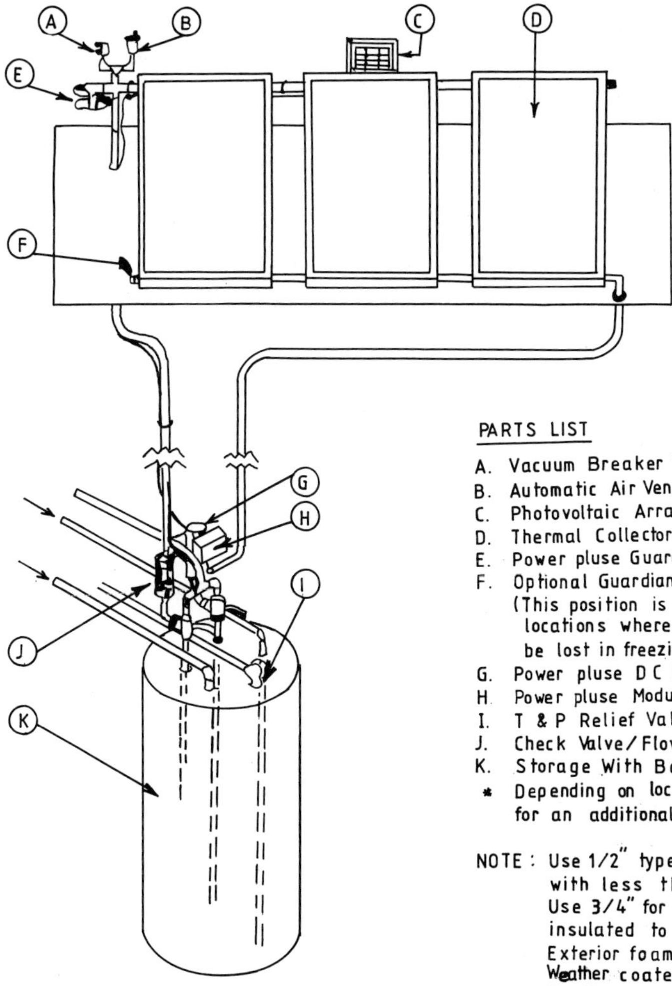

PARTS LIST

A. Vacuum Breaker Valve
B. Automatic Air Vent
C. Photovoltaic Array : up to 14 W
D. Thermal Collectors : up to 120 ft^2
E. Power pluse Guardian
F. Optional Guardian Location
(This position is recommended for locations where water pressure could be lost in freezing water.)
G. Power pluse D C Pump
H. Power pluse Module
I. T & P Relief Valve
J. Check Valve/Flow Indicator
K. Storage With Back-up
* Depending on local codes this position is for an additional T & P relief valve.

NOTE : Use 1/2" type L copper for installations with less than 50 ft. total run.
Use 3/4" for over 50 ft. All pipe must be insulated to a Valve of R-3.5.
Exterior foamed rubber insulation must be Weather coated.
Use 18 gauge weatherproof double stranded wire for up to 120 ft. runs.

SPECIFICATIONS

System Capacity:	UP to 120 gallons
Average Daily Energy Output:	UP to 110,100 BTU's
Electrical Back-up Max. Size	6 KW
Photovoltaic Array	Up to 14 W, 12 VDC
Pumping Rate:	UP to 3.5 GPM

FIG. 16 - DOMESTIC WATER HEATING SYSTEM WITH CONTROL VALVES AND PHOTOVOLTAIC POWER SOURCE

8. CONCLUSIONS

The necessary instrumentation in solar water heating system not only improves the efficiency of the solar energy collection but also makes it reliable and automatic. When a programmable logistics are developed for sequential operation, the system works automatically thereby reducing the maintenance cost. In the coming years, the water heating system runs on solar energy and collects and delivers solar energy.

REFERENCES

1. H.P. Jaiswal, A. Thomas, M. Ramakrishna Rao and J. Nagaraju, (1985), 'Instrumentation for a solar heating system', Proceedings of National Solar Energy Convention '85, January 1985, Bhopal.

2. Greg Peebles, (1984), 'How to compare SDHW systems using PV modules', Solar Engineering and Construction, pp 61-65, May/June 1984.

GALVANISED STEEL SOLAR ABSORBER - ITS CHOICE, CORROSION RATE AND EXPECTED LIFE

R.S. Soin
Hindustan Brown Boveri Ltd.
Baroda 390 001
Gujarat
India

1. INTRODUCTION

The choice of solar absorber compatible with solar water heating system is essential. A cost-effective solution requires highly thermal efficient collector at the lowest cost. Life cycle costing demands collector life of at least 10-15 years. Thus to design a solar absorber, one has to keep in mind thermal efficiency, cost and its expected life when subjected to actual operating conditions. Cost-effective solar absorber is easy to obtain. Expected life of an absorber can be estimated realistically by actually subjecting it to operating conditions. Accelerated ageing tests are to compare relative performance under identical conditions. The results must not be used to extrapolate linearly the life of an absorber.

A distinct feature of solar system is that it operates with different types of water, varying temperature in a day and in a year, different system configuration involving single metal/alloy or combination. Expected life estimation is thus complex. As experiences of actual installations are being accumulated, the corrosion problems in water heating systems overall and in absorber will help us to design better system. Meanwhile, one can draw inferences from other experiences in systems with nearly identical conditions.

It is the intention of this paper to illustrate the case of galvanized steel absorber. Its choice and expected life are discussed. Actual experiences are illustrated. Further for comparison, corrosion problems of copper collectors drawn from actual installations are also included. While corrosion studies is still being continued, the present work reviews status after 3 to 5 years of service of installed solar water heaters.

Any metal corrodes in water and the corrosion rate depends on many factors such as type of water (soft or hard), pH, dissolved gases like O_2, CO_2 etc., temperature, flow velocity etc. For solar water heating the most important parameters are temperature, type and flow velocities in an open system as described in Fig.1. Open system here implies that cold water enters directly into the solar collector or via storage tank. In a close loop system, a heat exchanger is used which isolates the

H. P. Garg (ed.), Solar Water Heating Systems, 199–230.

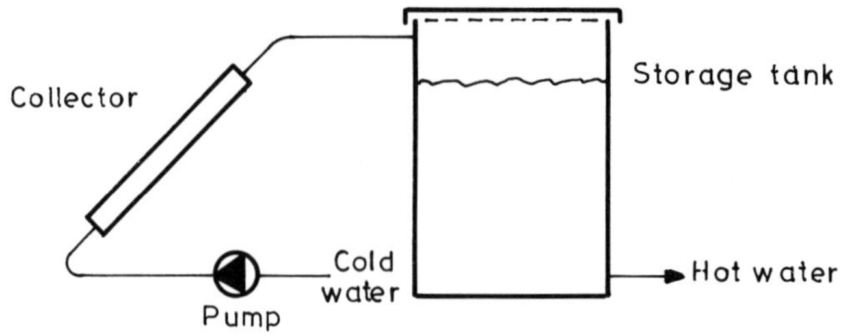

a. Single-pass Solar water heating system

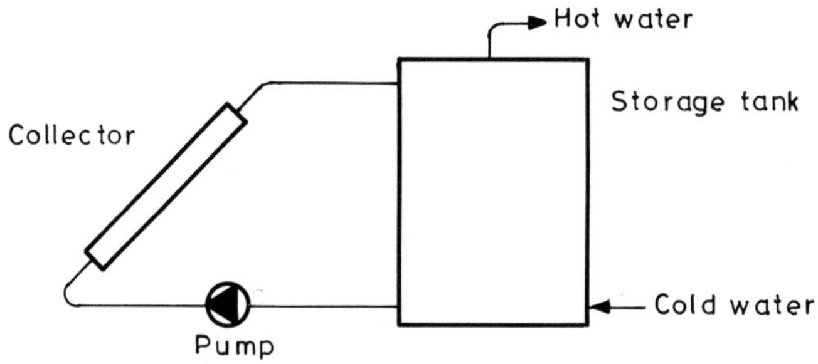

b. Multi-pass Solar water heating system

FIG.1 - CONFIGURATION OF OPEN-LOOP SOLAR WATER HEATING SYSTEM

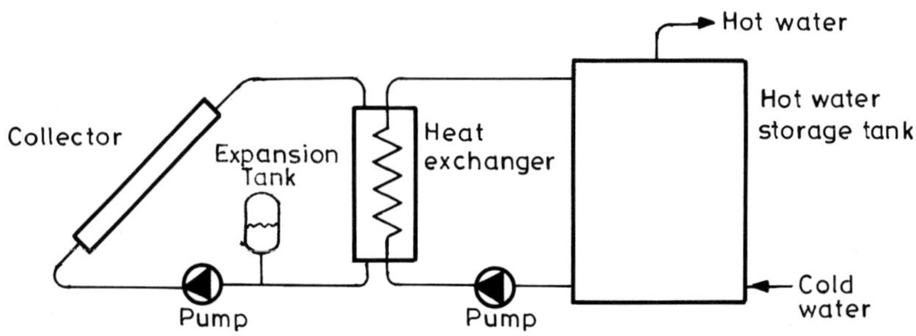

FIG.2 - CLOSE-LOOP SOLAR WATER HEATING SYSTEM

collector from cold water supply as shown in Fig.2. In such cases demineralised water is used, but even natural soft water can be used if the make-up rate is nominal. The close-loop system is used when cold water to be heated is highly scaling or in case of process industries could be corrosive also. In the present context, only open-loop system is of interest since for close-loop system the corrosion rate is practically small and selection of material is not as critical as in open-loop system.

2. CHOICE OF MATERIAL FOR SOLAR ABSORBER

For open-loop solar water heating, the practical choice of absorber material is limited to copper and galvanized steel. Aluminium has been ruled out (Ref.1) as a potential absorber material. The main impediments are that any extreme pH, dissolved oxygen increase the risk of corrosion. Flow velocity less than 90 cm/s increases corrosion, while more than 90 cm/s enhances erosion (Ref.1). In many of systems the life of absorber is not more than 6-12 months.

Copper, in general, is suitable for solar water heater. Actual installations prove that solar water heating systems with copper absorber are in service for more than 10 years and very few failures are reported. The failures are discussed in Section 5. In India, however, copper is an expensive material costing 7-8 times that of steel. Thus, to reduce the copper absorber cost, the tube spacing is normally kept at 100 mm while the fin is of 0.3 mm sheet. This results in collector efficiency, F', of 0.887 for 1-cover non-selective and 0.93 for 1-cover selective absorber. On the other hand, it had been established that steel absorber of integrated tube configuration has collector efficiency factor of 0.96 and 0.99 respectively (Ref.2). Thus, with proper geometry, steel absorber can be made highly efficient despite the fact that thermal conductivity is lower than copper, about one fifth. This is possible in case of steel absorber with integrated tube design since varying the tube spacing does not affect cost appreciably in contrast to copper absorber. Fig.3 shows the comparison of the two types of steel absorber with a copper absorber. The flexibility in the design of steel absorber is quite obvious without any change in cost of raw material.

However, bare steel absorber is normally not suitable for cold water and hot water systems and corrosion protection is necessary. The cheapest one and most reliable corrosion protection is by hot dip galvanizing. (For cold water pipes GI pipe is the best example of zinc as corrosion protection which is known to last for 12-20 years (see Section 4).) Here zinc electroplating is not sufficient. Firstly, due to absence of metallurgical bond and secondly, the electroplated zinc coating is porous. This is serious when the plating thickness is less than 15-20 microns.

There are two ways of making galvanised steel absorbers. Firstly, GI tubes can be welded together to form a pipe-matrix (Fig.4). The absorber sheet is either mechanically bonded or soft soldered as shown in Fig.5. This technique has several disadvantages, such as:

(a) The arc welding joint of GI riser and header is prone to corrosion since zinc is removed.

In hot water application, especially between 60°C to 90°C, zinc no longer acts sacrificeally and thus steel corrodes (Ref.3,4). This is illustrated in Plate 1, which shows corrosion at the welding joint inside the absorber panel where the galvanized steel header is arc-welded to zinc-sprayed steel absorber. This has been out from an absorber panel in service for four years. No pitting is observes so far, however. This joint corrodes even in the atmospheric condition as shown in the Plate 2. Table 1 shows how coupled zinc steel causes higher zinc corrosion rate compared to zinc alone.

TABLE 1 - Change in Weight of Coupled and Uncoupled Steel and zinc, g (Ref.4)

Environment	Uncoupled		Coupled	
	Zinc	Steel	Zinc	Steel
0.05 M $MgSO_4$	0.00	-0.04	-0.05	+0.02
0.05 M Na_2SO_4	-0.17	-0.15	-0.48	+0.01
0.05 M NaCl	-0.15	-0.15	-0.44	+0.01
0.005M NaCl	-0.06	-0.10	-0.13	+0.02

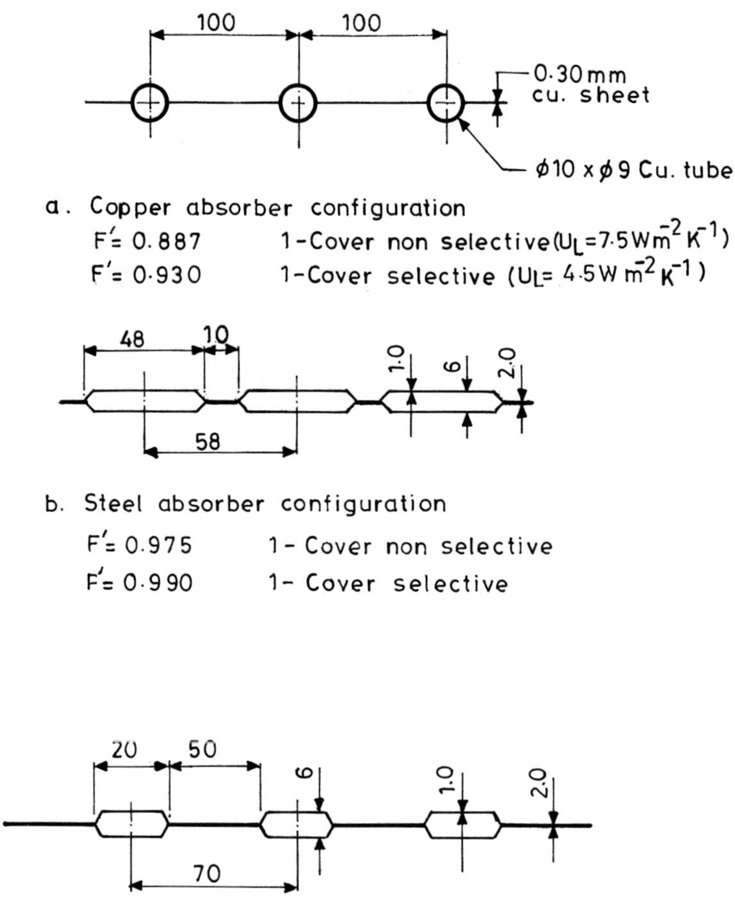

a. Copper absorber configuration
$F' = 0.887$ 1-Cover non selective ($U_L = 7.5 Wm^{-2} K^{-1}$)
$F' = 0.930$ 1-Cover selective ($U_L = 4.5 W m^{-2} K^{-1}$)

b. Steel absorber configuration
$F' = 0.975$ 1- Cover non selective
$F' = 0.990$ 1- Cover selective

c. Steel absorber configuration
$F' = 0.960$ 1- Cover non selective
$F' = 0.975$ 1- Cover selective

FIG.3 - DIFFERENT CONFIGURATION OF ABSORBER - COMPARISON OF COPPER AND STEEL

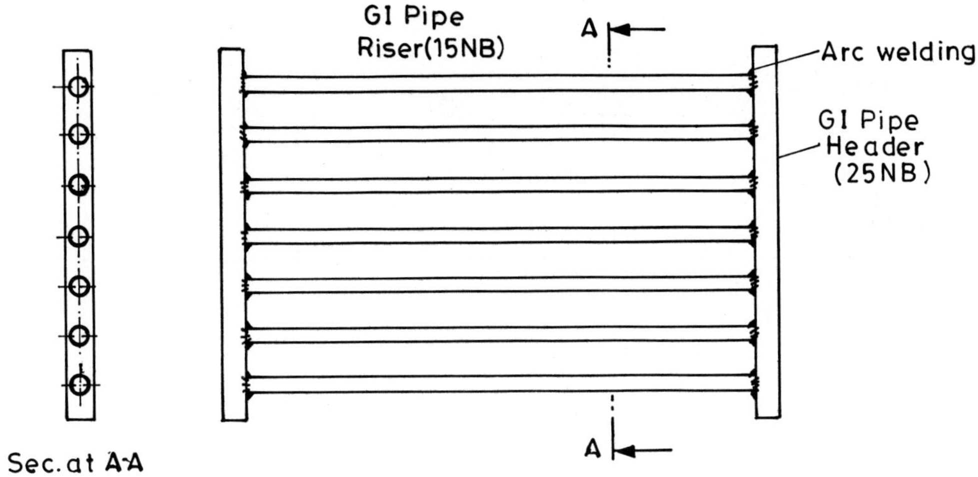

FIG.4 - G.I. PIPE MATRIX

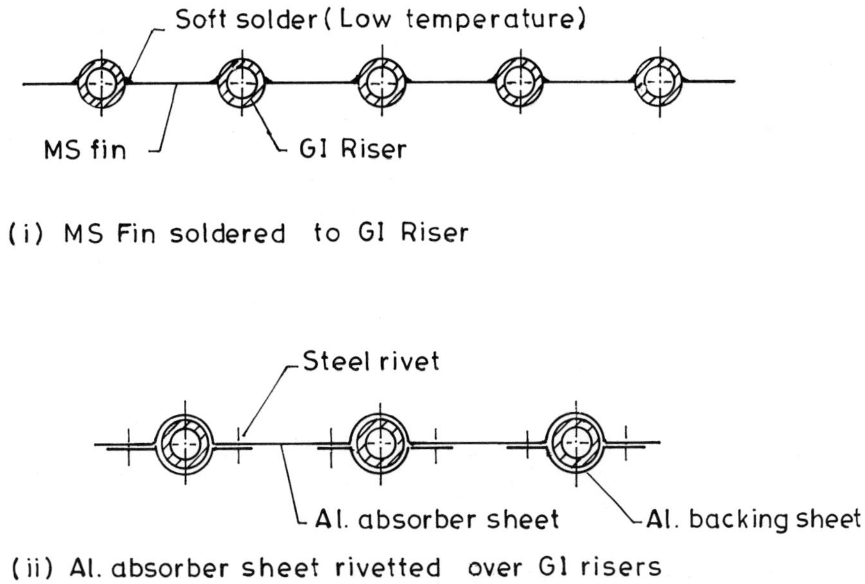

FIG.5 - METHODS OF FIXING ABSORBER SHEET OVER G.I. PIPE MATRIX (OF FIG.4)

PLATE 1 - CORROSION AT WELDED JOINT GALVANIZED HEADER AND ZINC SPRAYED STEEL PANEL; SERVICE LIFE 4 YEARS

PLATE 2 - ATMOSPHERIC CORROSION OF GI PIPE MATRIX AT THE JOINTS

The above indicates that normally zinc corrosion rate increases when coupled to steel. Thus for application where even zinc sacrificially corrodes in preference to steel, the life of zinc reduces considerably. In such cases only when non-flaking scale is formed the corrosion rate could be reduced. Such collectors have been used for more than 10 years, but no systematic data on corrosion aspects are available.

(b) Mechanical bonding of aluminium sheet gives very poor collector efficiency factor (less than 0.90) and actual measurements shows 10-15°C temperature gradient between the absorber hottest point and fluid in the tube (Ref.5).

Soft soldering can give away especially for selective absorber during stagnation condition, mostly when the collectors are not used for quite some time (Ref.6).

Second method of manufacturing is by cold pressing of sheet steel followed by resistance welding of two identical sheets to obtain integrated tube absorber as shown in Fig.6.

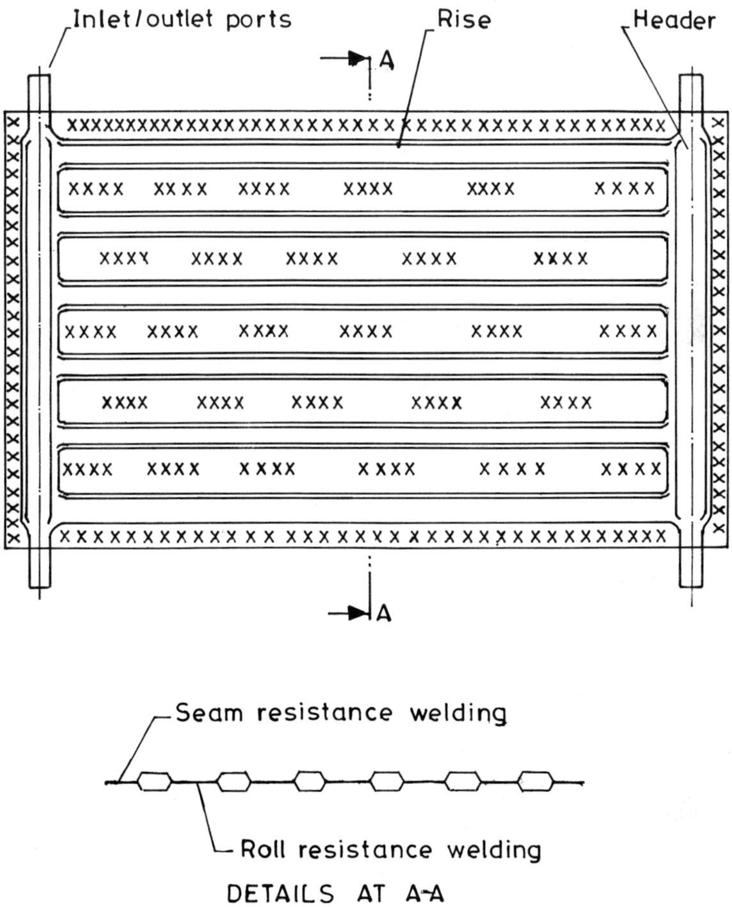

FIG.6 - INTEGRATED-TUBE GALVANIZED STEEL ABSORBER

After completion of welding of inlet/outlet ports and pressure testing, the whole absorber is hot dip galvanized (Ref.7). This process ensures complete coverage of zinc inside and outside and minimum thickness of 80-90 microns. Plates 3-5 shows the quality of galvanized steel obtained inside the absorber risers of two different designs. The uniform coating inside the riser ensures long life as will be discussed later in Section 4. Plate 5 shows how zinc is filled completely in crevices. The zinc coating is of 160-180 microns. In the next section the suitability of zinc for hot water application is discussed.

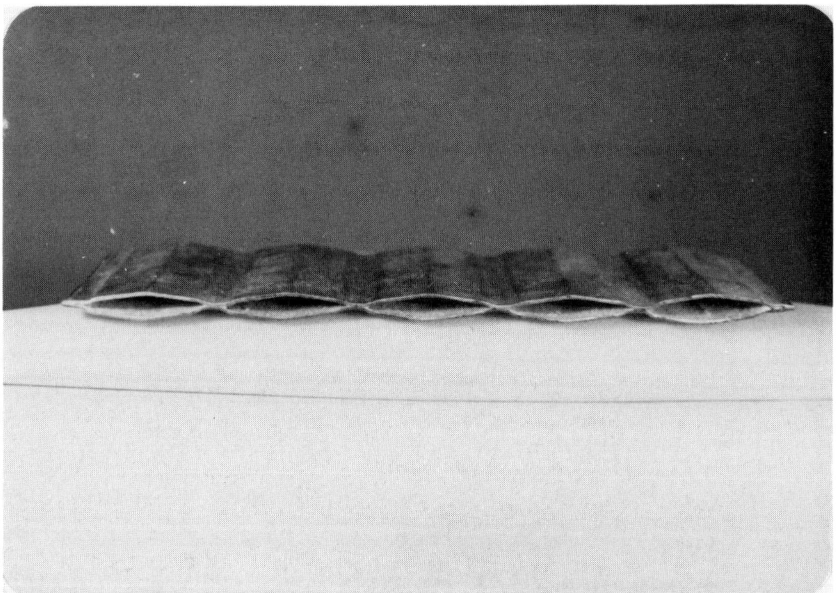

PLATE 3 - CROSS-SECTION OF GALVANIZED STEEL PANEL - INTERNAL GALVANIZING; UNUSED PANEL OUTDOOR AGEING(4 YEARS)

3. CORROSION RATE OF ZINC - ACCELERATED DATA

Before we proceed to discuss published data on corrosion rate of zinc, it may be cautioned that accelerated laboratory data on the effect of certain parameters need not necessarily reflect the life of actual system. While inferences can be made from laboratory data, experience of metal environment intereaction under actual operating conditions gives the most reliable indication on the suitability of chosen system-environment.

3.1 Effects of Parameters

In this section the effect of individual parameters or its combination

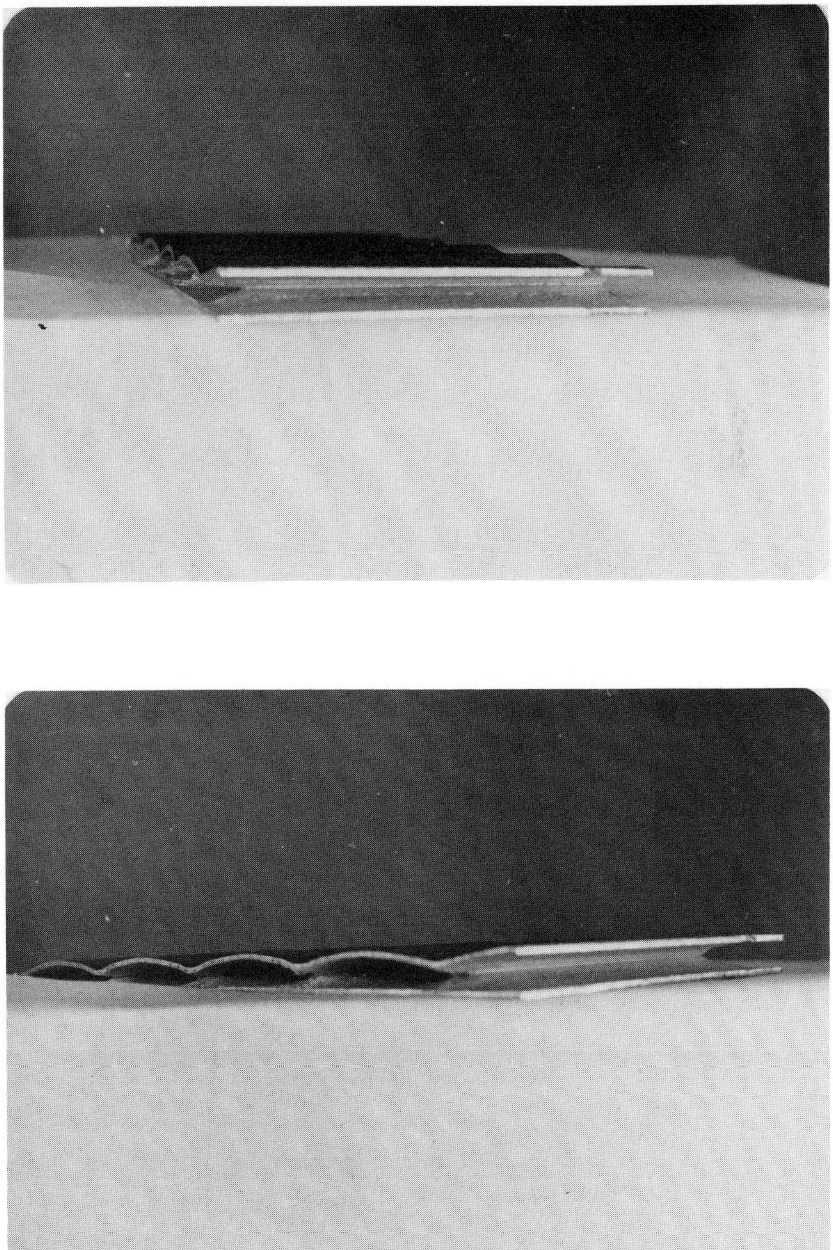

PLATE 4 - CLOSE-UP OF INTERNAL GALVANIZING SHOWING UNIFORM AND PROPER ZINC FILLUP AT THE RESISTANCE WELDING AREA

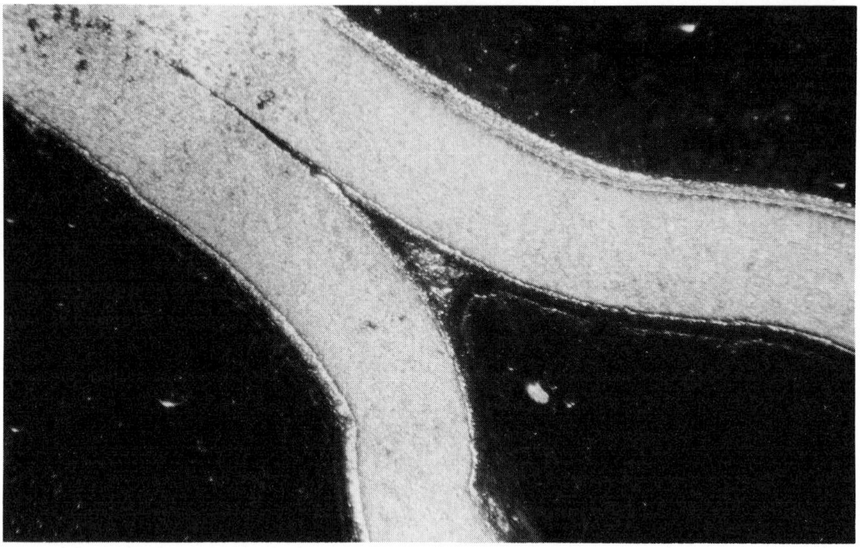

PLATE 5 - MICROSCOPIC CLOSE-UP OF ZINC FILLING IN A RISER (MAGNIFICATION 16X)

on the corrosion rate of zinc has been examined. For solar water heating applications the following parameters are important, viz.:

1. Temperature
2. Velocity
3. Dissolved gases, mainly oxygen
4. pH
5. Type of water

Table 2 illustrates types of water normally encountered with major properties of interest.

TABLE 2 - Typical Water Analysis, ppm

	Very soft lake water	Moderately soft surface water	Slightly hard water	Moderately hard water
pH	6.30	6.80	7.40	7.50
Total hardness (CaCO3)	10	53	120	230
Calcium hardness (CaCO3)	5	36	85	210

3.1.1 Effect of pH. In the range of 6-12, the corrosion rate of zinc is very small and lowest in the weakly alkaline water. The domestic water supply can have pH range of 5.8 - 8.5. The corrosion rate of zinc for the complete pH range is shown in Fig.7 (Ref.8).

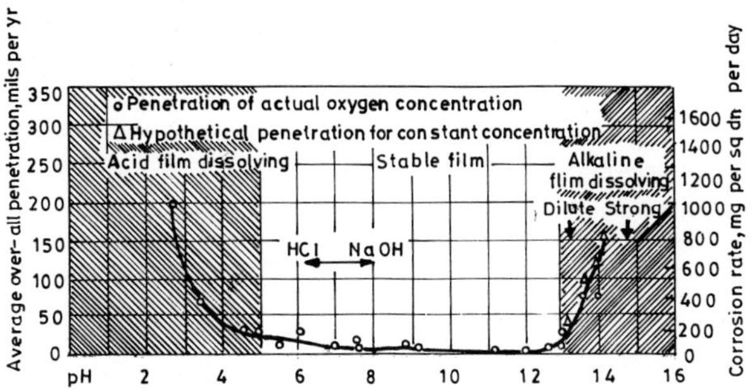

FIG.7 - EFFECT OF pH ON THE CORROSION OF ZINC (REF.8)

3.1.2 Effect of type of water. Mainly this may be classified in two groups for our discussion viz., soft and hard. Soft water, the extreme of which is distilled water, is more corrosive than hard water. In case of hard water (need not be very highly scaling, otherwise close-loop system is required) the deposit on zinc prevents or reduces corrosion rate.

Natural fresh water varies widely in composition and in corrosiveness to zinc. Typical data for a hard natural water and soft (distilled) water are given below in Table 3.

TABLE 3 - Corrosion of Various Grades of Zinc in Water (Ref.10)

- Test duration	30 days
- Temperature	room
- Aeration	none
- Corrosion criterion	change in weight after chemical removal of corrosion products
- Type of Test	sample immersion in open container

I

Grade of zinc	(Distilled water) Corrosion rate+ mdd	ipy
- Spectroscopically pure	12	0.0025
- High grade	11	0.0022
- Brass special	10	0.0020
- Selected	10	0.0020

* mdd : mg/dm^2/day; ipy : inch per year

II

	Corrosion rate			
	Distilled water		Tap water++	
	mdd	ipy	mdd	ipy
- High grade	27	0.0054	3	0.0007
- Die cast alloy XXV	28	0.0057	11	0.0023

++ Temperature hardness of 41 ppm and permanent hardness of 86 ppm.

Table 3 indicates that corrosion rate appears to be independent of grade of zinc in soft water. Further, the corrosion rate of zinc in tap water is much lower than that in distilled water. This is expected since deposition due to carbonate or other salts in water (e.g. sulphates and phosphates) reduces significantly the corrosion rate since the ionic movement is now by diffusion through the deposit. Further, corrosion rate is thus dependent upon nature of deposits from water and the zinc corrosion product.

3.1.3 Effect of dissolved oxygen. Out of dissolved gases, oxygen is the most important. Under conditions where the oxygen content cannot be replaced as quickly as it is consumed by the corrosion process, such as in stagnant water, zinc is attacked rapidly at local areas, creating pits. As more oxygen is made available the corrosion becomes more uniform. With further increase in oxygen content of water, the corrosion rate increases.

In flowing water, as in the case of water heating, localized attack disappears. Further, since the flow in absorber is in laminar region, the corrosion is normally constant over the practical flow rate (Ref.9). To illustrate the effect of oxygen, its combined effect with temperature and flow are given.

3.1.3.1 Effect of dissolved oxygen concentration and temperature.

The data on the corrosion rate of zinc in distilled water is reproduced from Uhlig (Ref.10) and is given in Table 4.

The following are observed:

1. Increase in corrosion rate with increase in O_2 content.
2. Increase in corrosion rate with increase in temperature.
3. Combined effect is again to increase corrosion rate.
4. With natural oxygen content and that available through air water interface the increase of corrosion rate is not as significant as that of due to intentional bubbling of O_2 (cases 3, 6, 9).

TABLE 4 - Effect of Dissolved Oxygen Concentration and Temperature

Test conditions:

- Water : distilled laboratory supply
- Temperature : room, 40°C and 60°C
- Duration of test : 7 days
- Specimen motion : none

Aeration	Temperature °C	Corrosion rate, mdd* Run 1	Run 2	Run 3
1. Water boiled-specimens inserted - flask sealed	Room	6	4	5
2. No aeration, except through air water interface	Room	10.5	10.5	10.5
3. Slow bubbling of O_2	Room	48	39	43
4. As in (1)	40	12	7	9.4
5. As in (2)	40	20	17	18.4
6. As in (3)	40	51.5	45	68.6
7. As in (1)	65	16	16.5	16.5
8. As in (2)	65	27	41	34
9. As in (3)	65	61	62.5	62

*determined after removal of corrosion products.

3.1.3.2 Effect of velocity and aeration (Ref.10).

The combined effect of velocity and aeration is given in Table 5.

TABLE 5 - Effect of Velocity and Aeration

Test conditions:

- Specimen size : 1 x 2 x 0.050 in
- Temperature : 20-25°C
- Test duration : 15 and 30 days
- Application motion: 6 rpm around center axis at 4 in radius plus 42 rpm independent rotation about specimen axis
- Aeration : 5-8 ml air per second when used
- Corrosion products removed chemically at the end of test.

TABLE 5 (cont'd)

Type of test	Corrosion rate, mdd Distilled water 15 days	30 days	3.5 NaCl solution 13 days	30 days
1 - No aeration; no rotation	24.8	24.0	20.2	25.9
2 - No aeration; rotation	46.4	32.1	70.2	47.0
3 - Aeration; no rotation	37.5	24.9	24.7	22.9
4 - Aeration; rotation	37.6	23.7	37.6	43.8

The following can be observed:

1. While rotation and aeration both increases corrosion rate, the combined effect need not increase corrosion rate as can be seen in cases (2) and (4) of Table 5. While both are rotated, the specimen with aeration shows lower corrosion rate in 3.5% NaCl medium and distilled water.

2. Except for cases (1) and (4) in 3.5% NaCl solution, the average corrosion rate of 30 days is lower than 15 days. This implies that corrosion product reduces the corrosion rate. In the long term one could expect reduction in corrosion rate of altogether prevention of further corrosion by the corrosion products if the latter forms uniform, stable, compact and adherent scale (Ref.12).

3.1.4 Effect of temperature. Table 6 illustrates the effect of temperature on the corrosion of zinc in distilled water. The test conditions specify that the specimen is rotated. Thus the results in Table 6 (Ref.10) are due to effects of temperature, flow and aeration. Fig.8 shows the same graphically (Ref.11).

Table 6 and Fig.8 indicate that the corrosion rate in distilled water is related to the change in the nature of corrosion product with temperature. From 20°C to 50°C, the rate approximately trebles and the product becomes less gelatinous and its adhesion is reduced. In the range of 50°C to 65°C there is a marked increase in rate, the product becomes decidedly granular, flaky and non-adherent. In range 70°C to 85°C, this type of product is accompanied by marked pitting. At higher temperatures the rate decreases owing to the decreased oxygen solubility and the product becomes compact and dense, and increases in adhesion.

TABLE 6 - Effect of Temperature, Aeration and Flow (Ref.10)

Test conditions:

- Material	:	Rolled high grade zinc
- Water	:	Distilled
- Duration of test	:	15 days
- Aeration	:	Unwashed air; other details not given
- Specimen motion	:	56 rpm on about 2 in radius; specimen horizontal

TABLE 6 (cont'd)

Temperature °C	Corrosion rate* mdd	ipy	Appearance of corrosion product
20	3.9	0.0078	definitely gelatinous; very adherent
50	13.7	0.0028	slightly less gelatinous; adherent
55	76.2	0.015	mostly granular; non-adherent
65	577	0.12	decidedly granular; becoming flanky and compact; non-adherent
75	460	0.092	decidedly granular; flanky and compact; non-adherent
95	58.7	0.012	compact, dense and flaky; adherent
100	23.5	0.0047	varies from greyish white to black, very dense resembling enabel; very adherent

*determined after removal of corrosion products. Average of two specimens. Values chosen from Fig.2 of original Reference 11.

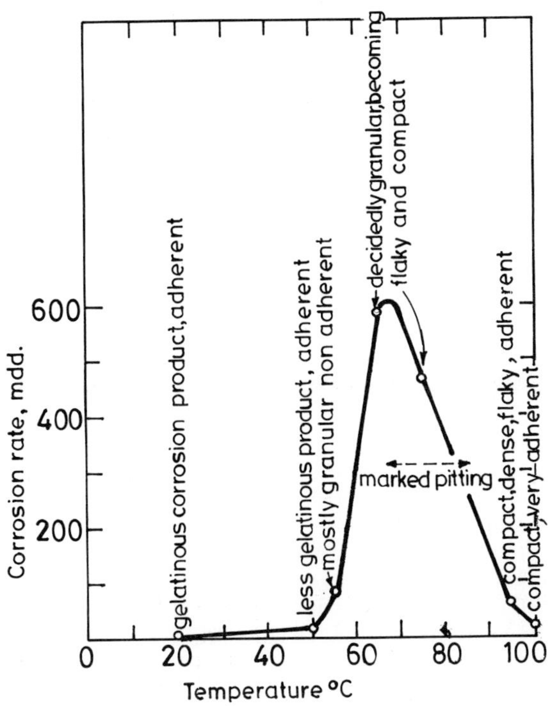

FIG.8 - INFLUENCE OF TEMPERATURE ON THE CORROSION OF ZINC IN DISTILLED WATER, 15-DAY ACCELERATED TEST (REF.11)

3.2 Discussion

It would appear from above that corrosion rate of zinc is very high in the range 60°C to 90°C. However, one should bear in mind that the test conditions specify that the corrosion test is done with distilled water and intentional aeration (with unwashed air). As pointed out earlier, the corrosion rate of zinc in tap water is much lower than in distilled water. In case of solar water heater source of O_2 is only due to the amount dissolved. It is also important to note that the corrosion rate deduced from the above experiment is based on 7 days duration, which is a very small period to extrapolate life of the zinc coating. From Table 5, it was observed that the corrosion rate decreases for longer duration. In actual condition corrosion rate would decrease as the ionic movement is restricted to diffusion through the corrosion product. If the data from Table 6 is used at 65°C where the corrosion rate is shown to be the maximum, then for 0.12 ipy corrosion rate, 3mm of zinc coating is required for expected life of one year. This is certainly absurd as we can see that it does not happen in practice. Therefore, in instances where the corrosion product is compact and adherent, then the rate of corrosion decreases with time and could even be prevented. HENCE ANY LINEAR EXTRAPOLATION CANNOT BE DONE AT ALL. In short, laboratory accelerated tests are not useful for service life prediction, but for comparison of relative corrosion behaviour of two materials under identical conditions.

Reverting to Table 4, the corrosion rate at 65°C in distilled water without motion is 62 mdd, while with motion it is 577 mdd. But comparing the corrosion rate of 48 mdd with slow bubbling of O_2 (Table 4) at room temperature and that of 37.6 mdd both with aeration and rotation (Table 5), one can be led to believe that the corrosion rate is lower in case of rotating specimen, thereby indicating that corrosion rate decreases with increase in velocity. This is against what is normally observed.

Therefore, from the above one can conclude that:

(a) The effect of each parameter or in combination as obtained in laboratory on the corrosion rate of zinc is applicable only for conditions identically similar to the ones existing during the laboratory tests.

(b) The corrosion rate obtained for a short period by accelerated test cannot be extrapolated linearly for service life. The corrosion rate deduced from the laboratory data is after removal of corrosion product.

(c) Corrosion rates obtained by different workers are under different conditions. Thus, the effect of each parameter or its combination by different workers will not necessarily give identical corrosion rate. Often the results can be misleading if test conditions are not specified (primarily by being not identical).

(d) Most of laboratory test results are for distilled water, while in actual practice of solar water heating system the nature of water varies from moderately hard to hard water.

In conclusion, it may be inferred from above that while the general behaviour of corrosion of zinc for different parameters individually or in combination are clear for short duration in distilled water, no inferences could be made as to the long term corrosion rate in tap water. Corrosion in an actual solar water heater is much more complex. In the next section, we shall examine the actual behaviour of galvanised steel solar absorber.

4. FIELD EXPERIENCE OF ZINC COATED STEEL

In this section we shall discuss two types of experiences. Firstly, the one observed by other people and need not necessarily be for solar water heating systems but for nearly similar applications. Secondly, the actual experience from solar water heating system.

4.1 Life Expectation of Zinc Coated Steel (Non-Solar System)

Extensive field data is available on the actual life of zinc coated steel. This has been compiled by Zinc Development Association, U.K. (Ref.12) and is reproduced here. Table 7 gives the specification for hot dip galvanizing of products. It specifies the minimum coating weight for various articles. As observed in practice and also pointed out in Table 7, actual thickness can be higher. In case of hot dip galvanized steel absorber with base steel thickness 1.00-1.25mm, minimum of 80-90 microns zinc coating thickness is obtainable (Ref.13). In actual practice, 125-150 micron is ensured for minimum expected life of 15 years (Ref.7).

Table 8 indicates the life expectation for typical locations and coating systems. The life is only for indicated zinc coating thickness.

For estimating the life of galvanized steel absorber, the coating thickness for reference is 80-90 microns or about 2 oz/ft^2. This falls in category A of Tables 7 and 8. Referring to "hot water supply" application, we can observe that the expected life ranges from 10 to 25 years. The range is due to varying water types and actual temperature or other conditions in the system. The "chemical characteristics of environment" indicates that the usage includes 60° to 90°C hot water supply. The need to have uniform coating is again stressed and the necessity of formation of non-flaking scale is obvious.

Comparing with other environment, the following are observed:

(a) Expected life for 2 oz/ft^2 (82.5 microns) in soft cold water supply is lower than for hot water supply. This is mainly attributed to high concentration of uncombined CO_2 in which case protective scale is not formed. The level of CO_2 in tap water is generally very low.

(b) Expected life in hard cold water supply is comparable to that for hot water supply for the same coating thickness. It is known that bicarbonate decomposition to carbonate and the corrosion product forms non-flaking and adherent protective layer for normal cold water supply temperature (20-30°C). This would lead to life of 20 years. On the other hand, any adverse conditions life copper ion deposition on zinc can cause severe pitting even at low temperature.

TABLE 7 - Specifications for Hot Dip Galvanizing of Products

For coatings (a) to (d) below the designer need only specify BS729:Part 1. Pickling is an inherent part of the normal pretreatment. Coatings (e) and (f) should be specified in full.

		Galvanized coating weight Minimum g/m^2	oz/ft^2	
A	Steel articles 5mm thick and iron castings	610	2.0	Standard finish universally obtainable. Structural steels usually have substantially greater coating weights than the minimum.
B	Steel articles 2-5mm thick	460	1.5	
C	Steel articles 1-2mm thick	335	1.1	
D	Threaded work and other articles which are centrifuged	305	1.0	Centrifuging is to ensure that threads are free-running. The reduced coating thickness may make extra protection, such as coating of zinc rich paint, desirable after erection.
E	Structural steel items Specify controlled shot blasting and $1200 g/m^2$ ($\sim 4\ oz/ft^2$) galvanized coating	1200	4.0	Where silicon-killed steel is used, shot blasting may not always be necessary.
F	As (a) to (d) plus two coats of paint	As (a) to (d)		Specified when colour or long total life is required (See 'Painting of Products Galvanized After Fabrication').

The thickness of the galvanized coating formed is largely self-regulating and variation is mainly due to differing grades of steel. For this reason coating weights normally range from the minimum quoted up to at least 50 per cent higher. The expected life figures quoted in Table 3 are based on the coatings normally obtained.

From Table 8 one can infer that hot dip galvanized steel with 125-150 microns coating thickness is suitable for hot water supply and a minimum life of 15 years can be expected. Longer life depends upon the actual system conditions.

TABLE 8 - Life Expectations for Typical Locations and Coating Systems (Ref.12)

Chemical characteristics of the environment	Location	Life in years before maintenance of protective system is needed (see Table 7)				
		A	B	C	D	E
In a warm dry atmosphere, stable zinc oxide forms	Indoors	Very long - probably centuries				
As above but with moisture present zinc hydroxide forms which is converted to basic zinc carbonate by atmospheric carbon dioxide.	Rural	25-50	18-35	12-25	12-25	50-100
	Coastal & Urban	12-25	8-15	6-12	6-12	25-50
Where sulphur dioxide is high, acidic rainwater on the zinc surface dissolves the protective film of zinc carbonate which reforms from the metal. This process slowly continues until all the coating is corroded away.	Chemically polluted (Industrial)	4-12	3-8	2-6	2-6	8-25
The corrosion rate of zinc can be high in acid (pH6) or alkaline solutions (pH12), but between these limits corrosion is much slower. In most mains water supplies calcium bicarbonate and the product of zinc corrosion produce a non-flaking scale which virtually stops corrosion. If the mains water supply is capable of dissolving copper, pitting corrosion may be caused in systems using copper pipes.	Hard cold water supply	Usually 12-20				
If the water is soft or has a high concentration of uncombined carbon dioxide, a protective scale is not formed and therefore the zinc dissolves slowly.	Soft cold water supply e.g. moorland	Usually 2-15				

TABLE 8 (cont'd)

In some waters between 60° and 90^oC, zinc does not act sacrificially when it is damaged and therefore it is important to obtain an early formation of non-flaking scale.	Hot water supply	Usually 10-25				
The high chloride content of salt water prevents a protective scale from forming, allowing the zinc to dissolve slowly.	Sea Water	5-12	3-9	2-6	2-6	10-20
In moist conductive soils sacrificial protection is given, but in inorganic moderately reducing soils and highly reducing organic soils extra protection is required. Clinker and ashes contain soluble sulphates and unburn carbon and give shorter lives than those quoted	Under-ground	5-20	3-12	2-10	2-10	10-40
Wet mortar, concrete and plaster have a slight etching action on zinc which stops soon after setting	Building materials	Indefinite				
Zinc is resistant over a wide pH range and particularly in moderately alkaline solutions but it should not be used with acids and very strong alkaline solutions.	Chemical plant					

4.2 Actual Field Experience with Zinc-Coated Steel Solar Absorber

Two types of zinc-coated steels are discussed here. Firstly, those coated with zinc spray internally and secondly those which are hot dip galvanized. In case of latter zinc coating is uniform internally as well as externally. In the former, the external surface is either coated with red oxide or zinc-rich paint followed by black paint. For the present discussion only the internal corrosion is of interest.

But, it may be mentioned that the corrosion of external surface with red oxide alone is insignificant over the observation period of 5 years in the absence of condensation inside the housing.

4.2.1 Experience with internally zinc sprayed steel solar absorber - Thermosiphon System.

PLATE 6 - 400 lpd 60°C THERMOSIPHON SOLAR WATER HEATING SYSTEM - INTERNALLY ZINC SPRAYED STEEL ABSORBER. COMMISSIONING DATE: JUNE 1979.

Plate 6 shows the system installed at Hindustan Brown Boveri's R&D Centre, Baroda. The system specifications are:

- Collector area : $3.75m^2$ (5 collectors of $0.75m^2$ in parallel).
- Absorber : Zinc sprayed internally, resistance welded.
- Glazing : Two cover - outer of glass, inner of polyester.
- Storage tank : 200 l, mild steel zinc rich painted.
- Piping : Galvanised iron.
- Thermosiphon flow rate : 50-100 $l\ h^{-1}\ m^{-2}$; linear velocity in absorber channel is 0.0031-0.0062 m/s or 0.31-0.62 cm/s (self-regulating).
- Commissioning date : June 1979.
- Type of water : Borewell water stored in overhead tank (see Table 10 for water analysis).

After 5 years of service there is no evidence of corrosion failure in the absorber. During the 5 years the absorbers had been undergoing temperature excursions of 45-70°C as shown below. However, this type of absorbers has been superseded by hot dip galvanized one due to the latter's superiority of zinc to steel metallurgical bond and longer expected life.

Time	Duration	Operating conditions
June 1979 - Dec. 1980	1½ years	No load test - fresh water change in the evening except Sunday; temperature range 55-70°C.
Jan. 1981 - Dec. 1981	1 year	Load test; hot water withdrawal 200 l - 300 l/day first month; 400 l/day for the rest of the year; temperature range 45-65°C.
Jan. 1982 - July 1984	2½ years	No load test; occasional water change; temperature range 55-70°C.

4.2.2 Field experience with hot dip galvanised steel absorber. To date there is no corrosion failure reported in solar water heating system employing galvanized steel absorber. Some of the installations are given in Table 9. The life of these installations to date varies from 6 months to 3 years.

The 100 lpd thermosiphon system (Plate 7) incorporating copper storage tank has been dismantled for examination. The temperature in this set up varies from 50°C to 75°C. It occasionally goes up to 80°C, since this set up was mainly constructed to study ageing effects on the solar water heater components under stagnation. The system pressure goes up to 2 Kg/cm^2. For the first two years the collector was provided with two covers (outer: window 3mm glass, and inner: 50 micron UV stabilised polyester film). Later inner cover was removed to study the window glass breakage pattern under stagnation. The maximum temperature later reduced to 70°C. The absorber panels after 3 years of service has been cut open and the observations are given below for the close-up shown in Plates 8-11.

Throughout, the zinc water reaction product on the upper surface (i.e. facing sun) is thinner with scattered dimples but uniform, while at the bottom it is rough, uniform and appears thicker. The reaction product throughout is compact and adherent. Plate 11 shows the difference in the nature of reaction product on the upper and lower surface of the absorber. There has been no pitting at both the surfaces. The microscopic examination of the galvanized steel surface is shown in Plate 12. The inner side in contact with water shows the free zinc layer and scale on top of it. There is clearly no evidence of pitting and uniform scale is observed. Plate 12 indicates that life of the absorber depends at what rate the reaction proceeds to consume inner zinc layer, steel and again outer zinc layer. The outer zinc layer as such has infinite life. Only in internal aggravating environment when the inner zinc layer and steel substrate have been consumed, the outer zinc layer will be affected.

Plate 12 indicates zinc reaction rate of 0.0010 ipy based on 3 years

TABLE 9 - Installed Solar Water Heating Systems (SWHS) employing ganvanized steel absorber

Location	Collector area, m^2	Type and capacity	Purpose	Date of commis-sioning	Status of absorber corrosion	Remarks*
1. R&D Centre, Hindustan Bown-Boveri Ltd.,Baroda	0.75	Thermosiphon; 100 lpd	Reliability++ test	Aug 1981	No failure	High temperature and high pressure test of associated compo-nents; stagnation allowed.
2. - do -	3.75	Forced circu-lation close-loop; 400 lpd 60°C	-do- +++	Oct 1981	No failure	For performance tests components reliabili-ty two-cover collect-or; differential tem-perature controller; initial filling DM water; make up by tap water.
3. Seshasayee Industries, Vadalur	300	Forced circu-lation 24,000 lpd, 60°C	Insulator cement joint curing	Aug 1982	No failure	Close-loop, DM water open expansion tank.
4. Traction Machine Workshop, Nasik	12.0	Thermosiphon 1000 lpd 60°C	Utensil washing	July 1982	No failure	-
5. Circuit House, Baroda	16.8	Thermosiphon 1250 lpd 60°C	Bathing	Dec 1983	No failure	City water supply.
	10.5	Thermosiphon 750 lpd 60°C				
6. Bajaj Auto Ltd.	190	Forced circu-lation 32,000 lpd, 45°C	Cooking+	Jan 1984	No failure	6,546,000 l water supplied to date.
	160	-do- 30,000 lpd, 45°C	Utensil washing		No failure	4,165,500 l water already supplied.

*Other components in the system, e.g. piping and fittings area of GI and tanks are of MS zinc-rich coated or galvanized; valves are of gunmetal or cast-iron. This is the same for all systems unless indicated. +Stainless steel tank for back-up heating up to 80°C, MS zinc-rich painted for 45°C storage.
++ Copper storage tank; +++ Copper tube heat exchanger.

service at 50-75°C, whereas in the same range of temperature reaction rate of 0.015 to 0.12 ipy (Table 6 and Fig.8) are observed. Further, the nature of scale observed in the accelerated test (Table 6) is flanky at specific temperatures of 55, 65 and 75°C while in the actual solar absorber (3 years service at 50-75°C) the scale is contrastingly adherent, compact and uniform. Reasons for such a difference are due to:

(i) 15-day accelerated test and actual 3-year data are not directly comparable because the reaction mechanism in case of solar absorber is self-healing, i.e. onset of reaction itself is inhibiting further reaction.

(ii) Chemistry of raw water in the solar absorber and distilled water in the accelerated test is very much different. It was observed from Table 3 (see 3.1.2) that raw water is significantly less reactive than soft (or distilled) water.

(iii) The solar absorber operates at variable temperature range of 50-75°C whereas accelerated test data are obtained at specific temperatures of 55, 65 and 75°C.

PLATE 7 - 100 lpd THERMOSIPHON SYSTEM - (GALVANIZED STEEL) RELIABILITY SET UP FOR STAGNATION TEMPERATURE AGEING OF COMPONENTS. COMMISSIONING DATE: AUGUST 1981.

TABLE 10 - Chemical quality of water sample

Particular of sample	pH	Suspended solid ppm	Total dissolved solids ppm	Alkalinity to phenophtta iem	Alkalinity to methyl orange ppm	Free CO_2 ppm	Chloride ppm	Silica ppm	SO_4 ppm	Fe ppm	Ca ppm	Mg ppm	Total hardness ppm $CaCO_3$	Temp. hardness ppm $CaCO_3$	Permanent hardness ppm $CaCO_3$
Well water sample	8,1	Nil	1072	Nil	640	8.0	142	17.0	8.2	Nil	30.4	57.1	312	312	Nil
				Total acidity as $CaCO_3$	HCO_3 ppm						Ca As $CaCO_3$	$CaCO_3$ ppm			
				120	780						76	236			

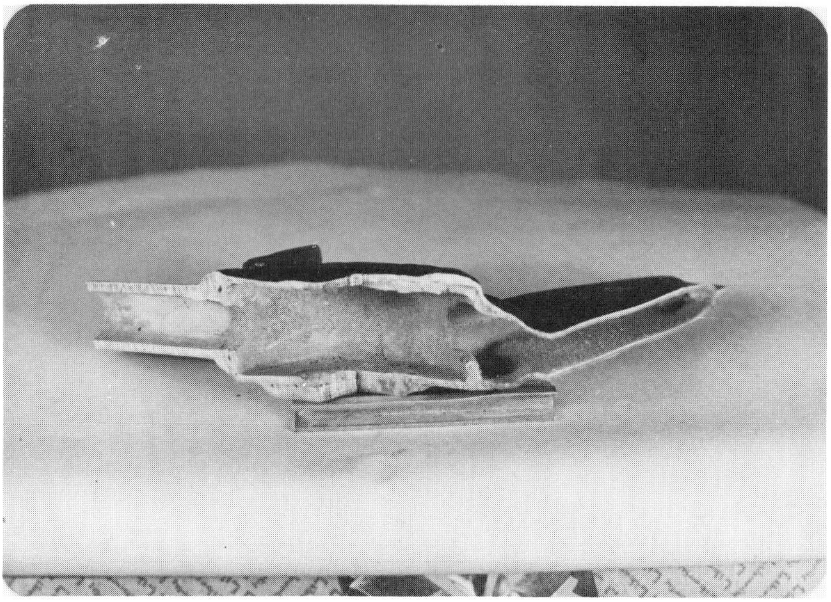

PLATE 8 - INTERNAL VIEW OF GALVANIZED STEEL ABSORBER OF PLATE 7; TOP MANIFOLD - AFTER 3 YEARS SERVICE AT 50-70°C

PLATE 9 - INTERNAL VIEW OF GALVANIZED STEEL ABSORBER OF PLATE 7; BOTTOM MANIFOLD - AFTER 3 YEARS SERVICE AT 50-70°C

PLATE 10 - INTERNAL VIEW OF GALVANIZED STEEL ABSORBER OF PLATE 7; EXTREME RISER - AFTER 3 YEARS SERVICE AT 50-70 ^{o}C.

PLATE 11 - INTERNAL VIEW OF RISER OF ABSORBER OF PLATE 7, SHOWING DIFFERENCE IN THE NATURE OF SCALE ON THE UPPER AND LOWER SURFACES. UPPER SURFACE IS THE ONE FACING THE SUN.

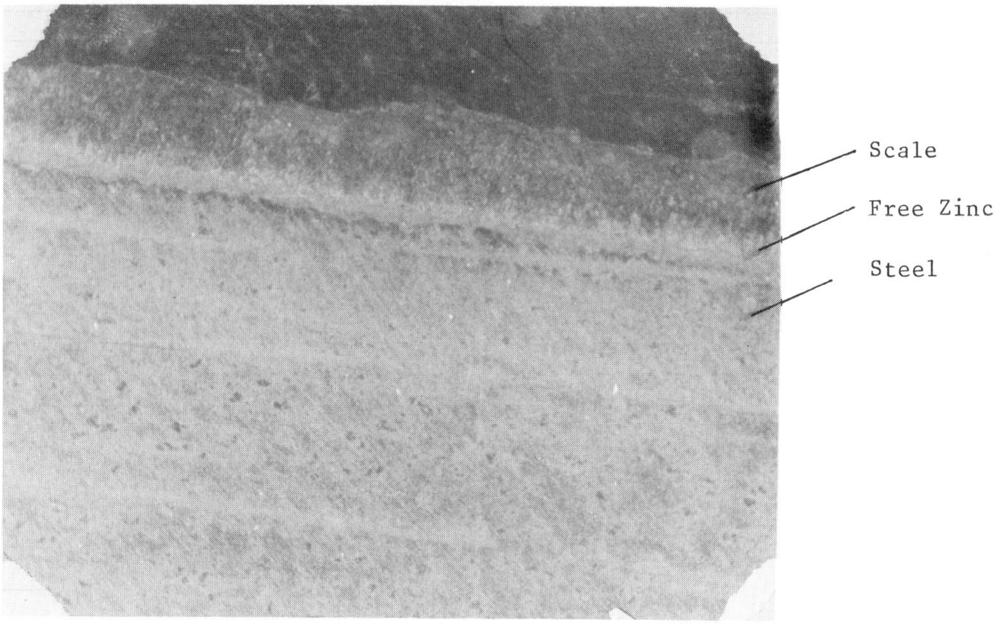

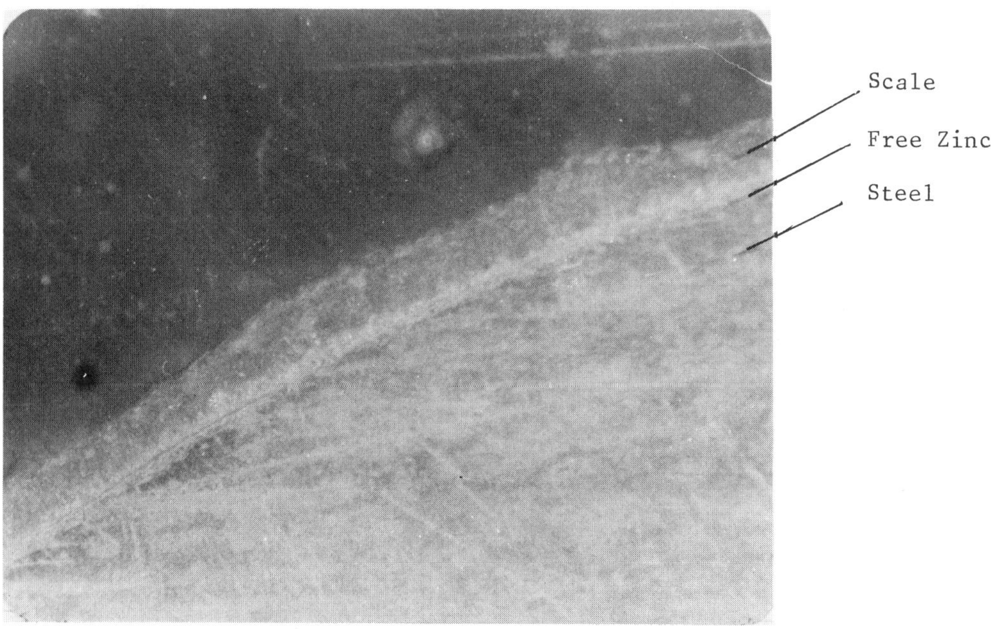

PLATE 12 - MICROSCOPIC EXAMINATION OF GALVANIZED STEEL ABSORBER'S RISER AFTER 3 YEARS SERVICE SHOWING FREE ZINC AND SCALE IN THE RISER (MAGNIFICATION 80X).

Based on the above observations IT CAN BE INFERRED THAT ACCELERATED TEST DATA CANNOT BE USED TO PREDICT WITH EXACTNESS THE BEHAVIOUR OF GALVANIZED STEEL WATER SYSTEM NOR CAN THEY BE USED TO EXTRAPOLATE SERVICE LIFE.

Further, comparing the reaction rate, what is OBSERVED IN ACTUAL 3-YEAR DATA is ONE HUNDREDTH OF PEAK VALUE INDICATED IN ACCELERATED TESTS. This rate is observed for aggressive water (chloride level of 142 ppm - see Table 10. Thus it can be expected that reaction rate will be much lower for normal water containing chloride level not exceeding 50 ppm.

There is no evidence of attack due to copper ions on the zinc. It is possible that the inner wall of the copper storage tank is passive and thus no significant copper ion goes into the solution. Further, the corrosion product on the zinc in absorber could have prevented direct attack, if any.

Type of water used for actual ageing tests:

The water analysis is shown in Table 10. Water is scaling type with high chloride content. Both have opposing effects. While cabonate deposits due to bicarbonate decomposition offers resistance to corrosion, chloride ions beyond 50 ppm is considered corrosive. The water used in water heaters of Plates 6 and 7 have 142 ppm chloride level and is, therefore, aggressive water. The actual ageing tests were, therefore, done under severe environment.

4.3 Suitability of GI Pipe-Matrix Absorber for Hot Water System

In Section 2, the disadvantages of GI pipe matrix were outlined. In the present section its suitability for solar water heating application is further discussed.

The GI pipes used for the pipe matrix are normally commercial pipes. There is no specific IS (Indian Standard) covering this product as far as hot dip galvanizing is concerned. Each GI pipe manufacturer has its own standards. These are mainly to cater to cold water application for which the product is meant. Actual measurement of commercial GI pipe indicated that zinc coating for 12 mm and 25 mm dia pipe are 40-50 micron and 70-80 micron, respectively. These coating thicknesses would be suitable, in general, for cold water application in cities, where water is slightly or moderately hard.

For hot water application, however, 12 mm pipe does not seem to be suitable for expected life of 10 years. This is because the zinc coating thickness is half of that indicated in Table 8, viz. 82.5 microns. 25 mm GI pipe just meets the minimum requirement.

Considering the internal and external welding joints of GI pipe matrix,where zinc is removed and much lower zinc coating thickness of the riser, it appears that GI pipe matrix is not a suitable choice for solar water heating application.

5. CORROSION PROBLEMS IN COPPER ABSORBER

The field experience of Australian copper collector has been reported by James recently (Ref.14). The following is an extract from his report:

"Absorber plates consisted of a silver soldered copper pipe manifold having an upper and lower header with smaller diameter riser pipes, either soldered to this copper plate or mechanically attached to aluminium plate with thermal paste to improve heat transfer. Those with copper plates usually had a selective surface treatment, and those with aluminium plate were painted. Corrosion in the form of copper oxides on the outside of the plates was occasionally observed but was slight. The domestic system manufacturer reported that internal corrosion of the pipe manifold in systems occurred in some parts of Western Australia after 4 to 7 years of service. This was the cause of 40 per cent of the failures in his product (0.6 per cent of total sales). The conditions singled out by the manufacturer were high chloride levels in water (250-300 mg/l) coupled with high radiation levels. The damage took the form of pitting corrosion, particularly just above any accumulation of sludge in the bottom of the absorber and in the top third of the absorber. According to Dillon (Ref.15), the water qualities which enhance the corrosion of copper are high acidity, softness, absence of organic matter, and high temperature. It is possible that chloride level is not the primary factor causing this corrosion but is merely a contributing factor."

"The collectors at Queanbeyan and those of the manufacturer occasionally developed leaks in the copper tube manifold. The manufacturer suggested rough handling during installation as the cause of the leaks in his collectors, as they underwent 100% pressure testing at the factory. The cause of the Queanbeyan leaks is unknown. Freeze bursting often occurred at the Canberia installation before the practice of draining the collectors on frosty nights was adopted."

The above two instances imply that it is not advisable to extrapolate experience from one system to another system of different working conditions. Thus, if copper fails it only means that "copper is suitable for water" has been stretched too far.

Particular/condition laid down by SUNSTRIP for usage of drawn copper tubes in open loop with water are (Ref.16):

(a) Fluid velocity should not exceed 0.9m/s.

(b) Water must not be acidic or containing corrosive chemicals.

(c) Chlorine content should not exceed 2 ppm.

Another serious failure, which has been experienced at our R&D Centre is due to stress corrosion. This occurred in a filtration unit after service of 3 years. The nitrate ions in water and stresses in the copper tube caused the failure.

6. CONCLUSIONS

(a) While accelerated test indicates the behaviour of zinc-water system under different environments, they cannot and must not be used to extrapolate service life. Accelerated tests are meant only to compare relative reaction rate of two different metals under similar operating conditions. The actual 3-year test data indicates reaction rate of one hundredth that of observed in accelerated test.

(b) Published data for long term performance of non-solar systems at 60-90°C hot water supply indicates suitability of hot dip galvanized steel. The minimum expected life for hot dip galvanized steel panel of 100-125 microns zinc is 15 years based on such long term experience.

(c) Actual experience of hot dip galvanized steel absorber panel after 3 years of service at 50-75°C indicates suitability of such absorber. Internal examination of hot dip galvanized steel absorber shows that the reaction product is compact, uniform and adherent. The reaction appears to be self-healing. From the reaction rate observed it can be inferred that actual reaction rate is one hundredth that of peak value indicated in accelerated test.

The fact that the above steel panel has withstood aggressive water used during the 3-year test with chloride level of 142 ppm shows that the reaction rate could be significantly lower in normal water with chloride level not exceeding 50 ppm.

(d) GI pipe matrix obtained by welding commercial GI pipes does not appear to be suitable for hot water application since not only the zinc is removed from the welding joints but the riser pipe has normally 40-50 microns zinc coating thickness. The latter cannot ensure even a minimum of 10 years life generally claimed.

REFERENCES

1. P.D. Thompson and M.B. Hayden, (1977), 'Corrosion problems in solar energy systems', Paper 63, National Association of Corrosion Engineers, Intl. Corrosion Forum, Mar 14-18, San Fransisco, Cal., USA, 1977.
2. R.S. Soin and V. Murali, (1982), 'Solar Hot Water Systems for Industry' HBB Technical Review, Vol. VI, No.2, Jan 1982.
3. Galvanizing Guide, Zinc Development Association, Galvanizers Association, 34 Berkeley Square, London W1X6AJ, UK, pp 11, Aug 1979.
4. M.G. Fontana and N.D. Greene, (1967), Corrosion Engineering, McGraw Hill Book Co., Inc.
5. S.S. Prasad and V.V. Rao, (1976), 'Studies on the performance of flat plate solar collectors for solar water pump', M.Tech. Thesis (Chem.Engg.), BITS, Pilani.
6. Author's witness of selective GI absorber at IIT Madras for solar refrigeration.
7. Author's Indian Patent 227/BOM/82.
8. Metals Handbook, 'Properties and Selection', Vol.1, pp 1163.
9. G. Butler and H.C.K. Ison, (1966), Corrosion and Its Prevention In Waters, Leonard-Hill, London, 1966, pp 136-37.
10. H.H. Uhlig, (1948), The Corrosion Handbook, John Wiley & Sons.
11. G.L. Cox, (1931), Ind. Eng. Chem. Vol.23, pp 902-904, 1931 (Quoted in Ref.9, p 141).
12. Galvanizing Guide, Zinc Development Association, Galvanizer Association, 34 Berkeley Square, London W1X 6AJ, Aug 1979.
13. Author's experience; compiled in R&D Report 'Flat Plate Solar Collector' (Project RD/ES/012). Final Report, R&D Centre, Hindustan Brown Boveri Ltd., Baroda, Sept 1983.

14. S.R. James, Sunworld, Vol. 7, No.1, Spring 1983, pp 2-5.
15. B.I. Dillon, (1976), 'Studies on solar hot water heaters', The Australian Mineral Development Laboratories, Report No.113, May 1976 (Quoted in Ref.14).
16. 'Specifications for fluids used in Sunstrip[R] Absorber Plates Fintubes Coils', SUNSTRIP International Inc., 26 Chapin Rd., P.O. Box 109, Pine Brook, N.J. 07058, USA.

WELDING TECHNOLOGY AND PROCEDURES

R.S. Parmar
Department of Mechanical Engineering
Indian Institute of Technology Delhi
Hauz Khas, New Delhi 110 016
India

1. INTRODUCTION

Almost all fabrications are now a days done by the use of welding and the allied processes which include soldering, brazing, braze welding, and adhesive bonding. The selection of process in a given situation depends upon the material to be joined, its thickness and the joint strength required.

Before selecting a joining method it is essential to understand the process capabilities of all of them.

1.1 Welding

It is the method of joining two or more components by the use of heat or/and pressure to achieve complete coalescence between them. Almost all metals in any thickness can be welded by one method or the other but the performance may differ. Welding of dissimilar metals and alloys as also the thermoplastics can be satisfactorily achieved. Welding of metals and non-metals is also feasible.

If welding is carried out properly the strength of the welded joint will be equal to the strength of the parent metal in case of metals and alloys, and about 60-80% of the parent metal strength in case of plastics.

1.2 Brazing

Brazing differs from welding in that the metals to be joined are not melted and deeply fused with the filler metal. Instead, brazing employs a filler metal which melts at a temperature below the melting point of the metal to be joined but above 450°C (and normally below 925°C), and is drawn into the close-fitting joint by capillary action. The surfaces must be clean so that a wetting action can take place. Wetting refers to the spreading of the filler metal and is due to the attraction that occurs between the molecules of the alloy and the base metal on properly cleaned surfaces at the right temperature.

H. P. Garg (ed.), Solar Water Heating Systems, 231–242.

1.3 Braze Welding

Braze (or Bronze) welding is similar to gas welding but the parent metal is not melted and the filler wire has a melting point above 450°C but below the melting point of the parent metal. It is different from brazing in that the filler is deposited rather than being made to flow into the joint by capillary action. The strength of the braze welded joint is similar to that of gas welding, but some of the forces effective in brazing, e.g. inter-allying and intergranular penetration are also at work in braze welding.

1.4 Soldering

Soldering like brazing is a process of joining two parts of metals or alloys by filling the joint space with a filler by capillary action. However, the melting point of the filler for soldering is less than 450°C and it involves no melting of the parent metal. A soldered joint is normally weaker than a brazed joint.

1.5 Adhesive Bonding

It is the process of joining metals by adhesives and finds considerable use in the manufacture of automobiles and aircrafts.

The mechanical strength of the adhesive bonded joint depends upon the joint configuration, the nature of adhesive, and its thickness. Generally speaking, the strength of a lap joint increases with the amount of overlap and decreases with increasing thickness of adhesive.

2. EQUIPMENT

The equipment for welding are of varied types depending upon the process used. It is therefore beyond the scope of this write up to describe all types of equipment used in the welding processes. However, brief description of some of the processes more suitable for fabrication of solar water heating systems is given here.

2.1 Gas Welding

This process uses oxy-fuel gas system to produce a flame which provides necessary heat for fusion. The main parts of the equipment include a welding torch, oxygen supply, fuel gas supply (usually acetylene), hoses for connecting torch to supply. A set up for gas-welding system is shown in Fig.1.

2.2 Shielded Metal Arc Welding (SMAW)

This is the most often used process in the world for welded fabrication and is more often known as manual metal arc welding or stick electrode welding. The equipment for the process consists of a welding power source which is an a.c. welding transformer, welding rectifier or a

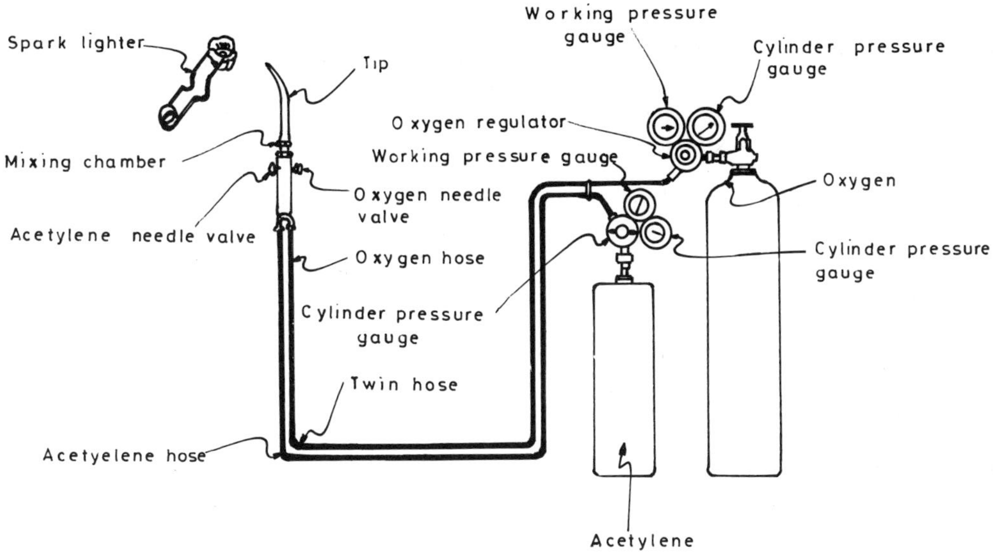

FIG.1 - OXY-ACETYLENE WELDING SET UP

motor-generator welding system. Whereas an a.c. transformer supplies a.c. power supply, rectifier and generator supply d.c. for welding. As solar heating systems usually require welding of thinner gauges of metal, it is better to use d.c. supply to achieve better control of heat and enhanced manouverability. The power source is connected to the electrode holder by cables and welding is achieved by the use of coated electrode. A set up for shielded metal arc welding system is shown in Fig.2. The power source can be any one of the above mentioned types.

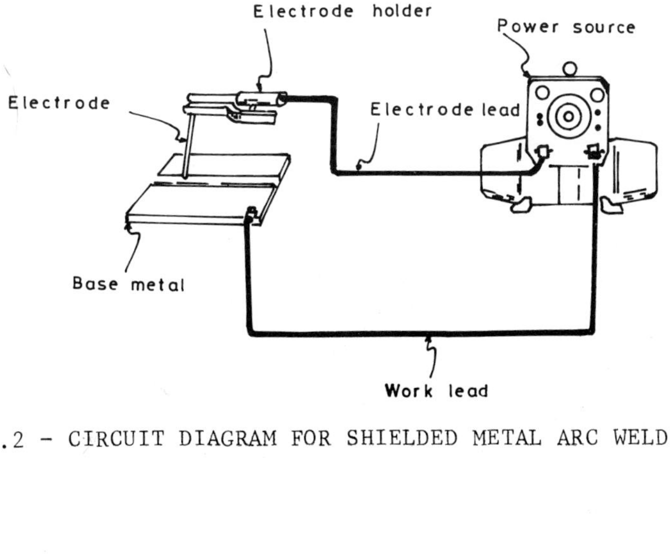

FIG.2 - CIRCUIT DIAGRAM FOR SHIELDED METAL ARC WELDING

These days more costly solid state power sources are also available which can supply the desired type power supply and also more accurate control on the welding current thus giving more uniform weld beads with better mechanical properties.

2.3 Gas Tungsten Arc Welding (GTAW)

Welding of aluminium and its alloys as well as stainless steel is more difficult by SMAW and hence the development of gas shielded tungsten arc welding. The arc is created between the non-consumable tungsten electrode and the workpiece to be welded. Filler metal is added, if required, from a separate filler rod just like in oxy-acetylene welding. The tungsten electrode and the weldpoint are given a shielding of inert gas, e.g. argon or helium. The power source used is usually of the a.c. type but when d.c. power source is used then electrode need to be connected to negative to reduce excessive heat at the non-consumable electrode. However, for welding aluminium, it is essential to connect negative to work which provides cleaning action due to movement of the cathode spot. This may necessitate water cooling of torch. A set up of GTAW system is shown in Fig.3.

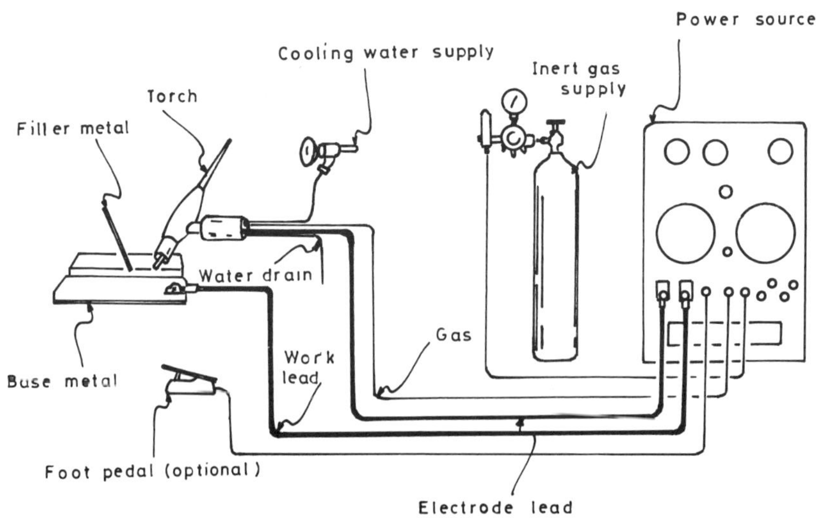

FIG.3 - CIRCUIT DIAGRAM FOR GTAW

2.4 Gas Metal Arc Welding (GMAW)

Gas metal arc welding followed the development of GTAW. This process uses a consumable electrode instead of non-consumable in GTAW.

Due to this change, the metal deposition rate is considerably increased and the process has an all-position welding capability. D.C. power source is invariably used with the electrode positive. The process finds extensive use for welding ferrous and non-ferrous metals. A schematic of GMAW process is shown in Fig.4.

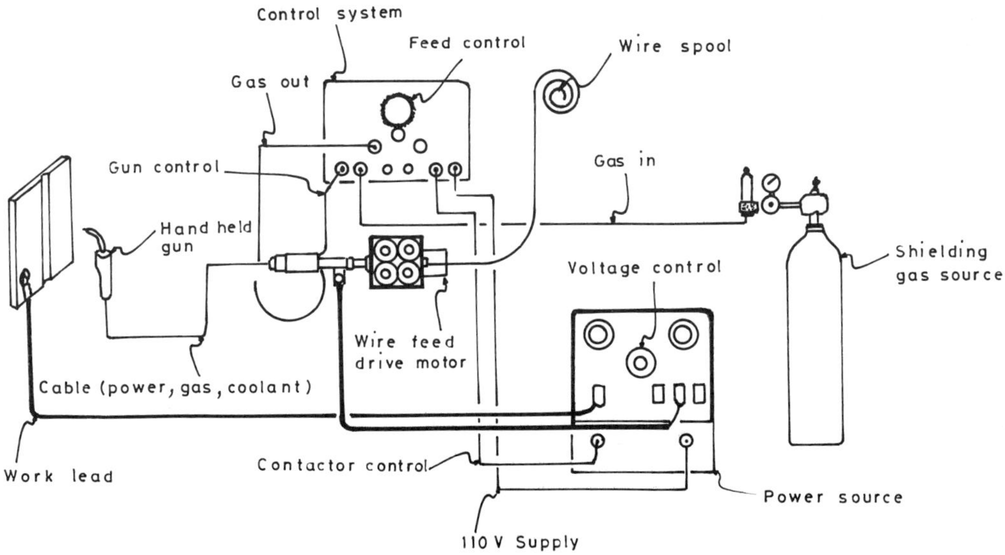

FIG.4 - BLOCK DIAGRAM - GMAW

Apart from the arc welding process, resistance welding processes like spot and seam welding can also be used in the fabrication of solar water heaters. A brief description of these processes follows.

2.5 Resistance Spot and Seam Welding

In the resistance spot welding process the pieces to be joined are placed between two rod electrodes and heavy current at low voltage (2-10 volts) is made to pass through the workpiece and joint is obtained. Time for which the current flows is controlled electrically and can be varied to deal with different thicknesses of matter. However, the heat generated is given by the relationship

$$H = I^2Rt$$

where H = heat generated, joules
I = welding current, amps
R = resistance, ohms
t = time , secs

This process is popular for joining sheet metal and is very effective where leak proof joints are not required. The set up for spot welding is shown in Fig.5.

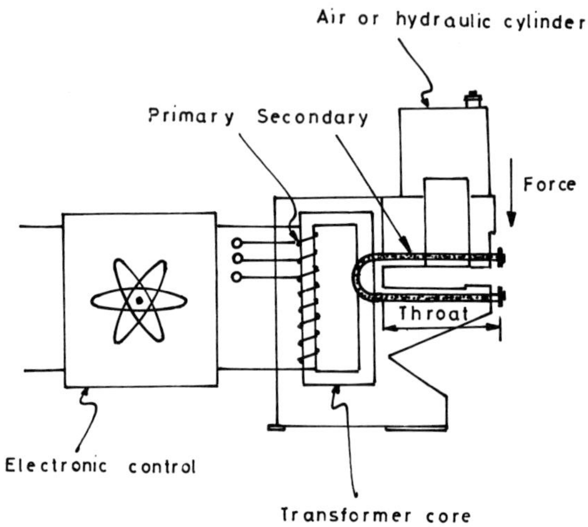

FIG.5 - PRESS TYPE RESISTANCE SPOT WELDING SET UP

When it is required to produce leak proof joint, the spot welding can be replaced by seam welding which is a continuous spot welding machine. Instead of rod electrodes of spot welding machine, seam welding has a set of copper wheels which act as electrodes.

Seam welding can be used to produce highly efficient water and gas-tight joints and the welds are often referred to as stitch welds due to its resemblance to stitches.

A part of the seam welding machine is shown in Fig.6.

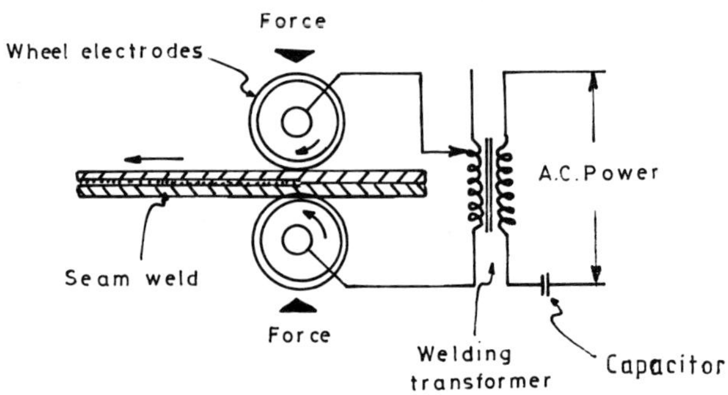

FIG.6 - RESISTANCE SEAM WELDING

Though many other welding processes may be used for the fabrication of solar water heating systems, but in most of the cases the above mentioned processes can handle almost all types of fabrications required.

2.6 Equipment for Soldering, Brazing and Bronze Welding

Rquipment for soldering is much simpler than the welding equipment and consists of a soldering iron, which may be electrically heated or may be by the use of other fuel like coal or kerosine if the work is at a site where electric power is not available. The filler used is called solder or soft solder and is an eutectiferous alloy of lead and tin. The joint to be soldered is cleaned and then soldered by filling the soft solder between the gap. Soldering fluxes are used to dissolve the oxide layer which may be present on the surface to be joined.

For brazing and braze welding of solar water heaters, the equipment commonly used is the gas welding torch. Brazing fluxes are used to dissolve the oxides on the surface to be joined and the joint is filled by molten brazing material which may be anyone of the following depending upon the metals to be joined and strength of the joint to be attained. In order of popularity these are: Silver, Copper and Copper-Zinc, Copper Phosphorous, Aluminium-Silicon, heat resisting materials, Copper-Gold, and Magnesium.

3. WELDABILITY OF METALS USED FOR SOLAR WATER HEATING

Generally solar energy collectors use pipes and sheets of low carbon steel, galvanised steel, copper and aluminium. These pipes and sheets are welded, soldered or brazed depending on the availability of equipment and/or the strength of the joint required. Brief description of methods employed for welding these metals in similar or dissimilar combinations are as follows:

3.1 Welding of Low Carbon Steels

Weldability of low carbon steel is very good. So joining sheets to plates create no problem provided proper precautions are taken before welding, brazing or soldering. These precautions include thorough cleaning of the joint by the use of wire brush.

FOR SMA welding proper sized electrode should be chosen depending upon thickness of sheet and wall thickness of tube or pipe. Usually the electrode employed are general purpose rutile electrode of 2.5 or 3.15 mm diameter. The pipes should first be tack welded then followed by regular continuous welding. All methods of welding described earlier can be used for welding this material. No serious difficulty is encountered in welding similar steels in the low carbon steel category.

If thin pipes and sheets are to be joined then brazing and/or soldering can be employed. Brazed joints being stronger it may be better to employ that rather than soldering. Torch brazing with copper-zinc filler rod with slightly reducing flame with AWS No.3B flux which contains boric acid, borates, fluorides, fluoborates and wetting agents. The flux can be applied in powder, paste or liquid form.

Steels can be soldered as well but the joint strength is low and the process as applied in the general fabrication may prove to be slow. However, where needed, torch soldering like torch brazing can be employed.

3.2 Welding of Galvanised Steel

Galvanised steel is popular for solar water heaters because of its corrosion resistant properties. However, zinc has a boiling point of 870°C and steel melts at 1535°C, thus zinc coating gets evaporated during welding leaving behind oxidation prone zone, the extent of which depends upon the heat input per unit length. The affected area is, therefore, greater with slower welding speed processes, e.g. oxyacetylene welding or gas tungsten arc welding.

When galvanised steel is resistance welded, the welding heat causes less disturbance of the zinc coating than the arc processes. The resistance to corrosion is not disturbed as the zinc forced from the spot will solidify adjacent to the spot weld and protect the weld nugget. Resistance welding of galvanised steel is, however, more of a problem than arc welding due to zinc pick-up by the resistance welding electrodes.

Zinc in gaseous state may get entrapped in weld metal causing proposity in the weldment. To eliminate this, the weld joint must be designed to allow the zinc vapours to completely escape from the joint. Fixturing backing straps, etc. should be arranged to allow for the zinc vapour to completely escape from the joint. However, the ultimate precaution would be to remove the zinc from the area to be welded.

When welding galvanized steel, positive ventilation should be ensured to avoid the ill-effects of poisonous fumes. For the same reason these steels should not be welded in confined areas.

When using coated electrodes, the electrode diameter and type will depend upon the thickness of the metal and the position in which welding will be done. Exx12 or 13 will be used for welding thinner material, the Exx 10 or 11 will be used for welding galvanized pipe and for welding hot-dipped galvanized parts of heavier thickness. Low-hydrogen electrodes can also be used on heavier thicknesses. The welding technique should utilize slow travel speed to permit degassing of the weld metal. The electrode should point forward to force the zinc vapour ahead of the arc. The quality of the welds will be equal to that of the bare metal.

GMAW process is being more widely used for welding galvanized steel. The shielding gas can be 100% CO_2 or 75% Argon and 25% CO_2 mixture. The amount of spatter in welding galvanized steel is slightly more than welding bare steel. This requires cleaning of torch nozzle more often.

The flux-cored arc welding can as easily be used for welding galvanized steel as GMAW. It is, however, recommended for heavy gauges and on hot-dipped galvanized parts. For this purpose highly deoxidised welding wire should be used.

GTAW is not popular for welding galvanized steel mainly because of its slow speed and contimation of tungsten electrode necessitating frequent redressing of electrode tip. In an effort to overcome this, extra high gas flow rates are sometimes used, which can be expensive.

Carbon arc welding process has been widely used for welding galvanized steel. Both the single carbon torch and the twin carbon torch can be used. The filler rod used with this process is 60% copper-, 40% zinc-alloy.

Apart from these welding processes, torch brazing is widely used for galvanized steel. The torch is directed towards the filler rod which melts and then fills the weld joint. A generous quantity of brazing flux is used to reduce the zinc loss adjacent to the weld.

For corrosion resistance of the weld it is sometimes advisable to use a corrosion resistant weld metal. This can be done by using a bronze deposit such as copper-zinc alloy, or a stainless steel electrode. In any case, when arc or acetylene welding is used the area adjacent to the weld will lose the protective zinc coating which must be repaired.

There are several ways of replacing zinc. One is by the use of zinc base paste sticks sometimes called zinc sticks or galvanized sticks. These sticks are wiped on the heated bare metal and the coating so obtained is blended with the original zinc coating. The coating should be two to two-and-one half times as thick as the original coating for proper corrosion protection.

3.3 Welding of Aluminium and its Alloys

Aluminium is an active metal and it reacts with oxygen in the air to produce a thin hard film of aluminium oxide on the surface. The m.p. of Al_2O_3 is about 2000°C which is more than three times the m.p. of aluminium (653°C). In addition, this Al_2O_3 film, particularly as it becomes thicker, will absorb moisture from the air. Moisture is a source of hydrogen which is the cause of porosity in aluminium welds. Hydrogen may also come from oil, paint and dirt in the weld area. Porosity will decrease weld strength and ductility depending on the amount.

The Al_2O_3 film must be removed prior to welding. If it is not all removed, small particles of unmelted oxide will be entrapped in the weld pool and will cause a reduction in ductility, lack of fusion, and may cause weld cracking.

Aluminium oxide can be removed by mechanical, chemical or electrical means. Mechanical removal involves scraping with a sharp tool, sand paper, wire brush (stainless steel), and filing etc. Chemical cleaning can be done either by the use of chemicals like NaOH or by the use of fluxes. Whenever etch cleaning or flux cleaning is used the cleaning material must be completely removed from the weld area to avoid future corrosion.

The electrical method is in the form of using d.c. power source and keeping the workpiece negative. The mobile cathode loosens the oxide layer which will float on top of the molten metal and form a part of slag. If a.c. is used then this cleaning action is effective for half the time when the workpiece is negative.

Since aluminium is so active chemically, the oxide film will immediately start to reform. The time of build up is not very fast but welds should be made after aluminium is cleaned within at least eight hours for quality welding.

Aluminium has high thermal conductivity, therefore preheat is often used for welding thicker sections. The preheat for aluminium should not exceed 200°C and the parts should not be held at the temperature longer than necessary. Because of the high heat conductivity procedures should utilize higher speed welding processes using high heat input. Both GTAW and GMAW are used for the purpose, the former mainly for thin sections and the latter for thicker sections. High thermal conductivity results in faster cooling of weld metal which helps in welding aluminium in all positions.

Aluminium does not exhibit colour change as it approaches its melting point. This is not of much consequence in GTAW and GMAW because the weld is quickly completed before the adjoining area would melt. However, precautions may have to be taken in brazing and soldering. When soldering or brazing aluminium with torch, flux is used and the flux will melt as the temperature of the base metal approaches the required temperature. The flux first dries out and then melts as the base metal reaches the correct working temperature.

Aluminium can also be joined by ultrasonic soldering in which high-frequency sonic waves are transmitted through molten solder to remove undesirable surface films and promote wetting of the base metal. Flux is not normally used.

3.4 Welding of Copper and Copper Alloys

Copper is widely used where heat conductivity plays a positive role, hence its use in solar water heaters. However, copper and its alloys possess properties that require special attention when welding. These are: high thermal conductivity, high thermal expansion coefficient, hot shortness, high fluidity of molten metal, high electrical conductivity, and that it gets annealed on heating.

High thermal conductivity requires that high heat input welding process should be used at the highest speed feasible.

High coefficient of thermal expansion requires proper control during welding to avoid undue distortion in the structure.

Copper has the highest electrical conductivity of the commercial metals and this is a big problem in resistance welding of copper and its alloys.

GMAW and GTAW processes are the most popular for welding copper and its alloys. GTAW is best for welding thinner gauges of copper and GMAW is used for thicker materials. GMAW is faster, has a higher deposition rate, and usually results in less distortion. It can produce high quality welds in all positions.

Several other welding processes can be used, e.g. plasma arc welding, cold-pressure welding, high frequency welding, and electron beam welding.

Brazing is widely used for joining copper. The copper-phosphorous filler material (BCuP) and some of the silver alloy (BAg) types are used. The copper-phosphorous is much less expensive but is not used for copper alloys containing more than 10% nickel.

Soldering is widely used for joining most of the copper alloys.

Both corrosive and resin type fluxes are used for soldering copper. However, solders containing more than 1.0% antimony or more than 0.02% arsenic should not be used to solder Cu-Zn alloys. They will produce brittle joints or have poor bonding. Soldering due to its low heat input does not cause annealing of the copper base alloy, hence the strength attained in cold working is retained.

3.5 Welding of Dissimilar Metals

There are three major obstacles to the joining of dissimilar metals: difference in physical properties such as melting point and thermal conductivity; the formation of brittle alloys in the fuse zone; and unsatisfactory behaviour in service for example,poor corrosion resistance.

Metals having a large difference in melting point may nevertheless be joined by fusion. An obvious example is tin and iron. When tin is melted on a solid iron surface, either in vacuum or in the presence of suitable flux, it wets the higher melting point metal and on cooling the resultant layer of tin is firmly bonded to the iron. This tinned iron can be fusion welded to tin.

If brittle phases are formed between two welded metals, e.g. Cu-Al, the joint will be unsatisfactory. In general, there are two ways of overcoming such problems:

i) to use a solid-phase welding process;

ii) to fusion weld with a filler metal which is compatible with both of the alloys to be joined.

Nickel and nickel alloys are particularly useful in this respect to join copper to steel. The technique is to deposit several layers of nickel alloy on the steel, i.e. buttering or surfacing the steel with a nickel weld metal deposit. The nickel deposit can be welded to the copper alloy using a nickel filler metal.

Flash butt welding also gives high quality welds between copper and aluminium. Other processes used for welding dissimilar metals include ultrasonic welding, friction welding, explosive welding, cold pressure welding, high frequency resistance welding, electron beam welding, and diffusion bonding. However, bulk of dissimilar metal joints are made by fusion welding, mostly with coated electrodes. GTAW and GMAW also find increased use. Resistance spot welding may sometimes be applicable, but this process is limited by the need to match electrical resistance and compression strength of the two materials as far as possible.

Finally, due consideration must be given to brazing and soldering, both of which have the advantage that little or no parent metal is melted. A very wide range of dissimilar metals may, for example, be joined by tin-lead solders, and, provided that the low melting point and morphology of the joint are acceptable. This is one of the most versatile methods. Adhesive bonding is also worth considering for light metals.

3.6 Adhesive Bonding

Generally speaking, the essential components of a metal adhesive is a thermosetting resin, although adhesives used in practice may be complex mixtures. One of the most successful adhesive techniques is Redux bonding, where the metal is first given a coat of phenol formaldehyde (taken up in a suitable solvent), polyvinyl formaldehyde powder is then scattered on the prepared surfaces, and subsequently they are brought together under pressure and cured. The polyvinyl resin is the main adhesive but the precoating of phenol formaldehyde is necessary to bond it to the metal. Other successful metal adhesives consist of combinations of phenol formaldehyde resin with nylon, nitrile rubber, or polyvinyl butyrate.

Adhesives may be applied in the form of liquid, the active ingredients being dissolved in a solvent which evaporates after application. Liquids are brushed or sprayed on the metal surface or applied by means of a roller.

When there is a solvent present in the adhesive it is necessary to allow sufficient time for it to evaporate before placing the surfaces together. The parts are then clamped with a moderate pressure, typically in the range 10-70 N/cm^2, in a suitable press. Complex parts are placed in a plastic bag which is then evacuated, allowing atmospheric pressure to apply clamping force. Surplus adhesive is removed when possible and the assembly is heated throughout the curing cycle, preferably in an oven although electric heating pads may be used for long items. A typical curing period is 30 minutes at 145°C, but shorter times at higher temperatures are applicable. Curing times are reduced (with some loss of bond strength) if an accelerator is added to the adhesive.

4. CONCLUSIONS

Fabrication of solar water heaters can be achieved by welding, brazing, soldering and adhesive bonding depending upon the metals to be joined, processes available and the strength of the joint required. Though some difficulties may be encountered but if proper precautions are taken as described in this write up, it is expected that the joints obtained will serve satisfactorily.

GLAZING MATERIALS FOR SOLAR COLLECTORS

N.K. Bansal and V.K. Sharma
Centre of Energy Studies
Indian Institute of Technology
Hauz Khas, New Delhi 110 016
India

1. INTRODUCTION

A cover over the absorber of a collector is required to trap the solar radiation through the so called 'green house effect'. An ideal cover should have maximum transmissivity for solar radiation and must be opaque to the long wave radiation emitted by the absorber plate. The important optical properties that determine the desired characteristics of cover are transmissivity, reflectivity and absorptivity, which are functions of incoming radiation, thickness of the cover, refractive index, and the extinction coefficient of the material. Apart from this the cover should be of low cost, durable, high abrasion resistant, weather resistant, etc.

2. REFLECTION BY COVER

Good covers should have as low reflectivity as possible. For smooth surfaces, the reflectivity of the glazing (defined as the fraction of the reflected radiation) can be calculated by using the expression

$$r = \frac{1}{2} - \left(\frac{\sin^2(\theta_2-\theta_1)}{\sin^2(\theta_2+\theta_1)} + \frac{\tan^2(\theta_2-\theta_1)}{\tan^2(\theta_2+\theta_1)}\right) \tag{1}$$

where θ_1 and θ_2 are angles of incidence and refraction respectively (Fig.1) since $n_1/n_2 = \sin\theta_2/\sin\theta_1$ for normal incidence, the reflectivity can be estimated by using the formula

$$r(0) = \left[\frac{(n-1)}{(n+1)}\right]^2 \tag{2}$$

The solar transmittance of nonabsorbing glass, having an average refractive index of 1.526 in the solar spectrum, has been calculated

H. P. Garg (ed.), Solar Water Heating Systems, 243–256.

for all incidence angles and the results for from one to four glass covers are given in Fig.2.

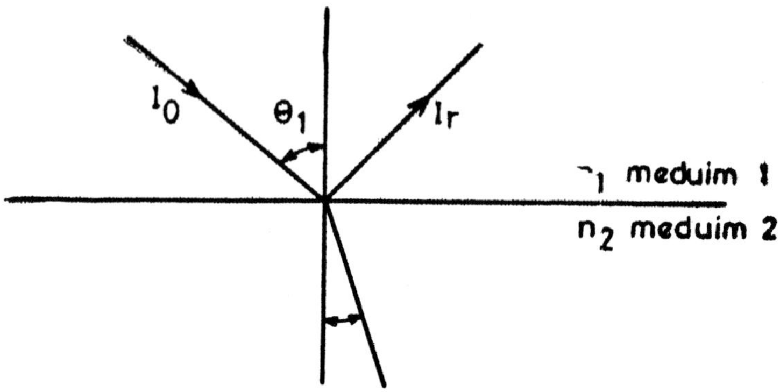

FIG.1 - ANGLE OF INCIDENCE AND REFRACTION

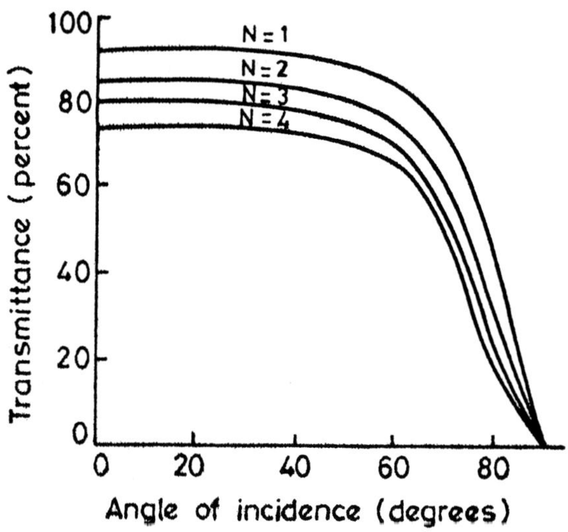

FIG.2 - TRANSMITTANCE WITH ANGLE OF INCIDENCE FOR NUMBER OF GLAZINGS N

The effectiveness of a cover plate as a transmitter of solar radiation is affected by absorption of energy in the glass as well as by reflection. As the angle of incidence increases the reflectivity increases. The amount of solar radiation absorbed in the glass is proportional to the exponent of the extinction coefficient (K) multiplied by the thickness of the glass (L). Each additional glass cover reduces thermal loss from the collector but at the same time it increases the optical attenuation. Figure 3 (Ref.1) shows the transmittance values considering reflection and absorption losses for different number of covers as a function of the angle of incidence.

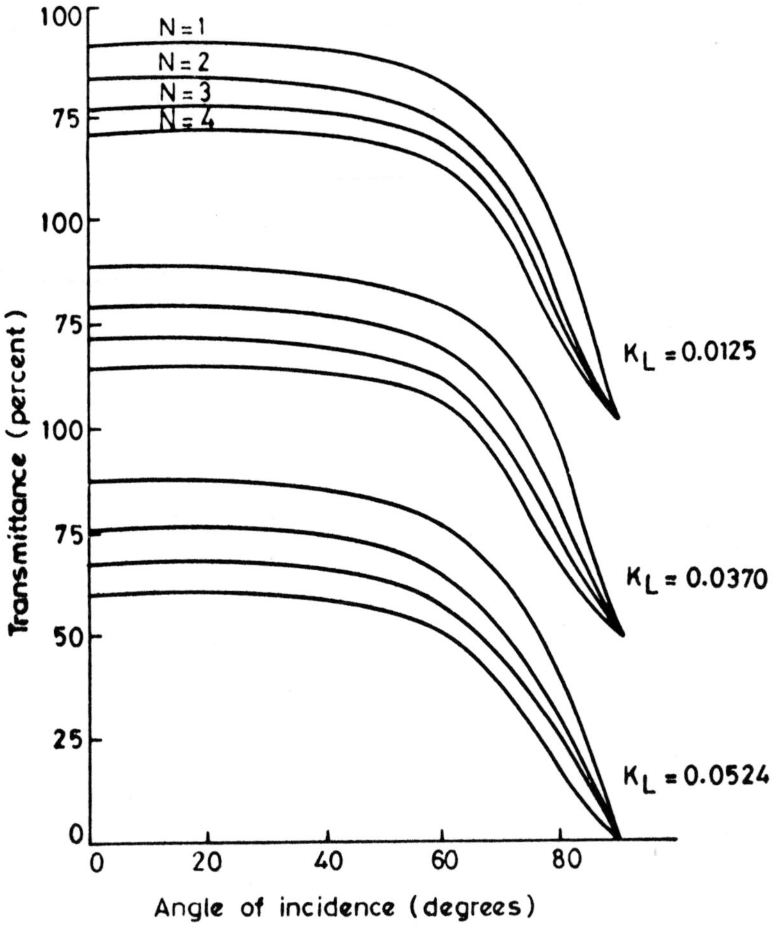

FIG.3 - TRANSMITTANCE FOR DIFFERENT TYPES OF GLASS PLATES WITH ANGLE OF INCIDENCE

The use of single or multiple glass covers depends on the compromise between optical and thermal losses in the collector operating range. The use of two glass covers modulates the heat loss coefficient (obtained from η versus $(T_p - T_a)/H$ plot), but it would be effective after certain temperature difference. However, optimum number of glass covers depends on the climatic conditions and more on outside air temperature and collector plate temperature.

Instantaneous collector performance is shown on Fig.4 for four different liquid heating collector designs: one and two covers, selective and nonselective absorber plate. These curves were generated with the incident radiation on the collector equal to 1000 W/m^2. This is a high level of radiation; if a lower radiation level had been used the curves would be even more nearly linear.

The important observation which can be made from this figure is that the selection of the "best" collector depends upon the type of application.

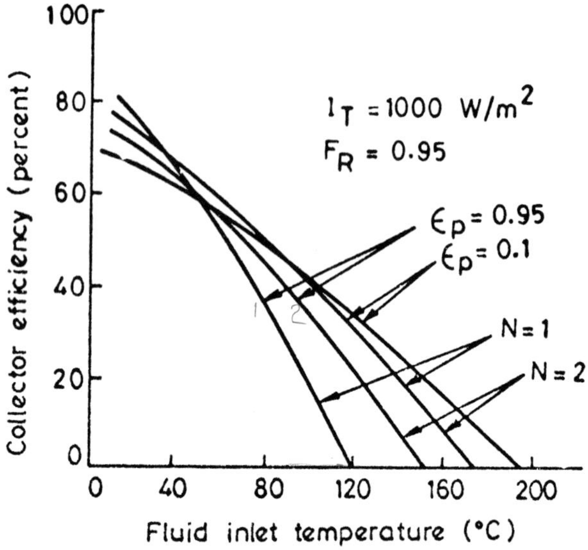

FIG.4 - COLLECTOR EFFICIENCY AT VARIOUS OPERATING TEMPERATURES

If low temperature water is supplied to these collectors, the least expensive collector, with one-cover, nonselective has the highest efficiency. As the temperature of the collector plate increases, the performance of the one-cover, nonselective collector falls off faster than the other collectors. When the inlet temperature exceeds 100°C, the most expensive collector, the two-cover, selective absorber has the highest performance.

An implication of the preceding paragraph is that a fluid delivery temperature is known for a particular application. This is generally not known. Most solar systems operate with a variable inlet temperature, therefore the selection of the best collector from a thermal performance viewpoint is not possible. Even if the thermal performance of one collector exceeds another collector over a wide range of temperature, cost should be the deciding factor.

3. PLASTIC MATERIALS FOR COVERS IN FLAT PLATE COLLECTORS (Ref.2)

Solar collector covers (glazing) are sheets or films that transmit nearly all incident solar radiation but block the passage of longwave radiant energy and convected heat from the absorber of the collector to outside. Glass owing to its easier availability and high desirable optical properties has been the most common material employed so far, to glaze flat solar collectors (Ref.3-7). Glass (3 mm) can transmit up to 91% of the incident shortwave radiation (from 0.3 to 3μm) at normal incidence, while virtually not allowing any longwave radiation (3.0-30 μm), emitted by the absorber, to escape outward. The weatherability of glass is also good. Glass has, however, very low impact strength which makes it very susceptible to breakage by hailstone, vandals or transit. This is in fact the most serious drawback. Other disadvantages of glass are its relatively high density and high conductivity as compared to the same thickness of plastics. For example, thermal conductivity of glass is about ten times that of PMMA; thermal losses to the environment are much higher in case of glass. Use of an IR reflective coating to reduce these thermal losses decreases the effective transmittance of the cover (for solar radiation) by 10% and increases its cost.

Plastics are generally lighter in weight and have better shatter resistance than glass. They are also not sensitive to oxidation, heat and moisture. Due to their ease of fabrication, low conductivity and good transmission of solar radiation, some plastics have been used as glazing in flat plate collectors and solar stills. Some plastics are more transparent to solar radiation than even glass. Table 1 summarises the transmission and other properties of some of the transparent plastics used in glazing or honeycomb structures. Most plastics, however, possess the undesirable properties of being partially transparent to longwave radiation emitted by the absorbing plate (Ref.6-9); the transmission of longwave radiation however depends on the thickness and molecular structure of the material. In Figs.5-10 (Ref.6), we illustrate the transmittance of radiation in the long wavelength region for a number of plastics of importance in solar energy applications; the thickness of the specimen of the material, chosen for transmission curves is typical of solar energy applications. As the thickness of the plastic sheet or film increases, the material becomes more opaque to IR radiation.

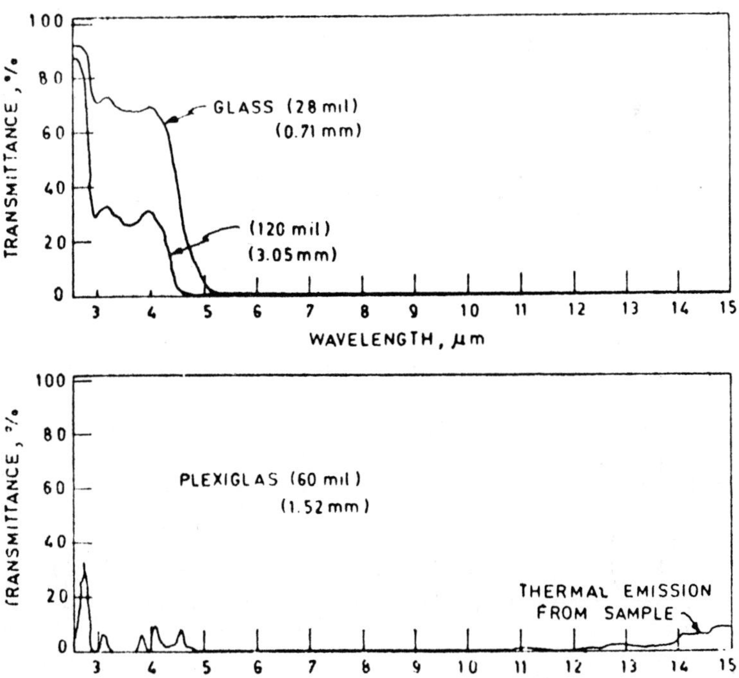

FIG.5 - INFRARED TRANSMITTANCE CURVES FOR GLASS AND PLEXIGLAS (Ref.6)

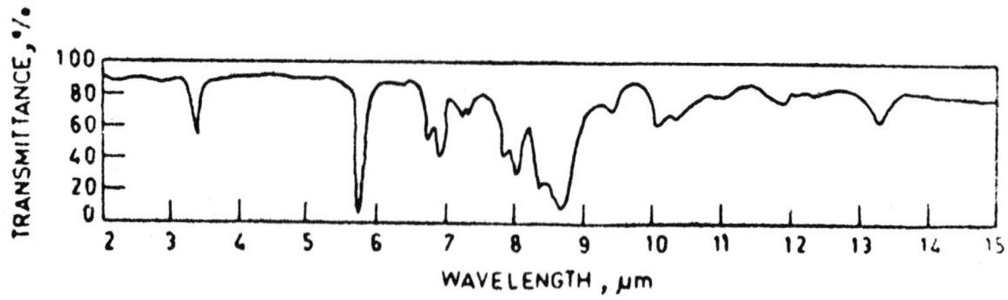

FIG.6 - INFRARED TRANSMITTANCE CURVE FOR POLYMETHYL METHACRYLATE FILM OF THICKNESS 0.012 mm (PMMA) (Ref.6).

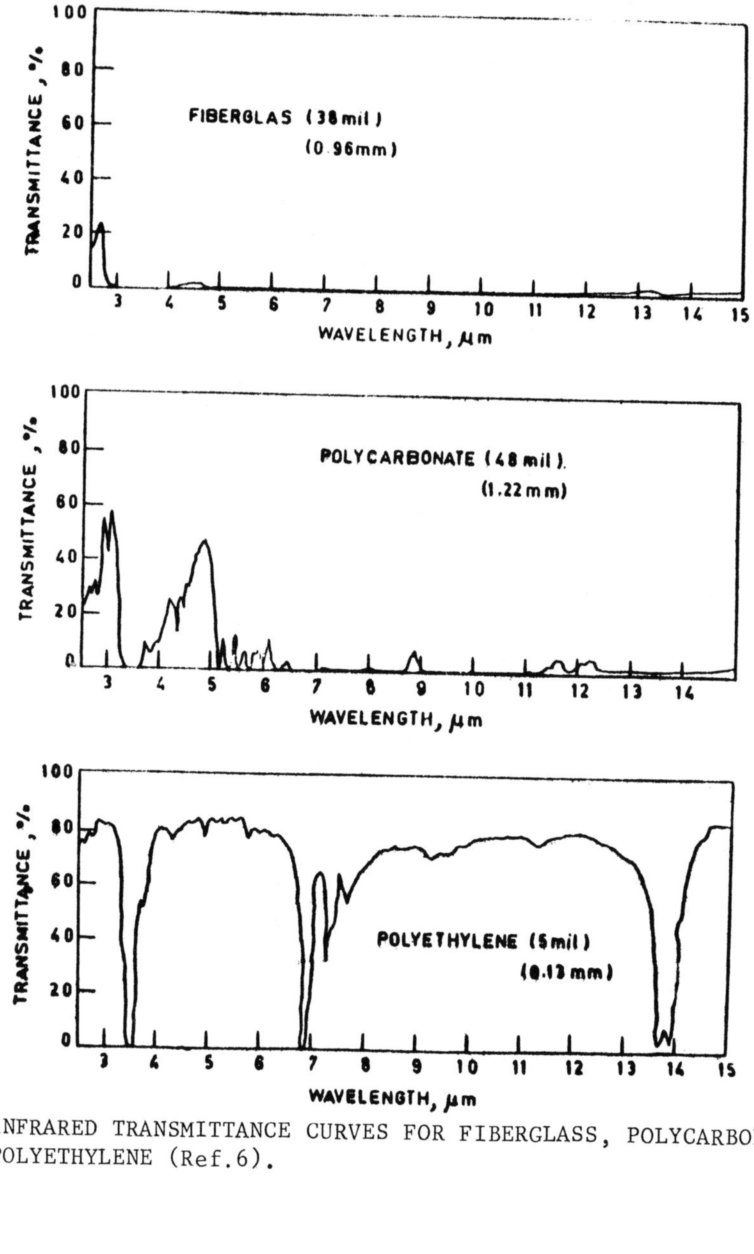

FIG.7 - INFRARED TRANSMITTANCE CURVES FOR FIBERGLASS, POLYCARBONATE AND POLYETHYLENE (Ref.6).

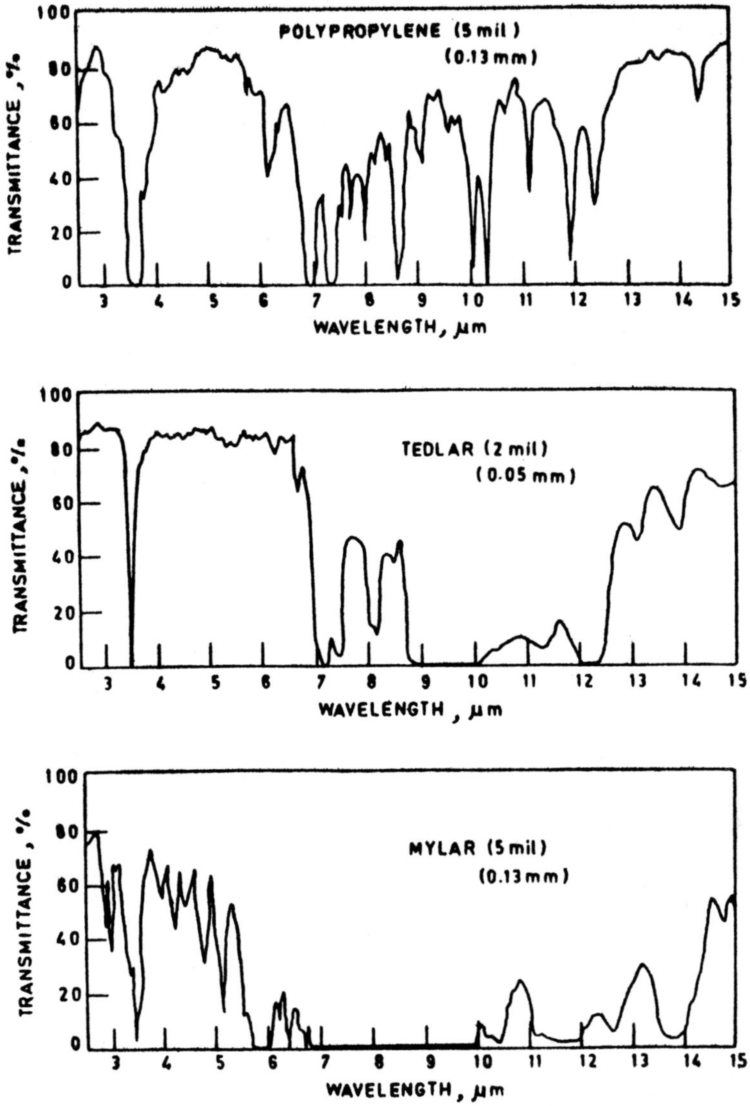

FIG.8 - INFRARED TRANSMITTANCE CURVES FOR POLYPROPYLENE-TEDLAR AND MYLAR (Ref.6).

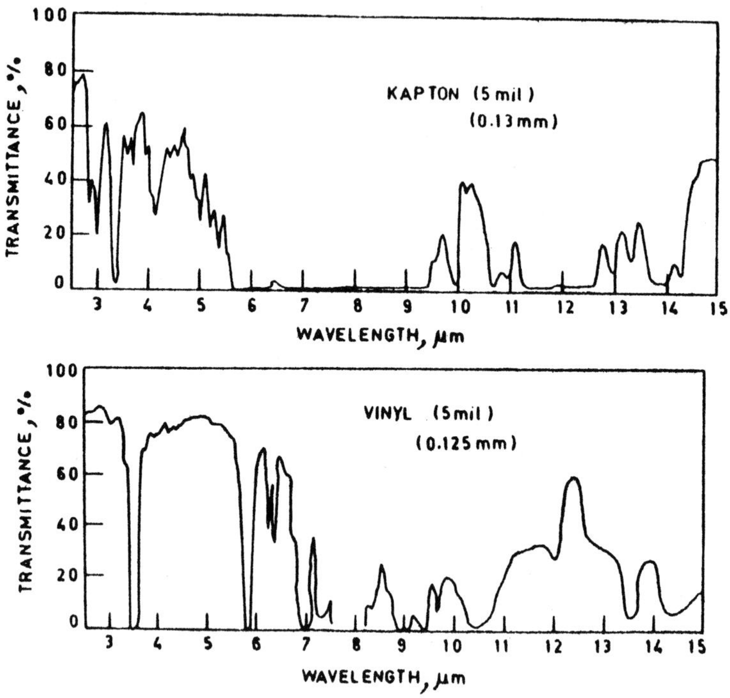

FIG.9 - INFRARED TRANSMITTANCE CURVES FOR KAPTON AND VINYL (Ref.6)

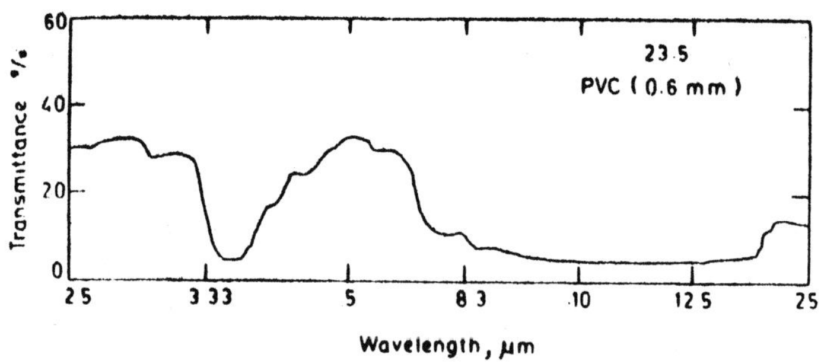

FIG.10 - MEASURED TRANSMITTANCE OF PVC COVER (Ref.6)

The plastics most commonly used for glazing in solar collectors are polymethyl methacrylate (PMMA), polycarbonate (PC), glass fiber-reinforced polyester (GRP) polyvinyl fluoride (PVF) and fluorinated ethylene propylene (FEP) copolymer.

Covers made from PMMA or PC sheet are used in institutional or industrial buildings (Ref.10); they weigh only half as much as those of similar thickness made of glass (Ref.3). PMMA and PC are good insulators and they transmit nearly 89 per cent of the incident solar radiation. Pure polymer of PMMA is not affected by UV light down to 0.285 μm. Acrylic plastic, which is very weather resistant and has a high transmittance (τ = 90%), is used as the glazing material in the solar-I house at the University of Delaware. Tests of acrylic plastic which had been exposed outdoors for 18 years revealed a slight increase in the absorptance (3% for 3 mm thick plastic) and a certain scattering of solar radiation at the surface. PMMA, however, has low softening point and thus it undergoes distortion when used as inner glazing material (Ref.11). In the experiments of Holz and Lueke (Ref.8), PMMA sheet used as inner-glazing lost its flatness at high collection temperatures. Surface treatments are available which can give the relatively soft acrylic plastic an improved resistance to wear. PC sheets have a slightly lower transmittivity for solar radiation than PMMA, but they have considerable higher heat resistance (Table 1). An UV-stabilised glazing was however found to develop microcracking at its surface after a short period of outdoor exposure (Ref.12). The surface deterioration, due to solar radiation along with moisture and/or temperature induced stress fatigue, resulted in gradual impairment of mechanical properties such as tensile (Ref.12,13) and impact strength (Ref.3). Existing commercial polycarbonate is thus still not adequately protected against the effect of radiation.

GRP, mainly suitable for sky lights and glazing for industrial community buildings, possesses desirable properties such as light weight, good impact resistance over a wide range of temperatures, high transmission for solar radiation and better heat resistance than other plastics. In outdoor environment GRP sheet, however, undergoes deterioration in properties (Ref.14); at the exposed surface fibres pop out and microcracks develop resulting in the deterioration of desirable properties (Ref.15). If the exposed surface of GRP is, however, covered with a film of PVF or Tedlar, the resistance to erosion by outdoor exposure considerably improves (Ref.16). GRP, being highly transmittive to solar radiation and nearly opaque to longwave radiation, is highly desirable material from solar energy collection point of view. The efforts are, therefore, being made to make it weather resistant. A GRP material highly stabilized to resist UV light degradation has been developed (Ref.17,18) for covers in solar energy collectors. The long term durability of this material under true service conditions is, however, yet to be assessed.

Plastics, based on fluorine containing polymers, have high thermal and chemical stability, low water absorption and very good weather stability (Ref.19,20). PVF and FEP copolymers, when used as film (up to 0.25 mm) have very good transmission and have been used as covers for solar collectors (Ref.21). PVF films are able to retain their appearance and physical properties even after their long exposure to outdoor weather; they have good resistance to abrasion. PVF film of 0.1 mm thickness,with an initial solar transmission of 92-94 per cent, are able to retain 95 per cent of its transmission even after five years of exposure in Florida weather (Ref.20). FEP films have transmission up to 97 per cent for solar radiation and are also able to retain their properties even after long outdoor exposure.

TABLE 1 - Guide to Properties of Sheet and Film Materials Used in Solar Energy Applications (Ref.3,22)

Material	Thickness (mm)	Sp.gravity (ASTM D792 g/cm^3)	Weight (Kgm^{-2}) for thickness of Col. 2	Transmission of solar radiation %	Thermal conductivity K (W/m^{o}C)	Water absorption %	Tensile strength (ASTM D 638 K Pa	Resistance to continuous heat Max. Temp. ^{o}C	Modulus of elasticity (ASTM D638) G Pa	Coefficient of thermal expansion (ASTM 696) x10^{-5}/^{o}C	Applications
CAB	0.15	1.15-1.22	-	90	0.167-0.334	0.9-2.2	18-48	60-90	-	11-17	(a)
PEP	0.05	2.12-2.17	0.10	97	0.245	$<$ 0.05	19-21	205	3.4	8.3-11	(b)
GRP	1.0	-	1.5	80-90	-	0.25	100-120	150-175	7.6	3.6-4.4	(c)
PC	3.2	1.2	3.8	82-89	0.187	1.15-0.18	65	120	2.1-2.4	6.6	
PET	0.025	1.38-1.4	0.034	85	0.86	0.01	170	104	3.8	3.0	(d)
PMMA	3.2	1.17-1.2	3.7	89	0.173-0.245	0.2-0.7	72.4	60-95	3.1	5.0-9.0	(e)
PVC	0.6	1.15-1.80	-	83	0.126-0.167	0.15-0.75	-	-	-	7_25	(f)
PVF	0.1	1.38-1.57	0.14	92-94	0.125	0.01	90	108	1.8	4.8	(g)
Glass	3.2	2.46-2.49	8.1	85-91	1.054	-	11	205	724	0.85	(h)
water white (10.10% iron)							(annealed) 4.4 (tempered)				

(a) Used for covers in flat plate collectors. (b) Glazing for collectors and solar stills, honeycomb structures, substrate in mirrors of concentrating collectors. (c) Glazing and honeycomb structures. (d) Honeycomb structures, substrate in mirrors of concentrating collectors. (e) Mainly as Fresnel lenses in concentrating collectors; as cover in flat plate collectors. (f) Used as covers in flat plate collectors and for the casing. (g) Glazing for collectors and solar stills, honeycomb structures, substrate in mirrors of concentrating collectors. (h) Glazing.

Investigation of the weather resistance of plastic films such as Teflon FEP, Tedlar, Mylar W and a number of other materials has been carried out in the moist and sunny climates of Florida (Ref.21). The life expectancy of Teflon FEP was set at 20 years, with corresponding times for Tedlar and Mylar W of 9 years and 4 years respectively. The expected lives of polyethene, PVC and ordinary Mylar are less than one year in this climate. The combination of moisture heat and solar radiation in a distillation plant reduces the life of Tedlar to 3-5 years.

4. OPTIMISATION OF THE AIRGAP THICKNESS BETWEEN THE COVER AND THE ABSORBER (Ref.24)

The effective Δt value across the cover and the temperature t_1 of the cover's surface on the inner side are dependent on the temperature t_2 of the absorber, the air gap thickness d between the cover and the absorber and the inclination angle of the system. With increasing d the conduction losses through air are reduced but the convection losses are increased. Thermal losses due to convection in the air gap can be calculated by the Fourier conduction equation by putting a correction factor X, i.e.

$$q_c = XK_a\left(\frac{t_2-t_1}{d}\right) \tag{3}$$

K_a being the thermal conductivity of air.

$$\text{With } K_a = f(\bar{t}), \tag{4}$$

$$\bar{t} = \frac{t_1+t_2}{2} \tag{5}$$

and

$$X = f(Pr,Gr) \tag{6}$$

Pr, Gr being the Prandtl and Grashoff Numbers (Ref.23).

The above equation shows increasing heat losses with decreasing d. Experiments, however, show that only below a definite air width the heat losses become critical. Measured performance of a few collectors(Ref.24) for various comibations of t_1 and t_2 and the inclination angles are given in Figs.11 and 12. For low temperatures ($t_2 < 60^oC$) an air gap of about 15 to 20 mm thickness is sufficient. For intermediate or higher temperatures ($t_2 > 160^oC$), an air gap of about 60 mm to 80 mm is necessary to limit the thermal losses.

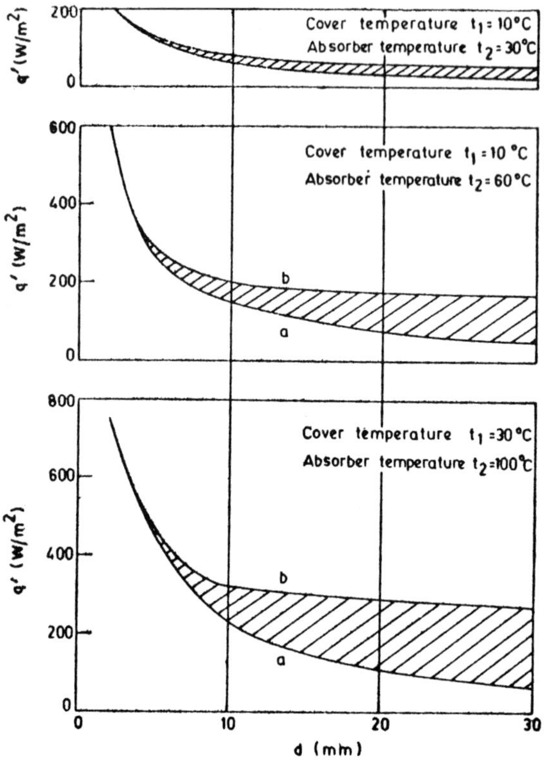

FIG.11 - HEAT LOSSES q' AS A FUNCTION OF AIR GAP THICKNESS d

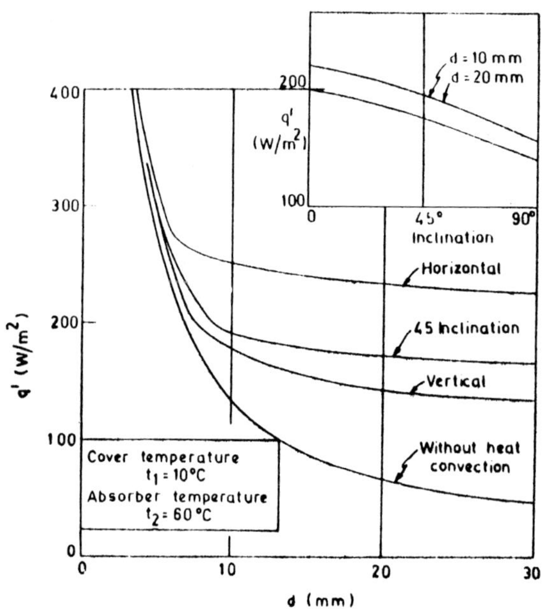

FIG.12 - THERMAL LOSSES AS A FUNCTION OF AIR GAP THICKNESS (INCLINATION 45°): (a)WITHOUT CONVECTION,(b)WITH CONVECTION

REFERENCES

1. J.A. Duffie and W.A. Backman, (1980), Solar Engineering of Thermal Processing, John Wiley and Sons, Inc.
2. N.K. Bansal, (1983), 'Plastic Solar Collectors', Chapter I, of Renewable Energy Resources, Vol.1, Wiley Eastern Ltd., New Delhi.
3. A. Blaga, (1978), 'Use of plastics in solar energy applications', Solar Energy, **21**, 331-338.
4. J.I. Yellot, (1974), 'Utilization of sun and sky radiation for heating buildings', ASHRAE Handbook and Product Directory, p. 59.1, Am. Soc. Heating Refrig. Air Cond. Engg., Inc., New York.
5. J.A. Duffie and W. Beckmann, (1974), Solar Energy Thermal Processes, John Wiley & Sons, Inc.
6. A.B. Meinel and M.P. Meinel, (1976), Applied Solar Energy: An Introduction, Addison Wesley, Reading, Massachusetts.
7. A.M. Zarem and D.D. Erway (Eds.),(1963), Introduction to the Utilization of Solar Energy, McGraw-Hill Book Co., New York.
8. P.R. Smith, M.H. Cobblem and L.L. Lukens, (1977), 'Parametric studies of the thermal trap flat plate collector', AICHE Sump. Ser., 73, No. 164, 169-170.
9. Teflon (R) FEP for Solar Collectors, E.I. du Pont de Numours and Co., Plastic Products and Resins Department (Inc.), Wilmington, Delaware (n.d.).
10. A.E. Sherr, (1972),'A bright future for glazing plastics', SPE J., **28**(ii), 24.
11. A.P. Saunders, (1972), 'Flexiglass DR thermal stability test', National Centre for Energy Management and Power, US National Technical Information Service, Tech. Dept., PB-238-580.
12. A. Balga and R.S. Yamasaki, (1976), 'Surface microcracking induced by weathering of PV sheet', J. Mat. Sci., **11**, 1513.
13. R.S. Yamasaki and A. Blaga, (1977), 'Degradation of PC sheeting on outdoor exposure - relationship between changes in molecular weight and tensile properties', RELEAM, Mat. Struct. **10**(58), 197.
14. B. Parkyn (Ed.),(1970), 'Glass Reinforced Plastics', Chapter 6, Hiffe Books, London.
15. A. Balga, (1975), 'Durability of GRP composites', Batiment International Build. Res. Prac., **5**(1), 10.
16. A. Balga and R.S. Yamasaki, (1977), 'Outdoor durability of a common type (Tetrachlorophthalic Acid Based) fire retardent GRP sheet', RELEM Mat. and Struct., **10**(58), 197.
17. J.S. White, (1977), 'Weatherability of fiberglass solar collector covers, Polym. News, **3**(5), 239.
18. R. Martino, (1976), 'Run for the sun: Solar heating opens a vast new construction market', Mod. Plast., **53**, 52.
19. J. Brydson, (1975), Plastic Materials, Butterworths, London.
20. H.F. Mark, N.G. Gaylord and N.M. Bidales, (1970-71), Encyclopaedia of Polymer Science and Technology, **13**, p. 654.
21. F.E. Edlin and D.E. Willaner, (1961), 'Plastics films for solar energy applications', Conference on New Sources of Energy, Paper 5/33, Rome.

22. C.A. Harper, (1975), Handbook of Plastics and Elastomers, McGraw-Hill Book Company, New York.
23. H.P. Garg, (1982), Treatise on Solar Energy, John Wiley & Sons, England.
24. G. Schenkel, (1977), 'Plastics for solar energy collectors', Construction Varients, Lueftung Khim, Houstech, CODEN HLHZA.

HEAT EXCHANGER OPTIMIZATION FOR HOT WATER HEATING SYSTEM

N.K. Bansal and Jugal Kishor
Centre of Energy Studies
Indian Institute of Technology
Hauz Khas, New Delhi 110 016
India

1. INTRODUCTION

A heat exchanger comprises of any of the following three components, namely (1) heating medium, (2) medium to be heated and (3) a partition between these two media. During the heat exchange process, the heating medium cools down and the other medium gets heated up as shown in Fig.1.

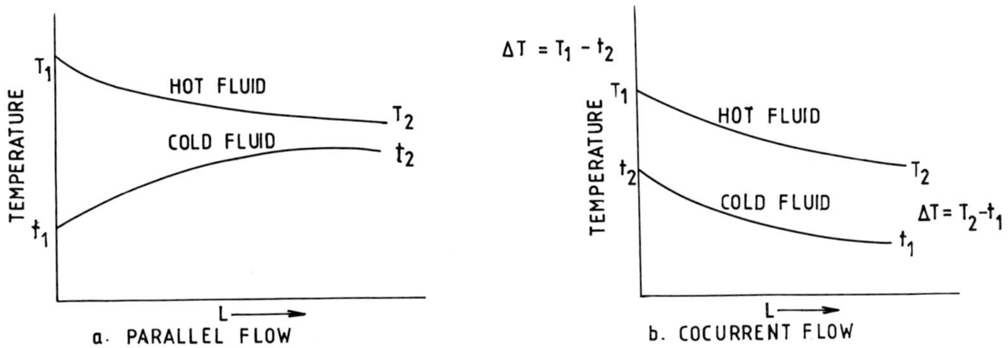

FIG.1 - TYPICAL FLUID TEMPERATURE CONDITIONS IN HEAT EXCHANGERS

Depending on the relative directions of fluid flow, the flow in a heat exchanger may be termed as parallel flow, counter flow and cross flow as shown in Fig.2. Clearly when the fluids flow in the same direction, the flow is a parallel flow. When the two fluid directions are opposite to each other, it is counter flow, and when the two fluids flow in directions perpendicular to each other, the flow in termed as cross flow.

The methodology of arriving at an optimum heat exchanger design is a complex one, not only because of arithmetic involved, but more particularly because of the many qualitative judgements that must be introduced. A schematic way of the design procedure is given in Fig.3. The inputs to the design theory procedure include, along with the problem statement or specifications, the surface heat transfer and flow

H. P. Garg (ed.), Solar Water Heating Systems, 257–277.

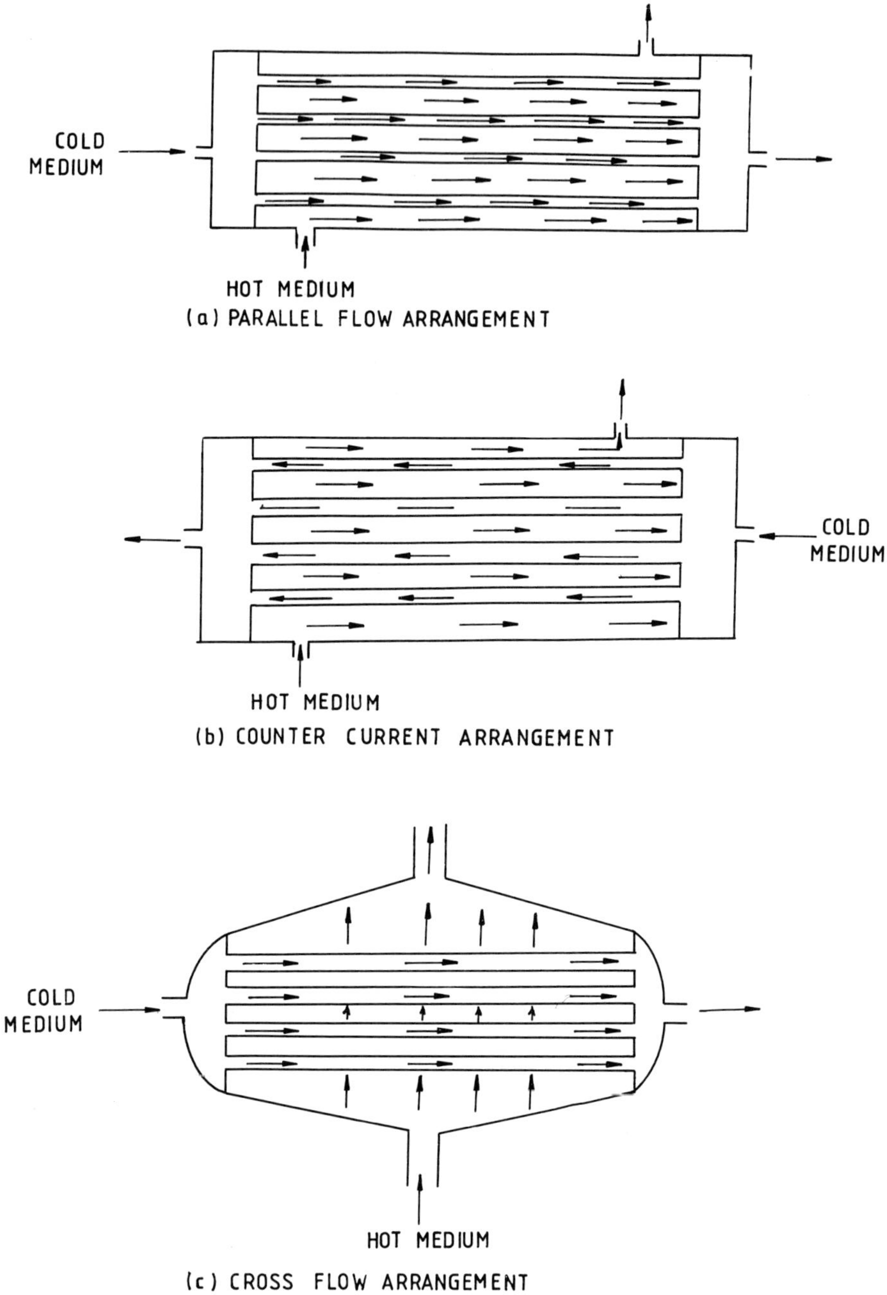

FIG.2 - DIFFERENT FLOW ARRANGEMENTS IN HEAT EXCHANGERS

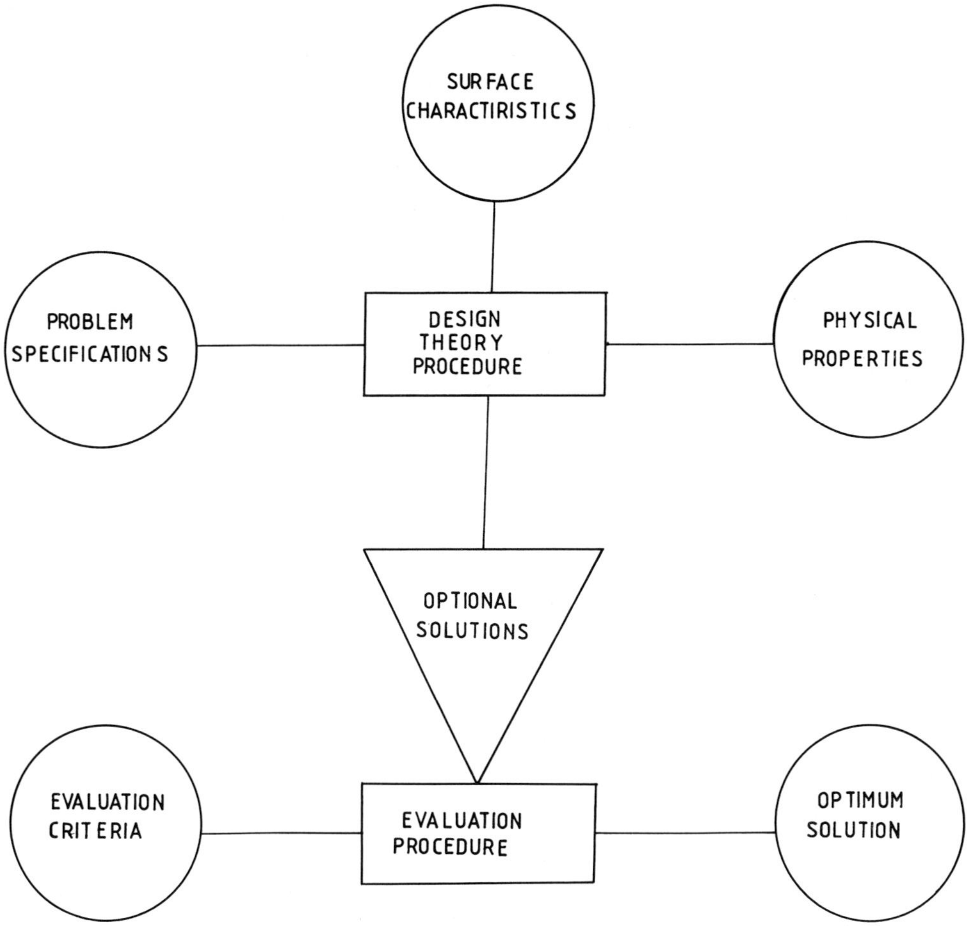

FIG.3 - METHODOLOGY OF HEAT EXCHANGER DESIGN

friction design characteristics and physical properties information. Additionally the statement may specify a consideration of different exchangers, for example shell and tube, double pipe, or coil type heat exchangers. The design theory procedure is the subject of this paper. The procedure described in this paper has been set up on a computer programme. The answer is a large number of optional solutions. Some of these optional solutions represent an estimate of what a competitor may offer; others may represent customer's suggestions. These optional solutions, plus evaluations criteria, serve as inputs to the evaluation procedure.

The evaluation procedure is in a large measure qualitative; shipping limitations, delivery dates, company policy, etc. are all examples of qualitative evaluation criteria. Additionally, one has to consider the trade off factors such as pressure drop, weight, heat transfer performance and leakages, etc.

The final output is an optimum design or possibly several such designs to be submitted to the customer. Alternatively, the final output may be used to formulate a new problem input statement for a parametric study leading to an optimum overall system rather than just an optimum heat exchanger based on somewhat arbitrary initial specification of rrequirements.

Clearly, a comprehensive design procedure of a general nature is out of scope of this paper. The paper contains quantitative inputs of surface characteristics and sampling of typical physical properties information, together with a systematic development of a design theory. Simple examples, wherever possible, have been included.

2. EXCHANGER VARIABLES

For a conventional two fluid heat exchanger, the parameters governing the performance of the exchanger are:

U = overall heat transfer coefficient ($W/m^2\,^oC$)

A = surface area on which U is used (m^2)

T_1, T_2 = hot fluid terminal temperature (^oC)

t_1, t_2 = cold fluid terminal temperature (^oC)

m_h = hot fluid mass flow rate (Kg/sm^2)

m_c = cold fluid mass flow rate (Kg/sm^2)

C_h = heat capacity of the hot fluid J/Kg^oC

C_c = heat capacity of the cold fluid J/Kg^oC

Flow arrangement - counter flow, parallel flow, or cross flow.

Once the heat capacities, the inlet temperatures and quantity of heat transferred are specified, principle of thermodynamics will immediately fix the outlet temperatures.

2.1 Overall Heat Transfer Coefficient

Overall heat transfer coefficient, U, determines, the overall heat transfer from hot fluid to cold fluid and may be assumed to be given by the following equation, which is similar to Ohm's law in electricity:

$$\frac{dq}{dA} = U(T_h - T_c) \tag{1}$$

where $\frac{dq}{dA}$ = heat flux per unit transfer area(W/m^2);

T_h = temperature of the hot fluid at a section where the heat transfer is being considered;

T_c = temperature of the cold fluid at a section where the heat transfer is being considered.

From equation (1) it is clear that U is an overall thermal conductance based on a temperature potential $(T_h - T_c)$ and a unit transfer area. The reciprocal of U is an overall thermal resistance which can be considered to be given by

$$R = \frac{1}{U} = \frac{1}{h_c} + \frac{t}{K} + \frac{1}{h_h} \tag{2}$$

where h_h = hot side film convection coefficient ($W/m^2\,{}^oC$);

h_c = cold side film convection coefficient ($W/m^2\,{}^oC$);

t = thickness of the conducting medium (m);

K = thermal conductivity of the conducting medium ($W/m\,{}^oC$).

A real pipe has different areas per unit length on its inside and outside surfaces, h_c and h_h must be referred to the same heat flow area. For a pipe with a thicker wall, equation (2) may be written as

$$R = \frac{1}{(Ac/A)h_c} + \frac{2.3D_o}{2K} \ln \frac{D_o}{D_i} + \frac{1}{h_h} \tag{3}$$

where D_o and D_i being the outer and inner diameter of the pipe through which cold fluid flows.

The convective film coefficients h_c and h_h are complex functions of the surface geometry, fluid properties and flow conditions. Except for some of the geometrically simple cases, the engineer generally relies on model experiments to establish these coefficients. The method of estimating heat transfer coefficients is given in standard text books (Ref.1-3). A simpe way of estimating these is given in Appendix I.

2.2 Fouling Factors

Over a time during which heat exchanger has been in use, dirt and scale deposit on the inside and outside of the pipe adding two more resistances that should be added to equation (3). In order to get the desired performance heat exchanger should be designed anticipating the deposition of dirt and scale by introducing a resistance R_f, called fouling factor. The overall heat transfer coefficient including the dirt factor then becomes

$$\frac{1}{U_o} = R + R_f \tag{4}$$

Example: Suppose in a double pipe heat exchanger h_h and h_c have been computed as 1700 $W/m^2\,{}^oC$ and 500 $W/m^2\,{}^oC$. Let us suppose that direct resistance is $1.7\times10^{-4}\,m^2\,{}^oC/W$ will deposite annually inside the pipe and $R_{do} = 2.56\times10^{-4}\,m^2\,{}^oC/W$ will deposit on the outside pipe. If the surface has to be cleaned annually, what is the value of U_o for which the surface area should be calculated?

$$R = \frac{1}{1700} + \frac{1}{500} = 0.00259 \ m^2 \, {}^oC/W$$

$$\frac{1}{U_o} = 0.00259 + 1.7\times10^{-4} + 2.56\times10^{-4}$$

$$= 3.016\times10^{-3}$$

$$U_o = 331.6 \ W/m^2 \, {}^oC$$

values of a few fouling factors for various types of water are given in Appendix II.

2.3 Logarithmic Mean Temperature Difference: Counterflow

When a liquid flows along the axis of a tube and absorbs or transmits heat, the temperature of the liquid varies over the entire length of the pipe. If the temperature of the inner circumference of a pipe is nearly constant over its entire length, as it might be when the fluid inside the tube is heated by steam, there will be two distinct temperature differences at each end: one between the pipe wall and the inlet liquid and one at the other end between the pipe wall and the heated liquid. A simple analysis (Ref.2) shows that the heat transferred is given by the expression (referring to Fig.4).

$$\dot{Q} = \frac{h_c A_c (\Delta t_2 - \Delta t_1)}{\ln \frac{\Delta t_2}{\Delta t_1}} \tag{5}$$

The expression $(\Delta t_2 - \Delta t_1)/\ln(\Delta t_2/\Delta t_1)$ is the logarithmic mean temperature difference abbriviated as LMTD.

For a counter flow (Fig.1b), LMTD is written as

$$LMTD = \frac{(\Delta t)_{big} - (\Delta t)_{small}}{\ln \frac{(\Delta t)_{big}}{(\Delta t)_{small}}} \tag{6}$$

where Δt is the difference in the temperature of the two fluids at the inlet and outlet ends and whichever difference is higher, that is desigated as (big) and the other (small).

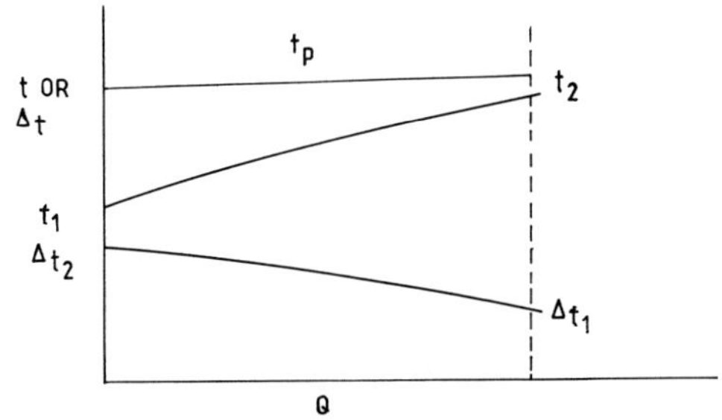

FIG.4 - TEMPERATURE DIFFERENCE BETWEEN A FLUID AND A PIPE WALL

For parallel flow also, equation (6) is valid and it may appear that from the two flow arrangements there is little to choose. The example below, however, demonstrates that except where one fluid is isothermal (such as condensing steam), there is a distinct thermal disadvantage to the use of parallel flow.

Example 1 - Calculation of LMTD: A hot fluid enters a concentric pipe at a temperature of 150°C and it is to be cooled to 90°C by a cold fluid entering at 35°C and being heated to 65°C. Shall they be directed in parallel flow or counter flow?

(a) Counter flow

$$\text{LMTD} = \frac{85-55}{\ln \frac{85}{55}}$$

$$= 70^{\circ}\text{C}$$

(b) Parallel flow

$$\text{LMTD} = \frac{115-25}{\ln \frac{115}{25}}$$

$$= 59^{\circ}\text{C}$$

LMTD is more in case of counter current flow than parallel flow arrangement. It is therefore recommended to have the counter current flow arrangement.

Example 2 - Calculation of LMTD with equal outlet temperature: A hot fluid enters a concentric pipe apparatus at 150°C and is cooled down to 90°C by a cold fluid entering at 60°C and heated to 90°C.

(a) Counter flow

$$\text{LMTD} = \frac{60-30}{\ln \frac{60}{30}}$$

$$= 43.3^{\circ}\text{C}$$

(b) Parallel flow

$$\text{LMTD} = \frac{90-0}{\ln \frac{90}{0}}$$

$$= 0$$

From the above example it is clear that the lowest temperature theoretically attained by the hot fluid is the outlet temperature of the cold fluid. If this temperature were attained LMTD would be zero and from Fourier conduction equation, i.e. $\dot{Q} = UA\Delta t$, since $\dot{Q}$ and U are finite, one requires infinite surface area, A, which is a physical impossibility.

3. ANALYSIS OF HEAT EXCHANGERS

Rate of heat transfer $\dot{Q}$ from hot fluid to cold fluid can be expressed in terms of overall heat transfer coefficient, surface area of the heat exchanger, and inlet and outlet temperatures of the hot and cold fluid, i.e.

$$\dot{Q} = \dot{m}_h C_h (T_1 - T_2) \tag{7}$$

$$\dot{Q} = \dot{m}_c C_c (t_2 - t_1) \tag{8}$$

$$\dot{Q} = UA\,(LMTD) \tag{9}$$

The use of these equations to calculate the rise in the temperature of cold water and the required heat exchanger area is illustrated in a few examples below.

Example 3: In a shell and tube type heat exchanger water at temperature $70^{o}C$ (T_1) comes out at temperature $50^{o}C$ (T_2) to heat water at temperature $20^{o}C$ (t_1). What is t_2 and how much heat exchange area is required if $\dot{m}_c = 2000$ Kg/h, and $\dot{m}_h = 1000$ Kg/h. Overall heat transfer coefficient $U = 200\ W/m^2 {}^{o}C$.

Using equations (7) and (8),

$$\dot{Q} = \frac{1000 \times 4200 \times (70-50)}{3600} = \frac{2000 \times 4200\ (t_2 - 20)}{3600}$$

Yielding $t_2 = 10+20 = 30^{o}C$

In counter current flow

$$LMTD = \frac{40-30}{\ln \frac{40}{30}}$$

$$= 34.8^{o}C$$

Yielding

$$\frac{1000 \times 4200 \times 20}{3600} = 200 \times A \times 34.8$$

$$A = 3.35\ m^2$$

For 25 mm pipe required length = 44.5 cm.

Example 4: In this example we consider a case in which only the inlet temperatures of the hot and cold fluids are given. In practice, however, one needs to know the cooling suffered by the hot fluid and rise in cold fluid temperature for given flow rates. For the following data

$\dot{m}_h = 10{,}000$ Kg/h $\qquad T_1 = 200^{o}C$

$\dot{m}_c = 3000$ Kg/h $\qquad t_1 = 20^{o}C$

$U_o = 300\ W/m^2 {}^oC$ and $A = 0.487\ m^2$

$C_h = 1900\ J/Kg^oC$ $C_c = 4200\ J/Kg^oC$

Solution: From equations (7) and (8)

$$10{,}000 \times 1900 \times (200 - T_2) = 3000 \times 4200 \times (t_2 - 20)$$

Hence $$t_2 = 200 - 1.5\ T_2$$

$$LMTD = \frac{(200-t_2)-(T_2-20)}{\ln \dfrac{200-t_2}{T_2-20}}$$

Equating

$$10{,}000 \times 1900 \times (200 - T_2) = 300 \times 0.487 \times LMTD$$

Hence $$t_2 = 125.5^oC$$

In equation (9) the LMTD is the true temperature difference for a double pipe heat exchanger. The same is not valid for multipass shell and tube heat exchanger because the heat exchangers having more than one pass consist of both the flows counter current and parallel flow as shown in Fig.5.

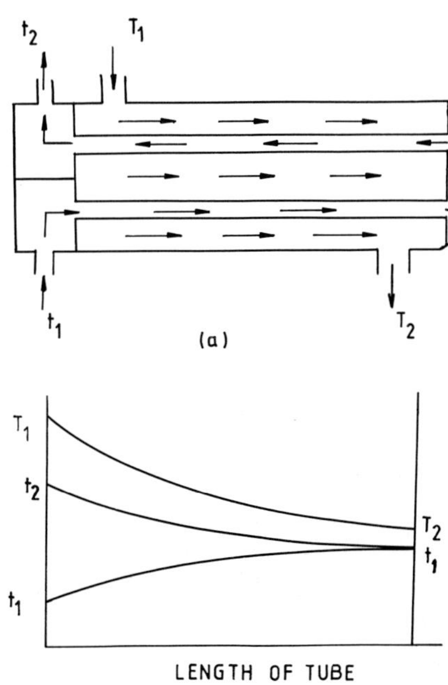

FIG.5 - TEMPERATURE PROFILE IN A 1-2 HEAT EXCHANGER

Therefore, to obtain the true temperature in case of multipass shell and tube heat exchanger, the LMTD is multiplied by a factor F.

To evaluate the value of F, the method was suggested by Underwood (Ref.4), Nagle (Ref.5) and Bowman et al. (Ref.6) have derived the following expression for F.

$$F = \frac{\sqrt{(R^2+1)}\ \ln(1-S)\ (1-RS)}{(R-1)\ \ln \dfrac{2-S\ (R+1-\sqrt{R^2+1})}{2-S\ (R+1+\sqrt{R^2+1})}} \tag{10}$$

where $R = \dfrac{T_1-T_2}{t_2-t_1}$ and $S = \dfrac{t_2-t_1}{T_1-t_1}$

To avoid the computation every time, a graph (as shown in Fig.6) has been plotted between F and S. Finally equation (9) can be written as

$$\dot{Q} = UA\ (F \times LMTD) \tag{11}$$

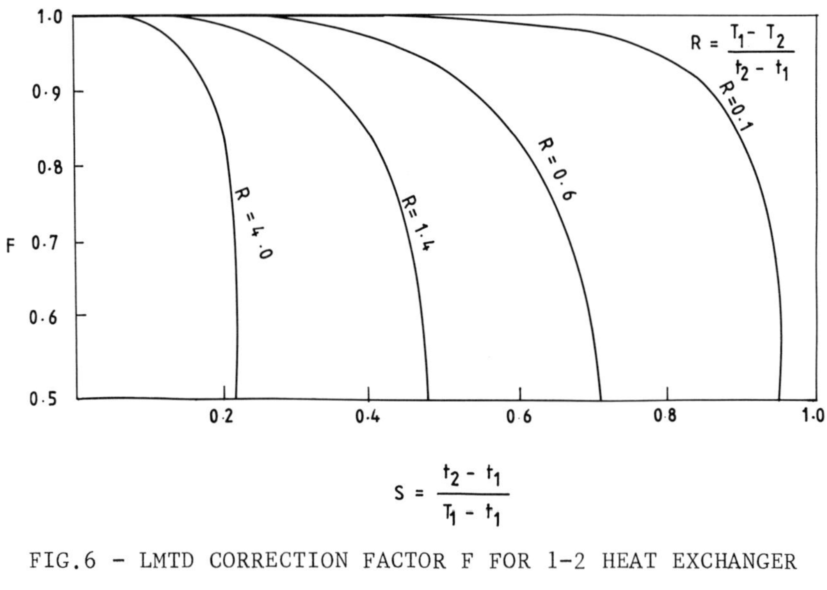

FIG.6 - LMTD CORRECTION FACTOR F FOR 1-2 HEAT EXCHANGER

Different charts are available for different types of multipass shell and tube heat exchangers. In the 1-2 heat exchangers, it is advisable that the factor F should be greater than 0.75 for practical purposes.

4. PRESSURE DROP

When the fluid flows on any surface or any object is moving in a fluid, the fluid exerts a force on the solid surface. The friction of force (related with kinetic behaviour of the fluid and the object) acted on the surface area A and characteristic kinetic energy per unit volume K. Therefore the pressure drop in terms of head is

$$\Delta F_a \propto A \times K$$

or $$\Delta F_a = f \times A \times K$$

where f = friction factor (dimensionless quantity).

This friction factor f is called fanning friction factor. For pipes or tubes with radius R and length L

$$\Delta F_a = f \times (2\pi RL) \times (\frac{1}{2}\rho\langle V\rangle^2) \tag{12}$$

where ρ = density of the fluid ,

$\langle V\rangle$ = average velocity of the fluid.

Generally ΔF_a is measured experimentally in terms of pressure drop (Δp) across the length L of pipe or tube. ΔF_a can be written in term of(Δp) as given below.

$$\Delta F_a = (\Delta p) g_c \pi R^2 \tag{13}$$

Equating both the equations (12) and (13)

$$f(2\pi RL)(\frac{1}{2}\rho\langle V\rangle^2) = (\Delta p) g_c \pi R^2$$

or $$\frac{p}{\rho} = \frac{f\langle V\rangle^2 L}{g_c R}$$

By putting $\Delta F_a = \frac{\Delta p}{\rho}$ and $\langle V\rangle = \frac{G}{\rho}$ (where G is the mass flow rate per unit flow area), we obtain

$$\Delta F_a = \frac{2fG^2L}{g_c \rho^2 D} \tag{14}$$

For more than one tube pass, the R.H.S. must be multiplied by the number of tube passes.

There are other losses in pressure while the fluid enters or leaves the hairpin. These losses (ΔF_1) can be computed as one velocity head per hairpin.

$$\Delta F_1 = \frac{\langle V\rangle^2}{2g_c} \tag{15}$$

In case of 1-2 heat exchanger (ΔF_1) is given as below.

$$\Delta F_1 = \frac{4n}{s^2} \frac{\langle V\rangle^2}{2g_c}$$

where n = number of passes,

and s = specific gravity of the fluid.

Therefore, tube side total pressure drop

$$\Delta F_t = (\Delta F_a + \Delta F_1) \qquad (16)$$

In any shell and tube type heat exchanger, the shell side pressure drop depends upon the number of crosses M + 1 (i.e., $M = L_1/B$; where L_1 is the shell length and B is the baffle spacing). Therefore total length L travelled by the fluid in the shell side is $D_s x(M+1)$ (where D_s is the inside diameter of the shell). Thus shell side pressure drop is

$$\Delta F_s = \frac{2fG^2 D_s (M+1)}{gc\,\rho^2 D_e} \qquad (17)$$

where D_e is the equivalent diameter of the shell.

The fanning friction factor can be determined with the following correlations given by Hagen-Poiseuille and Blasius respectively:

for laminar flow behaviour

$$f = \frac{16}{Re} \qquad (18)$$

for turbulent flow behaviour

$$f = \frac{0.0791}{Re^{0.25}} \qquad (19)$$

5. TYPES OF HEAT EXCHANGERS

A number of heat exchangers with different designs are employed in solar water heating systems. Some of them are discussed below:

(a) Coil type
(b) Shell and Tube type
(c) Double Pipe

(a) Coil type heat exchanger: In closed loop solar water heating systems, generally the coil type heat exchangers (as shown in Fig.7) are used because it is one of the cheapest and convenient means of obtaining heat exchanger area. The tubes are rolled in the form of helixes or double helical coils of either type is placed near the wall in vertical cylindrical vessels. Another type of coil, which is called pancake coil, is rolled in the spiral shape. This coil is placed in horizontal tanks in a single plane near the bottom to transfer heat by free convection.

(b) Shell and tube type heat exchanger: Shell and tube type heat exchangers (Fig.8) provide more heat exchange surface. They contain a tube bundle placed in a shell. They can be employed in horizontal or vertical positions. Vertical position is preferred in case of free convection and horizontal position in forced circulation. In case of

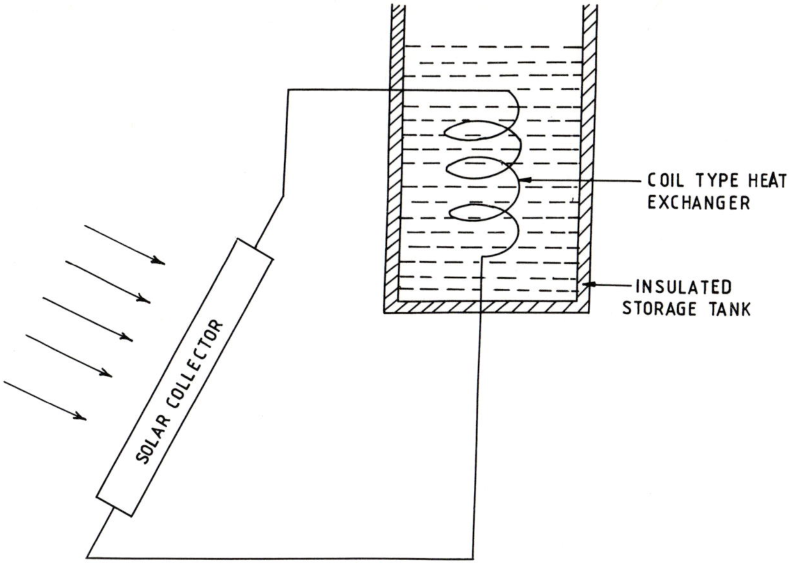

FIG.7 - COIL TYPE HEAT EXCHANGER EMPLOYED IN
CLOSED LOOP SOLAR WATER HEATING SYSTEM

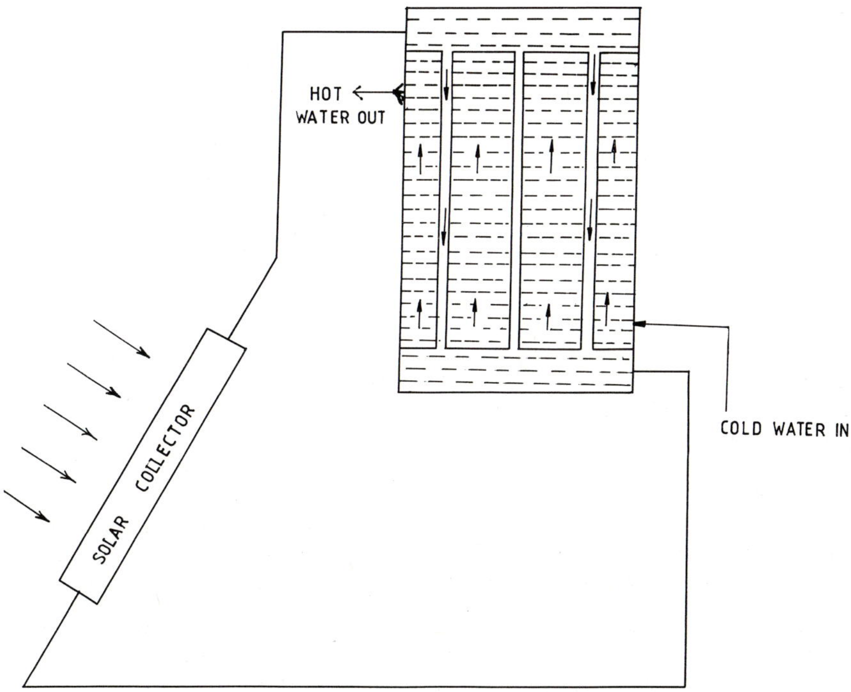

FIG.8 - VERTICAL SHELL AND TUBE TYPE HEAT EXCHANGER
EMPLOYED IN SOLAR WATER HEATING SYSTEM

natural or free convection, it is advisable to employ 1-1 heat exchanger. In case of the forced circulation, the shell and tube type heat exchangers with multipasses can be employed depending upon the requirement.

(c) Double pipe heat exchanger: In the above mentioned heat exchangers the water can be stored for heating and later on it is used. But for continuous withdrawal of hot water the double pipe of heat exchanger, as shown in Fig.9, will be suitable. This type of heat exchangers can be successfully used in industries for preheating the liquids or gases. To preheat the gases the fin type double heat exchanger must be used in order to achieve higher efficiency. These heat exchangers consist of a tube in another tube. They can be rolled in helical form but it is difficult to fabricate. Therefore, these exchangers are put horizontally in straight tube form. They can be connected to each other in the following three ways:

(1) one series hot stream and one series cold stream;
(2) one series hot stream and n-parallel cold stream;
(3) one series cold stream and n-parallel hot stream.

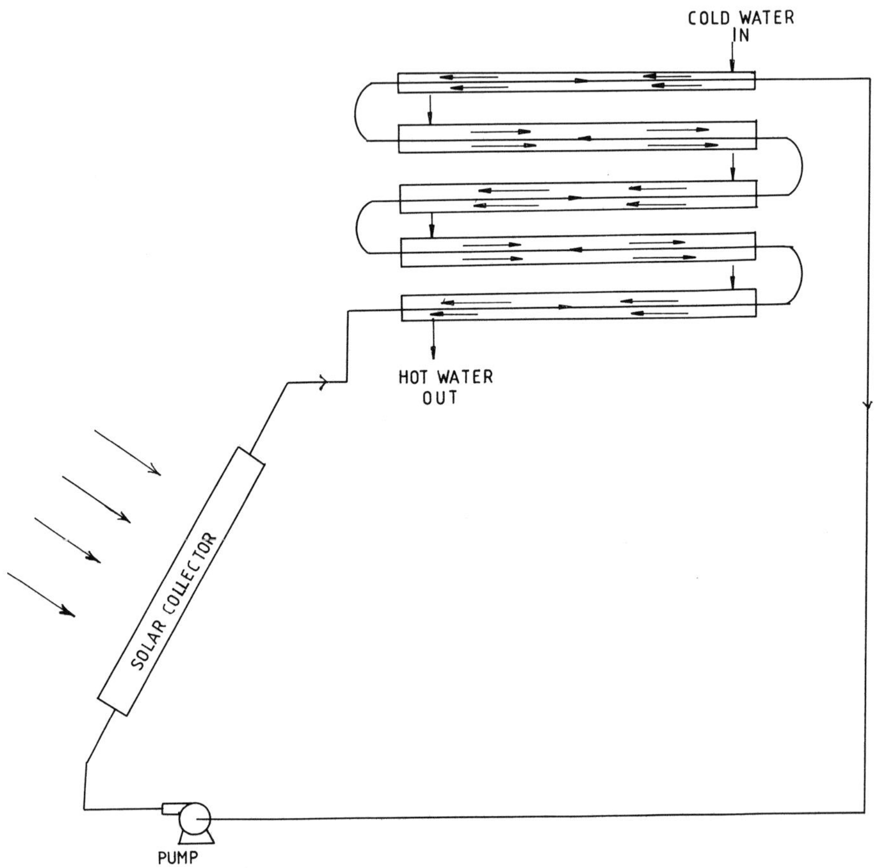

FIG.9 - DOUBLE PIPE HEAT EXCHANGER EMPLOYED IN SOLAR WATER HEATING SYSTEM

This combination of systems is decided mainly on the basis of the heat exchange area, i.e. length of heat exchanger and pressure drops across the heat exchanger. In any heat exchanger, one has to be careful about the pressure drops in both the streams. It helps in determining the pumping power in forced circulations and in designing the heat exchangers.

APPENDIX I

Calculations of h_h and h_c

(1) Coil type heat exchanger

In this type of heat exchanger, McAdams (Ref.7) has reported that the inside heat transfer coefficient is (1+3.5 d/D) times the heat transfer coefficient inside the straight tubes (where d is the tube diameter and D the coil diameter). The correlations for heat transfer coefficient inside the straight tubes are given below:

(a) In laminar flow conditions:

$$\frac{hd}{k} = 1.86 \left[\left(\frac{dG}{\mu}\right)\left(\frac{c\mu}{k}\right)\left(\frac{d}{L}\right) \right]^{1/3} \left(\frac{\mu}{\mu_w}\right)^{0.14} \tag{20}$$

(b) In turbulent flow conditions:

$$\frac{hd}{k} = 0.027 \left(\frac{dG}{\mu}\right)^{0.8} \left(\frac{c\mu}{k}\right)^{1/3} \left(\frac{\mu}{\mu_w}\right)^{0.14} \tag{21}$$

where
d = tube diameter,
L = length of the tube,
G = mass flow rate per unit area,
k = thermal conductivity,
c = specific heat,
μ = viscosity,
μ_w = viscosity at wall temperature.

Inside the storage tank the heat from the coil is being transferred by free convection (due to the absence of stirring the tank fluid). Therefore outside heat transfer coefficient can be evaluated by the expression, given by Chilton et al. (Ref.8):

$$h_c = 116 \left[\left(\frac{k^3 \rho^3 c\beta}{\mu}\right) \left(\frac{\Delta t}{d}\right) \right]^{0.25} \tag{22}$$

where β is the coefficient of thermal expansion and Δt is the temperature gradient.

(2) Shell and Tube type heat exchanger

Shell side heat transfer coefficient depends upon the tube arrangement and baffle spacing. It is therefore necessary to know equivalent diameter of the shell, d_e and flow rate per unit area in order to calculate the shell side heat transfer coefficient.

(a) For square pitch tube arrangement:

$$de = \frac{4(P_T^2 - \frac{\pi d_o}{4})}{\pi d_o}$$

where P_T = tube pitch

and d_o = outside diameter of the tubes.

(b) For triangular tube pitch arrangement:

$$d_e = \frac{4(0.5P_T x 0.86P_T - 0.5\pi d_o^2/4)}{0.5\pi d_o}$$

(c) Shell side flow rate:

$$G_s = \frac{W_s}{a_s}$$

where W_s = mass flow rate per unit time,

a_s = fluid flow area in the shell.

a_s can be evaluated as follows:

$$a_s = \frac{IDxC'xB}{P_T}$$

where I_D = inside diameter of the shell,

C' = clearance,

B = baffle spacing,

P_T = tube pitch.

(d) Tube side flow rate:

$$G_t = \frac{nW_t}{N_t a_t}$$

where W_t = mass flow rate per unit time,

a_t = flow area of one tube,

n = number of tube passes,

and N_t = number of tubes.

To compute the heat transfer coefficient in any shell and tube type heat exchanger for either side using the equivalent diameter and flow rate per unit area, the expressions (20) and (21) from Sieder and Tate (Ref.9) are employed.

(3) Double pipe heat exchanger

In this type of heat exchanger, the coefficients are calculated with the same expressions (20) and (21). The equivalent diameter of the annulus

$$d_e = \frac{D^2-d^2}{d}$$

where D = inside diameter of the annulus

and d = outside diameter of the inner tube

is used to obtain the annulus side heat transfer coefficient.

In all the heat exchangers the properties are generally evaluated at the arithmatic mean temperatures of inlets and outlets. In fluid-fluid heat exchange the hot fluid possesses a viscosity on entering which becomes greater as fluid cools. The cold counterflow fluid enters with a viscosity which decreases as it is heated. The variation in heat transfer coefficient at inlet and outlet was found to be large.

Hence with the assumptions:

(a) the overall heat transfer coefficient varies linearly with the temperature;
(b) constant mass flow rate;
(c) constant specific heats of the fluids;
(d) no partial phase change occurs during heating or cooling of the fluids.

The following expression could be derived for caloric temperatures of the fluids at which the properties of the fluids are determined in order to have the heat transfer calculations:

$$T_c = T_2 + F_c(T_1-T_2) \tag{23}$$

$$t_c = t_1 + F_c(t_2-t_1) \tag{24}$$

where T_c = caloric temperature of hot fluid,

t_c = caloric temperature of cold fluid,

F_c = factor.

F_c can be directly evaluated from the graph as shown in Fig.10 with the help of another factor K_c and $\Delta t_c/\Delta t_h$. The expression for K_c is given below.

$$K_c = \frac{U_h-U_c}{U_c}$$

where U_h = overall heat transfer coefficient at hot end,

U_c = overall heat transfer coefficient at cold end,

Δt_c = temperature difference at cold end,

Δt_h = temperature difference at hot end.

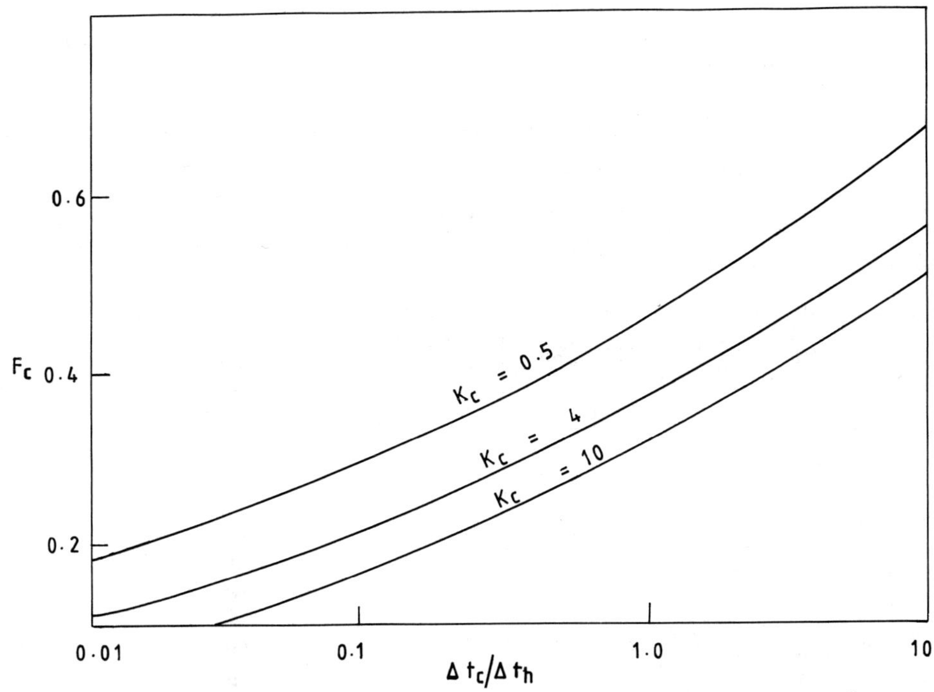

FIG. 10 - THE CALORIC TEMPERATURE FACTOR F_c VERSUS $\Delta t_c / \Delta t_h$

APPENDIX II

TABLE - Foulding Factors in $m^2 {}^oC/Watt$

Temperature of heating medium	Up to 115.6°C		115.6 to 204.4°C	
Temperature of water	51.7°C or less		Over 51.7°C	
Water	Water velocity m/s		Water velocity m/s	
	0.9144 m and less	Over 0.9144 m	0.9144 m and less	Over 0.9144 m
Sea water	0.000088	0.000088	0.000176	0.000176
Brackish water	0.000352	0.000176	0.000529	0.000352
Cooling tower and artificial spray pond				
Treated make-up	0.000176	0.000176	0.000352	0.000352
Untreated	0.000528	0.000528	0.00088	0.000704
City or well water	0.000176	0.000176	0.000352	0.000352
Great Lakes	0.000176	0.000176	0.000352	0.000352
Muddy or silty	0.000528	0.000352	0.000704	0.000528
Hard (Over 0.2138 Kg/m^3)	0.000528	0.000528	0.00088	0.00088
Engine jacket	0.000176	0.000176	0.000176	0.000176
Distilled	0.000088	0.000088	0.000088	0.000088
Treated hoiler feed	0.000176	0.000088	0.000176	0.000176
Boiler blow down	0.000352	0.000352	0.000352	0.000352

REFERENCES

1. D.Q. Kern, (1980), Process Heat Transfer, International Student Edition, McGraw-Hill Kogakusha Ltd.
2. H.Y. Wong, (1977), Heat Transfer for Engineers, Longman, London.
3. C.O. Bennett and J.E. Myres, (1974), Momentum, Heat and Mass Transfer, 2nd edition, McGraw-Hill Book Company.
4. A.J.V. Underwood, (1934), J. Inst. Petroleum Technology, 20, 145-158.
5. W.M. Nagle, (1933), Ind. Eng. Chem. 25, 604-608.
6. R.A. Bowman, A.C. Muller and W.M. Magle, (1940), Trans. ASME, 62, 283-294.
7. W.H. Mcadams, (1942), Heat Transmission, 2nd edition, pp 177-184.
8. T.H. Chilton, T.B. Drew and R.H. Jebens, (1984), Ind. Eng. Chem., 36, 510-516.
9. E.N. Sieder and G.E. Tate, (1936), Ind. Eng. Chem., 28, 1429-1436.

PUMPS FOR USE WITH HOT WATER

D.P. Agrawal
Department of Mechanical Engineering
Indian Institute of Technology
Hauz Khas, New Delhi 110 016
India

1. INTRODUCTION

In modern practice, machines for delivering liquids are known as pumps. During the flow of fluid through it, liquid head increases due to absorption of the external work in it. Pumps find innumerable applications in engineering areas and it is difficult to think of any application where principle of pumping is not used in some form or the other. Over the period of time these machines have been designed to cater to maximum efficiency, complicated systems and special duty requirements. The main task, therefore, faced mainly by the user, is the selection of proper pump out of data available and to satisfy himself of its performance when used along the intended system. This emphasises the need for clear understanding of basic pump concepts, testing procedures and component intricacies. The present article deals with this objective.

2. CLASSIFICATION OF PUMPS

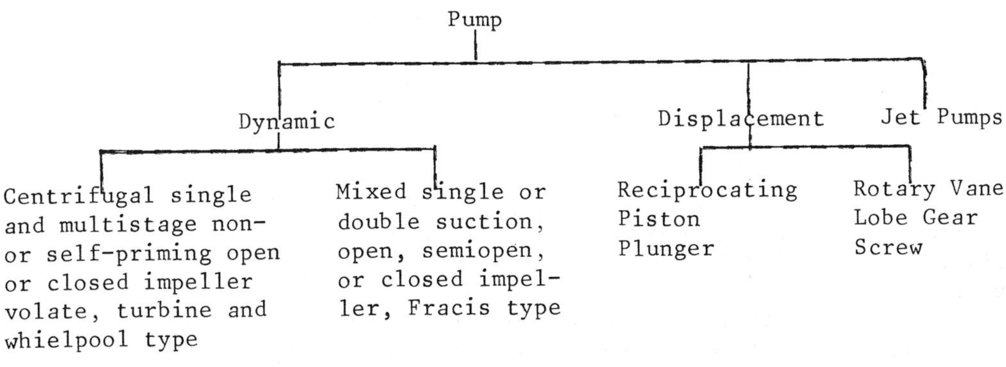

Figure 1 shows the shape of impeller and flow path on it for dynamic pumps.

The main parameters of any pump are the capacity and head it develops. Each type of pump is appropriate for specific discharges and

H. P. Garg (ed.), Solar Water Heating Systems, 279–289.

heads, so, for instance, reciprocating and rotary pumps are used where high head and small quantity are required, while axial flow pumps are specifically designed to deliver large flow quantities at lower heads. Figure 2 shows a head-capacity range for different pumps.

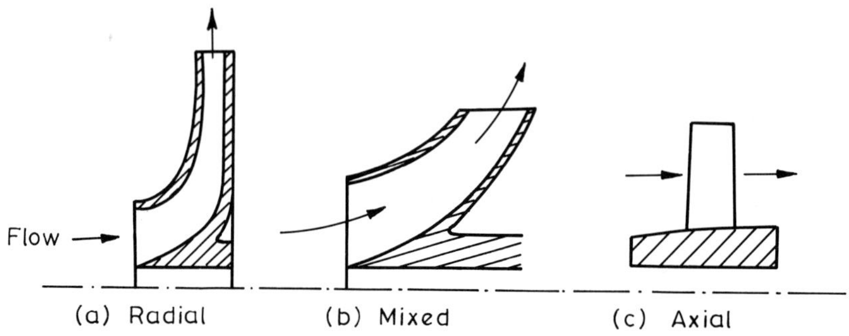

FIG.1 - IMPELLER SHAPES AND FLOW PATH

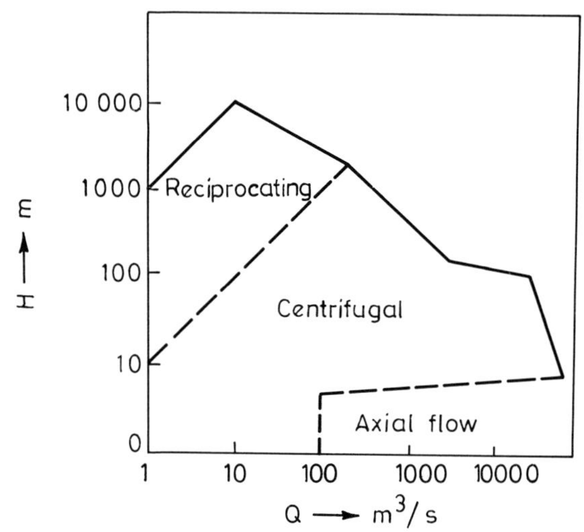

FIG.2 - HEAD-CAPACITY RANGE FOR PUMPS

Pumps are called upon to handle variety of fluids; large number of them handle clean and normal water. Hot water, oils, chemicals, viscous liquids, sewage and liquid with solid particles are some other fluids in use.

Pumps in solar water heaters are mainly centrifugal type used for circulation of water and use small power 0.05 to 0.1 K.W. of energy. These pumps are either in line type or end suction type. For larger requirements, propeller pumps are also used. It has been recommended that though the fluid properties affect the pump performance, but there are not major changes in design and performance for water temperatures up to 120°C. However, beyond this, proper selection is needed in view of cavitation etc.

2. HEAD OF A PUMP

(i) Static Head or Lift: The difference between the liquid levels in pump (suction tank) and storage reservoir is called static head. This consists of a suction and delivery heads, as shown in Fig.3. This figure also depicts the fact that the definition of static head is not affected by the placement of the two tanks or the way the water is delivered. Further it should be appreciated that the suction head may be positive or negative.

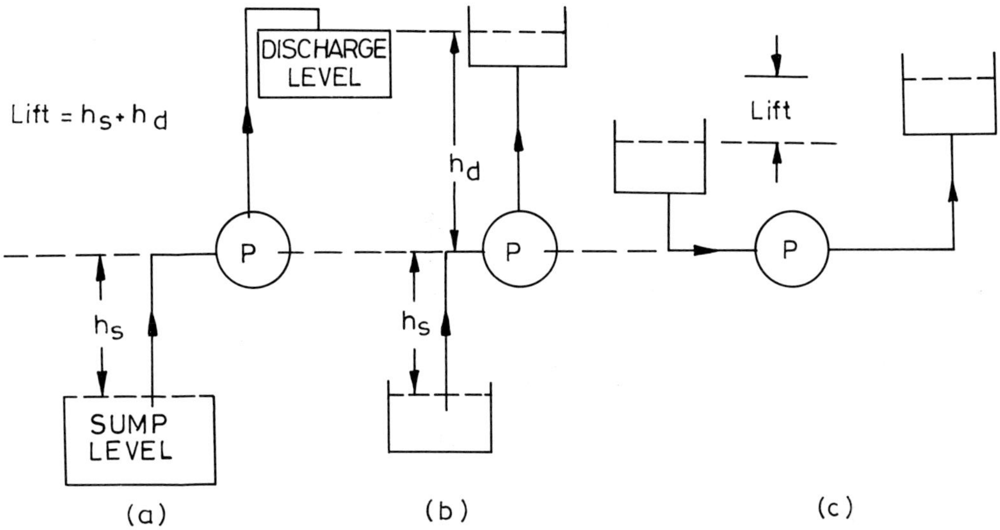

FIG.3 - STATIC HEAD FOR A PUMP

(ii) Total Head (H): A pump is called upon to develop pressure energy equal to the static head (h), frictional head (hf) in pipes and fitting and velocity head ($Va^2/2$) at the delivery end.

$$H = h + hf + \frac{Va^2}{2} \tag{1}$$

This represents the total energy per Kg of the liquid and hence is used in designing the pump.

(iii) Manometric Head: This indicates the difference of pressure head between the inlet and exit flange of a pump and is related to total head by

$$H_m = H + \frac{Vs^2}{2} - \frac{Vd^2}{2} \tag{2}$$

where Vs is the suction velocity in pipe.

3. ENERGY TRANSFER IN A PUMP

Total energy transfer in any rotodynamic pump is obtained from Euler's turbomachines equation:

$$\text{Work done}/\dot{m} = Vu_2u_2 - Vu_1u_1$$

$$= \frac{1}{2}\,\Delta V^2 + \frac{1}{2}\,\Delta U^2 - \frac{1}{2}\,\Delta W^2 \tag{3}$$

Each component refers to the way the energy transfer to the fluid has manifested on the fluid. Using velocity triangles (Fig.4) at impeller inlet and outlet, theoretical total head developed

$$H_{th} = \frac{U^2}{g}\,(1-\phi\cot\,\beta_2) \tag{4}$$

This is an equation of straight line and is plotted in Fig.5 for different values of exit impeller blade angle (β_2) and is designated as the Ideal pump characteristics. The variation of β_2 result in three different types of impeller shapes, viz. forward curved, radial tipped and backward curved. The back curved impeller is most commonly employed because of its highest efficiency. Fig.6 shows the modification of the ideal curve to the actual one by considering the effect of finite number of blades, losses and off-design conditions. Figure 7 presents the shape of performance curves of centrifugal, mixed and axial flow pumps.

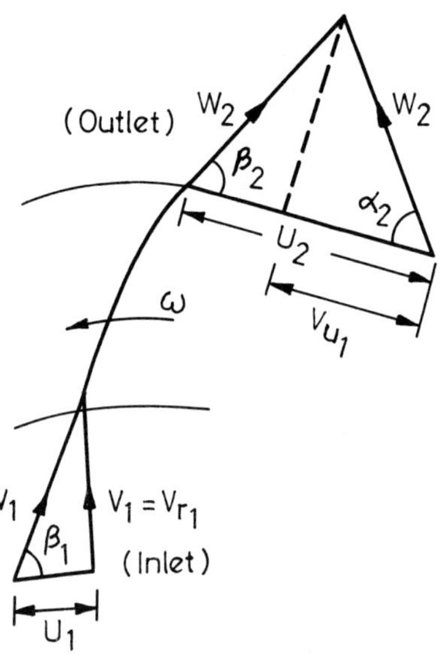

FIG.4 - VELOCITY TRIANGLES

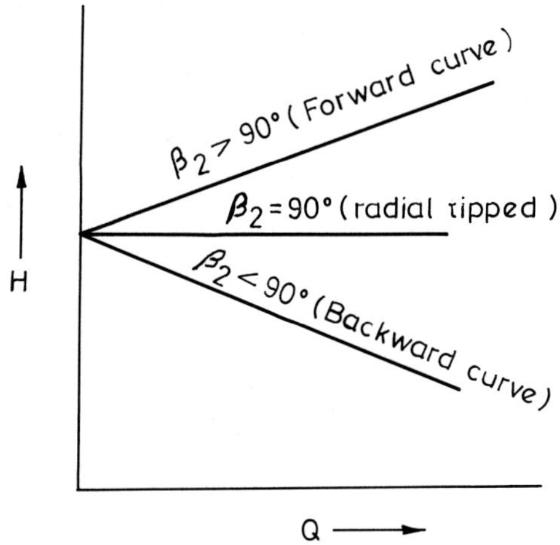

FIG.5 - IDEAL CHARACTERISTIC OF A PUMP

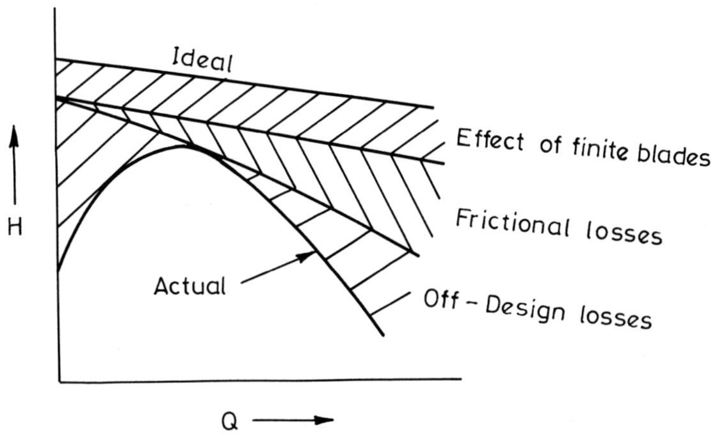

FIG. 6 - MODIFIED PUMP CHARACTERISTICS

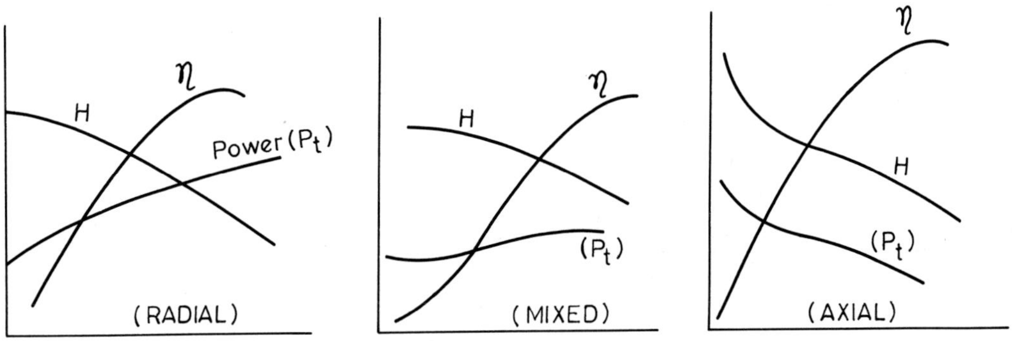

FIG.7 - PERFORMANCE CURVES FOR PUMPS

4. SUCTION LIFT

A pump is to suck liquid from its suction reservoir, the height through which the fluid is sucked could be +ve or -ve. For negative suction lift, if the pump surface is open to atmosphere, there will be reduction in static pressure at the pump inlet and in limit this pressure approaches liquid vapour pressure, and due to the formation of bubbles the pump is not able to lift the liquid. Theoretically, the maximum allowable suction lift (MSLA) is equal to atmosphere pressure. However, various energy losses in suction pipe reduce this value. Table 1 gives the practical values of MSLA.

TABLE 1 - Practical Values of MSLA

Pump type	MSLA m of water	practical value m of water
Single stage (CF)	8.25	4.5
Multi stage (CF)	8.25	4.5
Mixed flow	4.5	1.5 - 3.0
Reciprocating	8.25	6.7
Gear pump	8.25	1.5

Figure 8 shows the effect of temperature on suction lift.

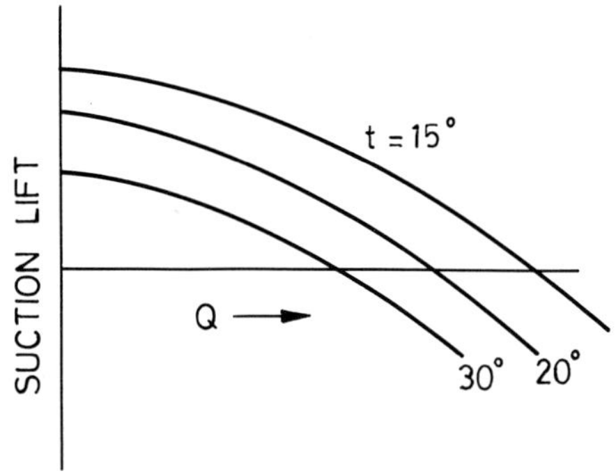

FIG.8 - EFFECT OF TEMPERATURE

5. SPECIFIC SPEED

Since the pump performance depends on number of inter-relating parameters, it is necessary to find one parameter which can help categorise various pumps more scientifically and may act as a common basis for comparison and classification of various pumps. A commonly used such parameter is called specific speed and is defined as the speed of a pump to develop unit head when using unit power.

$$N\,s = \frac{-NQ^{1/2}}{H^{3/4}}$$

the value of specific speed (Ns) for various pumps is:

Rotary and Reciprocating pump	Ns ⩽ 40
Centrifugal pump	Ns ⩽ 40-300
Mixed flow pump	Ns ⩽ 300-600
Axial flow pump	Ns ⩽ 600-1000

Figure 9 presents the variation of pump efficiency with this parameter.

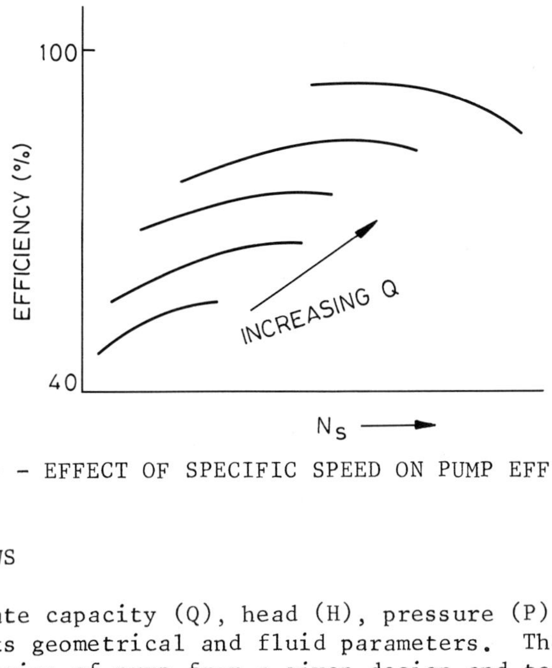

FIG.9 - EFFECT OF SPECIFIC SPEED ON PUMP EFFICIENCY

6. AFFINITY LAWS

These laws relate capacity (Q), head (H), pressure (P) and power (P_t) of a pump to its geometrical and fluid parameters. They are helpful in developing a series of pump from a given design and test a small model of an actual pump. The relations are given in Table 2.

Speed of the drive could vary continuously and is only limited by commercial consideration of the drive. The impeller exit diameter is changed by cutting it at its periphery as per the following:

For	Low specific speed	15-20%
	Mixed flow	3-4 %
	Axial flow	No cut

TABLE 2 - Variation of Q, H, P , P_t

With D, N, ρ	With ρ	With D	With n
$Q_1 = Q_2(\frac{D_1}{D_2})^3 \frac{n_1}{n_2} \cdot \frac{\eta_{vol_1}}{\eta_{vol_2}}$	$Q_1 = Q_s$	$Q_1 = Q_2(\frac{D_1}{D_2})^3$	$Q_a = Q_s \frac{n_1}{n_2}$
$H_1 = H_2(\frac{D_1}{D_2})^2 (\frac{n_1}{n_2})^2 \frac{\eta h_1}{\eta h_2}$	$H_1 = H_2$	$H_1 = H_2(\frac{D_1}{D_2})^2$	$H_1 = H_2(\frac{n_1}{n_2})^2$
$P_1 = P_2 \frac{\rho_1}{\rho_2}(\frac{D_1}{D_2})^2 (\frac{n_1}{n_2})^2 \frac{\eta h_1}{\eta h_2}$	$P_1 = P_2 \frac{\rho_1}{\rho_2}$	$P_1 = P_2(\frac{D_1}{D_2})^2$	$P_1 = P_2(\frac{n_1}{n_2})^2$
$P_{t1} = P_{t2} \frac{\rho_1}{\rho_2} (\frac{D_1}{D_2})^5 (\frac{n_1}{n_2})^3 \frac{\eta_2}{\eta_1}$	$P_{t1} = P_{t2} \frac{1}{2}$	$P_{t1} = P_{t2}(\frac{D_1}{D_2})^5$	$P_{t1} = P_{t2}(\frac{n_1}{n_2})^3$

7. PUMP EFFICIENCY

Pump power output $= \rho QgH$ Watts

Pump input power (electrical) $= \sqrt{3}\, V_L I_L \cos \phi \; \eta_{motor}$

$$\text{Pump efficiency } (\eta) = \frac{\text{Power output}}{\text{Power input}}$$

$$= \frac{\rho QgH}{\sqrt{3} V_L I_L \cos \phi \; \eta_{motor}}$$

$$\eta_{pump} = \eta_{overall} = \eta_{hydraulic} . \; \eta_{vol} . \; \eta_{mech.}$$

$$\eta_{vol.} = \frac{Q}{Q + AQ}$$

$$\eta_{hydraulic} = \frac{\text{H manometric}}{\text{Electric work}}$$

8. PUMP DUTY POINT

In any given pumping systems, it is essential to understand both the pump and the system characteristics, the intersection of these two curves, decides the duty point of the pump. The pump duty changes considerably by the change in the system through which the fluid is supplied. Each system imposes its own pressure restrictions on the pump, and sometimes in extreme makes the pump inoperative. Figure 10 shows the duty point for two system cases.

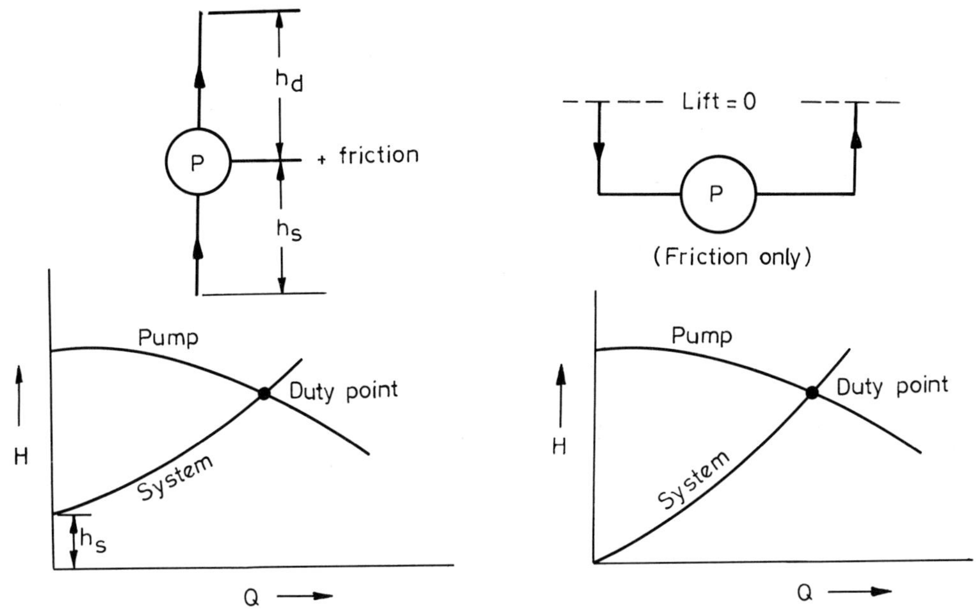

FIG.10 - PUMP DUTY POINT FOR DIFFERENT SYSTEMS

9. MATERIALS FOR CONSTRUCTION

Use of common materials is quite prevalent for temperatures lower than 100°C. For elevated temperatures with no impurities and corrosiveness grey cast iron, spheroidal graphite cast iron, cast steel and cast chrome steels are used for impeller and casings. It is necessary that the casing must be sealed with '0' rings, flat gaskets and spiral wound asbestos ropes.

10. TESTING OF PUMPS

Pumps are tested following the codes existing in each country for:

(i) acceptance test at Maker's Works;
(ii) acceptance test on site;
(iii) performance test on site.

The following indicate the type of instruments used for these tests.

Quantity to be measured	Acceptance Test	Site Test
Capacity, Q	Measuring tank, flowmeters (venturi, orifice Dall) measuring weir for large quantities	Flowmeters, integrating and recording current meters, pitot tube
Head, H	Dial type pressure or vacuum gauge, mercury column	Similar to acceptance test
Speed, n	Revolution counter and stop watch, Tachometer, stroposcope	Similar to acceptance test
Power, P_t	Tortional or eddy current dynamometer, electric motor with swing yoke	Single or double water-meter method

11. SHAFT SEALS

If the pressure in the pump interior is above or below atmospheric, suitable seals are installed at the points where the shaft passes out through the pump casing. Common seals are:

(i) Contact friction seals or stuffing boxes, stuffing boxes are soft packed and are made water cooled for hot water pumps.

(ii) Mechanical seals.

(iii) Floating seals.

12. CAPACITY REGULATION

In actual practice, a pump is required to operate at variable capacity depending on system requirements. There are three methods of capacity regulation:

(i) Throttling at exit: It reduces machine capacity and efficiency.

(ii) Speed regulation: The best method as the pump characteristics is automatically maintained.

(iii) Inlit guide blades at inlet: A cumersome method.

13. CONCLUSIONS

In conclusion it may be mentioned that for working fluid temperature below 100°C, the commercially available mono-block pumps could be used without any loss of efficiency in solar water heaters. However, at elevated temperatures care is needed because of their effect on suction lift and thereby cavitation. It is further recommended that at high temperatures, the seats and the materials be selected appropriately.

FLOW AND HEAT TRANSFER OF WATER IN PIPES

Ashok Malhotra
Department of Mechanical Engineering
Indian Institute of Technology
Hauz Khas, New Delhi 110016
India

1. INTRODUCTION

A pipe represents a convenient equipment for transport of water from one part of an equipment to another and for its heating or heat exchange. Due to its widespread use the flow and heat transfer in pipes has been the subject of numerous studies. Vast amounts of information both analytical and experimental is available on the topic. A short note like the present one can therefore only serve to focus attention on the topic. Only single phase flow of water flowing in a smooth circular pipe is considered here. This specific configuration is the starting point for studies in the area and at times results for rough pipes as well as those of different cross sections are often derived by simple extensions.

The thermophysical properties of interest are density ρ , specific heat C_p, thermal conductivity K and viscosity μ or alternatively the kinematic viscosity (μ/ρ). In the range of temperatures from 0 to 100^oC kinematic viscosity exhibits a sixfold decrease. The variation of other properties is less severe and often suitably defined averages are used. The Prandtl number, Pr, of water varies in the range of 13.6 to 1.74. The non-dimensional combination of properties represented by the Prandtl number tells more about the type of heat transfer that occurs in a fluid than the magnitudes of the individual properties. Considering that the Prandtlnumber of liquids can vary from several hundreth parts of unity to several thousands, the Prandtl number of water lies in a very convenient range. Because of this special effects such as axial conduction (in liquid metals) and viscous heating (in oils) are normally not significant in water. Another importance of the Prandtl number being close to unity is that analogies between momentum and heat transfer which are often used for convenience are more nearly applicable for water than they would be for other liquids.

2. TERMINOLOGY

In order to study the topic of flow and heat transfer in pipes, it is

H. P. Garg (ed.), Solar Water Heating Systems, 291–298.

useful at the outset to know the meaning of terms commonly used by engineers working in the area.

The fluid velocity during flow in a pipe varies over the cross section falling to zero at the walls. To simplify calculations much work is therefore carried out with the help of mean velocity, u_m. It is the velocity which multiplied by the pipe cross sectional area and fluid density provides the mass flow rate through the pipe. The mean temperature t_m, also variously called the bulk temperature or cup-mixing temperature is similarly defined. This is the temperature which will result if the fluid at any cross section is gathered and mixed. If variations in density or specific heat are neglected which is reasonable for water then mean temperature can be related to local temperatures at a pipe cross section mathematically through

$$t_m = \frac{2}{u_m R^2} \int_0^R r u t \, dr \qquad (1)$$

When fluid enters a pipe and makes contact with the internal surfaces a boundary layer starts to develop (Ref.1) characterised by a layer in which velocities vary steeply. The width of the boundary layer grows as the fluid progresses in the pipe. At some distance, the boundary layer growing all around the inside surface of the pipe will merge possibly near the centre of the pipe (if the situation is not unsymmetric because of gravity or another force). The distance at which the merger occurs is called the hydrodynamic entrance length. This is typically of the order of 0.05 Re diameters in case flow is Laminar and between 10 and 60 diameters in case of turbulent flow. The subsequent region of pipe is called the fully developed region. This region is characterised by vanishing axial gradients of velocity. In the preceding entrance region axial gradients of velocity vary rapidly in the boundary layer due to its growth. If fluid temperature changes along the length of pipe due to heat transfer in the fully developed region then axial gradients will never become zero due to expansion and contraction caused by density variations. However, these latter axial variations are normally an order of magnitude less than in the entrance region.

When the fluid enters a pipe, if the pipe surface temperatures are different from the fluid temperature then heat transfer takes place and a thermal boundary layer begins to develop. The thermal boundary layer for water is thinner than the hydrodynamic boundary layer and will exist within it. Therefore the thermal entrance length is larger and typically of the order of 0.05 RePr diameters in Laminar flow. The thermally fully developed region is defined analogously, with a difference though, that axial gradients of suitably non-dimensionalised temperature approach zero rather than temperature itself. The temperature must go on changing in a pipe as long as heat transfer takes place.

The fully developed regions are characterised by constant or nearly constant friction factor and heat transfer coefficients. This eases the task of design calculations in the fully developed region or in long pipes where the entrance region is ignored. The entrance region on the other hand is characterised by large changes (decrease) in friction factor and heat transfer coefficient. In short pipes, therefore, it becomes necessary to account for this region during calculations.

The heat transfer coefficient mentioned in the previous paragraph is defined locally through the equation

$$q_w = h\ (T_w - T_m) \tag{2}$$

where q_w is the local heat flux at any point on the pipe surface where it is desired to know the heat transfer coefficient and T_w is the temperature at that point. From time to time there have been voices that equation (2) is not a very sound basis for heat transfer calculations. Some of the reasons for this area that situations have been found where h itself is a function of q_w (e.g. supercritical fluids) or q_w depends nonlinearly on $(T_w - T_m)$ (e.g. free convection). In spite of these differences the convenience of using 'h' as defined by equation (2) cannot be debated and its use is likely to continue long into the future.

Two excellent texts that throw light on the fundamentals and practices in this area and will give details of definitions are due to Kays (Ref.2) and,Eckert and Drake (Ref.3).

3. THEORETICAL DESCRIPTION

A theoretical description of the phenomenon is based on the physical laws of conservation of mass, momentum and energy; the laws of stokes and Fourier; thermophysical property equations and turbulence hypothesis. The mathematical problem based on these laws poses two problems: the first is, in many cases an adequate description of turbulence is not available; the second problem is purely a mathematical or numerical one, that of solving the resulting simultaneous differential equations. However, considerable progress has been made in recent decades to overcome the difficulties. An understanding of this progress can be acquired from a text due to Arpaci and Larsen (Ref.4). Presently let us consider the relatively simple problem of Axi-symmetric flow in a smooth circular pipe. It can be described by the following governing equations.

$$\frac{\partial}{\partial x}(\rho u r) + \frac{\partial}{\partial y} + \frac{\partial}{\partial y}(\rho v r) = 0 \tag{3}$$

$$\rho u \frac{\partial u}{\partial x} + \rho v \frac{\partial u}{\partial y} = -\frac{\partial p}{\partial x} + \frac{1}{r}\frac{\partial}{\partial y}[r(u \frac{\partial u}{\partial y} - \rho \overline{u'v'})] + F_x \tag{4}$$

$$\rho u C_p \frac{\partial t}{\partial x} + \rho V C_p \frac{\partial t}{\partial y} = \frac{1}{r}\frac{\partial}{\partial y}[r(K\frac{\partial t}{\partial y} - \rho C_p \overline{V't'})] \tag{5}$$

F_x is the body force. For vertical up flow this is $-\rho g$. The coordinate system is depicted in Fig.1. u and v are the velocity components in the x and y direction. These equations are applicable to both laminar and turbulent flow with the understanding that if flow is laminar than the fluctuating correlations $\overline{u'v'}$ and $\overline{v't'}$ will be zero and if the flow is turbulent the velocities u and v and temperature t are time averaged quantities. These equations along with boundary conditions will yield a solution for velocity and temperature variations in the pipe which can in turn be used to assess bulk quantities such as heat transfer coefficients.

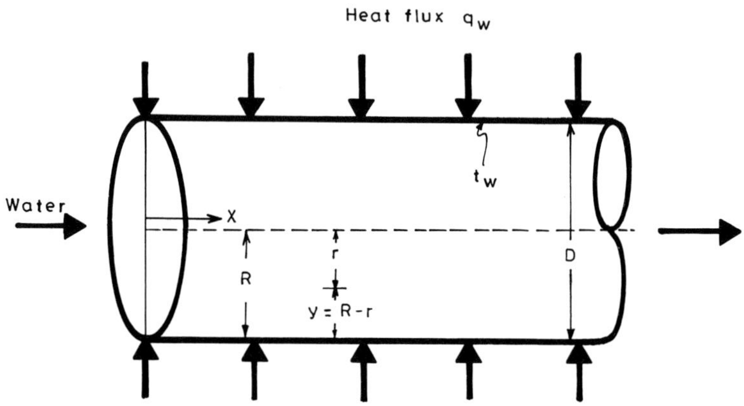

FIG.1 - SCHEMATIC DIAGRAM OF THE PROBLEM

A turbulent hypothesis is required to close these equations and a simple algebraic hypothesis will suffice as long as the flow remains parabolic (non-recirculating) and buoyant influences are moderate to vanishing. For example, the expressions used in reference (Ref.5)

$$-\overline{u'v'} = k^2 y^2 [1-\exp(-y^+/28)]^2 \qquad (\frac{\partial u}{\partial y})^2 / (1+3\,\frac{Y}{R} + 3\,\frac{y^2}{R^2})$$

and $$\overline{v't'} = \frac{1}{P_{r_t}}\, \overline{u'v'}\, \frac{\partial T}{\partial y} \Big/ \frac{\partial u}{\partial y}$$

where $$Pr_t = 1.01 - 0.09\, Pr^{0.36}; \quad 1 < P_r < 145$$

It may be noted that the present equations apply only for the symmetric case. Symmetry will be violated whenever the boundary conditions are unsymmetric or when the pipe is not vertical and buoyant influences are present.

An example of the numerical results of this model is given in Fig.2, where calculated temperature profiles are exhibited at three different Prandtl numbers. The results at a Prandtl number of 5.7 are compared with the experimental results of Gowen and Smith (Ref.6). The boundary conditions and geometry used for calculations corresponded closely to that of the experiments. The Nusselt number found from these temperature profiles also matched very well with previously available data (Ref.5).

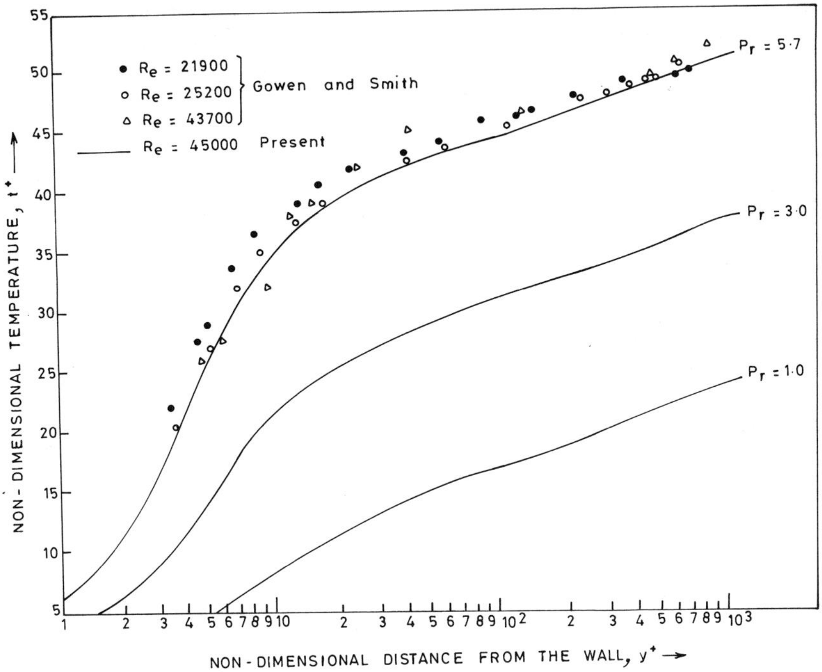

FIG.2 - TEMPERATURE PROFILES IN CIRCULAR PIPES COMPARED WITH EXPERIMENTAL DATA AT P_r = 5.7

4. ENGINEERING CALCULATIONS

The designer of equipment consisting of piping is not as concerned with the details of the flow as much as with overall effects such as heat transfer coefficients and pressure drops. He is not usually in a position to undertake a numerical solution of the governing equations. The best approach for design is therefore to use available correlations which have been previously determined either experimentally or theoretically. Care, however, needs to be exercised in choosing the appropriate correlations for friction factors or Nusselt number.

Two things need to be ascertained before the most applicable correlation can be found. First, the boundary conditions and secondly the heat transfer regime. Fortunately when flow is turbulent thermal boundary conditions do not play a significant role. Laminar flow on the other hand can be quite sensitive to boundary conditions. A majority of laminar flow heat transfer correlations have been ascertained for two standard conditions of uniform wall temperature or uniform heat flux. The actual situation may not correspond to either of these. In that case it is probably wisest to choose the boundary condition which will lead to the safest design.

To demarcate the flow regime it is a good approach to calculate the relevant non-dimensional parameters such as Reynolds number Re, Prandtl number Pr, Eckert number Ec, Grashof number Gr, Rayleigh number Ra, as well as the non-dimensional pipe length X/D and relative roughness at the outset. The Reynolds number will indicate if the flow is laminar or turbulent. Eckert number will indicate the presence of viscous heating. A calculation of Pr and X/D will indicate how much of the pipe is covered by the entrance length. Calculation of (Gr/Re^2) will indicate the extent of buoyant effects. This latter parameter does not fully indicate the effect of buoyancy on the turbulent fluctuations which requires the estimation of another parameter the Richardson number. However, for water, this latter sophistication will rarely be necessary. If the flow turns out to be free convective then Rayleigh number, Ra will help in further demarcating the free convection regime.

After the flow situation is demarcated with the help of non-dimensional parameters and boundary conditions the required correlations may be sought from literature, i.e. texts, research publications and handbooks, e.g. (Ref.7). It is useful for designers to maintain a record of the most convenient correlation when several are available for the same type of flow. For example, designers often make use of the Implicit Colebrook and While formula for friction factor in turbulent pipes whereas now much simpler explicit formulae are available,e.g. (Ref.8). Another example is the often used Dittus-Boelter equation (Ref.9) for turbulent forced convection. The careful measurements of Allen and Eckert (Ref.10) have shown that Nusselt number for water can be up to 20 percent, higher than predicted by this equation. An alternative to achieve accuracy is the equation due to Petukhov and Popov (Ref.11). Another equally accurate but simpler equation is due to Notter and Sleicher (Ref.12) later modified by Sleicher and Rouse (Ref.13). The effect of property variations can be separately accounted for by the property ratio method. Recently a theoretical justification to the property ratio method has been provided by Herwiz (Ref.14). Similarly roughness effects are sometimes separately accounted for through correction factors (Ref.3) whereas now convenient correlations are beginning to appear which include relative roughness (Ref.15).

Finally it must be mentioned that the effort put into ascertaining the heat transfer coefficients must eventually depend on its role in the overall design. For example, if the pipe is exposed to an atmosphere where the outside coefficient is much lower than that of the water in the pipe then the design will be controlled by the outside coefficient and even gross approximations for the coefficient inside the tube will be acceptable.

In conclusion, the present note has left out much more than it has covered about the flow and heat transfer of water in pipes. It only serves to encourage further studies in a direction suited to the needs of the interested reader.

LIST OF SYMBOLS

C_p	constant pressure specific heat
D	pipe diameter
Ec	Eckert number, $u_m^2/C_p(T_w-T_m)$

f	friction factor $\Delta P / \left[(L/D)\, (\rho u_m^2/2) \right]$
Gr	Grashof number $\rho^2 g \beta (T_w - T_b)\, D^3/\mu^2$
g	acceleration due to gravity
h	heat transfer coefficient, $q_w/(t_w - t_m)$
K	thermal conductivity
L	pipe length
Nu	Nusselt number, $h\, D/K$
P	pressure
Pr	Prandtl number, $C_p \mu/k$
Pr_t	turbulent Prandtl number
R	pipe radius
Ra	Rayleigh number, $Gr\, Pr$
Re	Reynold number, $\rho VD/\mu$
r	radial coordinate
t	local temperature
t_m	mean temperature
t'	fluctuating component of temperature
t^+	non-dimensional temperature, $(t_w - t) \rho C_p \sqrt{\tau_w/\rho} / q_w$
t_w	wall temperature
u	local velocity in x direction
u'	fluctuating component of velocity
u_m	mean velocity
v	local velocity in Y direction
v'	fluctuating component of velocity
x	co-ordinate
y	co-ordinate
y^+	non-dimensional coordinate $y\sqrt{\tau_w/\rho_w}\ \ (u_w/\rho_w)$
ρ	density
κ	Von-Karman's constant
μ	viscosity
τ_w	wall shear stress

REFERENCES

1. H. Schlichting, (1960), Boundary Layer Theory, McGraw Hill Book Co., Inc.
2. W.M. Kays and M.E. Crawford, (1980), Convection Heat and Mass Transfer, McGraw Hill Book Co., New York.
3. E.R.G. Eckert and R.M. Drake, (1972), Analysis of Heat and Mass Transfer, McGraw Hill Book Co., Inc.
4. S. Vedat Arpaci and Poul S. Larsen, (1984), Convection Heat Transfer, Prentice Hall Inc., New Jersey.
5. A. Malhotra and S.S. Kang, (1984), International Journal of Heat and Mass Transfer, 27(11), 2158.
6. R.A. Gowen and J.W. Smith, (1967), Chem. Engg. Sci., Vol.22, 1703.
7. Heat Exchanger Design Handbook, Hemisphere Publishing Corporation, 1983.
8. U. Schacham, Ind. Eng. Chem. Fundamentals, Vol.19, p.228, May 1980.
9. F.W. Dittus and L.M.K. Boelter, Univ. of Calif. Publications in Engr. 2(13), p.443, Oct. 17, 1930.
10. R.W. Allen and E.R.G. Eckert, (1964), Trans. ASME, Series C,Vol.86, p.301.
11. B.S. Petukhov and V.N. Popov, (1963), Transactions in High Temperature, 1(1).
12. R.H. Notter and C.A. Sleicher, (1972), Chem. Engg. Sci., Vol.27, p.2073.
13. C.A. Sleicher and M.W. Rouse, (1975), Int. J. Heat Mass Transfer, Vol.18, p. 677.
14. H. Herwiz, (1985), Int. J. Heat and Mass Transfer, 28(2), p.423.
15. J.J.J. Chen, (1985), Int. Comm. Heat Mass Transfer, Vol.12, p.219.

DOMESTIC THERMOSYPHON WATER HEATING SYSTEMS

S.S. Mathur and N.K. Bansal
Centre of Energy Studies
Indian Institute of Technology Delhi
Hauz Khas, New Delhi 110 016
India

1. INTRODUCTION

Tall chimneys releasing hot gases is a common sight in power plants and other industries. This phenomenon of fluid flow without any external fan or blower is called natural or free convection. The movement of the fluid is caused by the difference in densities at different heights because of the temperature gradient that exists along the chimney. This phenomenon is often made use of in the design of solar water heaters, particularly of small capacities. Water heated in a flat plate collector rises up to the storage tank by virtue of its low density and is replaced by colder water from the bottom of the tank. A continuous circulation is thus established between the collector to the tank which helps in heating up the entire water of the storage tank.

The efficient functioning of a thermosyphon system depends upon the unobstructed circulation of water in the system. It is strongly influenced by the shapes and sizes of its various components. A number of studies have been made to investigate the relationship between the design and performance of such water heaters. These are briefly discussed here.

One of the main problems in thermosyphon system is due to the reverse flow from the tank to the collector at night. When the collector cools down, hot water from the top of the tank enters the collector, flows down and re-enters at the bottom of the tank after losing its heat to the surroundings. In this paper some design considerations to prevent this reverse flow are also presented.

2. THERMOSYPHONIC FLOW

A simple schematic thermosyphon solar water heater is shown in Fig.1. It consists of a collector, a storage tank installed at a level higher than the collector and the connecting pipes. As the collector gets heated up in the sun, temperature gradients cause density variations giving rise to the fluid flow through the collector. In the stationary conditions, the pressure due to buoyancy forces balance the pressure losses due to friction.

H. P. Garg (ed.), Solar Water Heating Systems, 299–326.

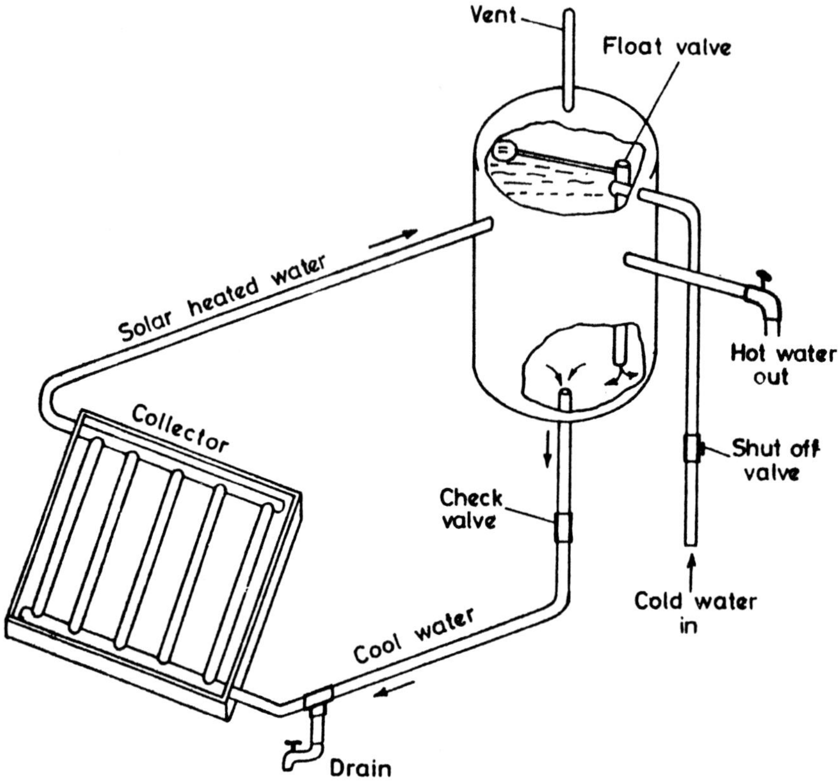

FIG.1 - A BASIC THERMOSYPHON SOLAR WATER HEATER

As a simple case, we consider that the cold water enters the collector at the bottom at a temperature t_1 and leaves it at a higher temperature t_2 after collecting some heat from the collector. The effective pressure Δp_t responsible for this flow can be considered to be made up of two parts, one representing the pressure drop across the collector (Δp_1) and the other between the top and bottom of the tank (Δp_2) such that

$$\Delta p_t = \Delta p_1 + \Delta p_2 \tag{1}$$

If $\rho(y)$ be the density at a distance y from the inlet of the collector, ρ_1 the density at the inlet and ρ_2 the density at the outlet of the collector (which are the same as at the bottom and top of the tank), then

$$\Delta p_1 = g \operatorname{Sin} \theta \int_0^L [\rho_1 - \rho(y)]\, dy \tag{2}$$

and

$$\Delta p_2 = (\rho_1 - \rho_2) gH \tag{3}$$

where θ is the inclination and L the length of collector, and H is the height of the storage tank.

The variation of density with temperature can be presumed to be

$$\rho(t) = \rho_o(1 - \beta T) \tag{4}$$

where T is the temperature in celcius, ρ_o the density at 0^oC and β the coefficient of volume expansion of the fluid. Δp_t can, then, be expressed as

$$\Delta p_t = g\beta\rho_o [\sin\theta \int_0^L \{t(y)-t_1\} dy + H(t_2-t_1)] \tag{5}$$

In the stationary state, this pressure is equal to the total pressure loss in the tubes and tank etc. on account of viscosity. Denoting the pressure drop across the collector as Δp_c and in the connecting tubes as Δp_2 the total pressure drop on account of viscosity Δp_s can be written as

$$\begin{aligned}\Delta p_s &= \Delta p_c + \Delta p_2 \\ &= \Delta p_c(1+r_p)\end{aligned} \tag{6}$$

where $r_p = \Delta p_2/\Delta p_c$ and is a measure of the relative flow resistance in the connecting tubes and the collector. It may be noted that the flow resistance in the tank has been ignored as it is much smaller in comparison.

Further, assuming the temperature to increase linearly along the length of the collector, we may write

$$t(y) - t_1 = \frac{(t_2-t_1)}{L} y \tag{7}$$

and equating (5) and (6), we get

$$g\beta\rho_o(t_2-t_1) [\frac{L}{2} \sin\theta + H] = \Delta p_c(1+r_p)] \tag{8}$$

The difference of temperatures (t_2-t_1) can be calculated from the energy balance equation of the collector namely,

$$\dot{Q}_u = \dot{m}_c C_f(t_2-t_1) = F' A_c [I_a - U_L(T_m - T_a)] \tag{9}$$

where $\dot{Q}_u$ is the useful power from the collector, $\dot{m}$ the mass flow rate in the thermosyphonic loop, A_c the collector area, I_a the effective absorbed radiation, U_L the total heat loss coefficient from the collector, T_a the ambient temperature and T_m the mean collector temperature which can be approximated as

$$T_m = \frac{t_1+t_2}{2} \tag{10}$$

The flow resistance per unit length ($\Delta p_c/L$) of the collector can be estimated depending upon the design of the collector. There are two simple geometries which are often used, namely (i) parallel tube thermosyphon and (ii) parallel plate thermosyphon (Fig.2).

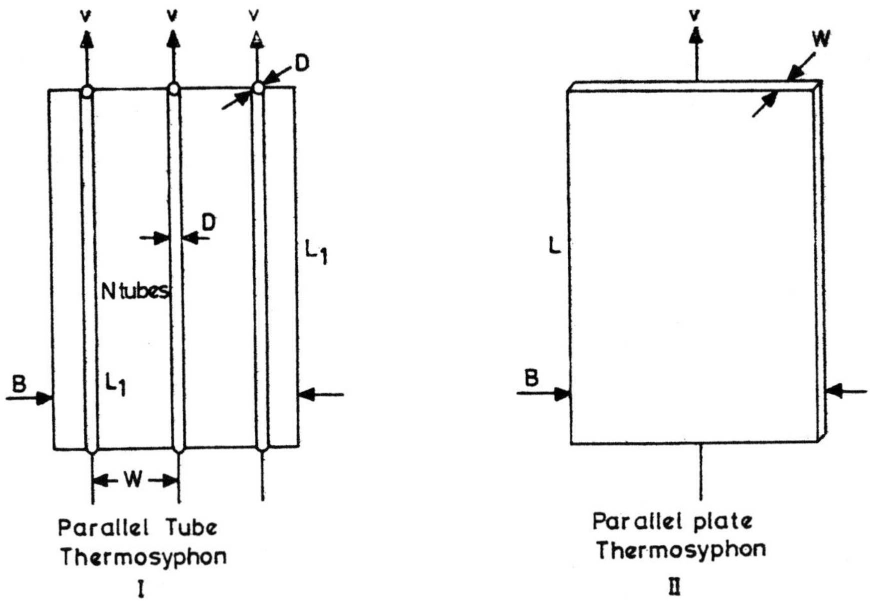

FIG.2 - PARALLEL TUBE AND PARALLEL PLATE THERMOSYPHON ABSORBERS

The expressions are:

$$\frac{\Delta P_c}{L} = \frac{32\,\nu\rho v}{gD^2}$$

and (11)

$$\frac{\Delta P_c}{L} = \frac{12\,\nu\rho v}{gw^2}$$

respectively, where ν is the viscosity, v the velocity and D, the diameter of the tube. For the parallel plate thermosyphon system, w is the separation between the two plates. The mass flow rates, for the two cases, are related to the flow velocity v by the relations

$$\dot{m}_c = N\rho v \frac{\pi D^2}{4} \quad \text{(parallel tube)}$$

and (12)

$$\dot{m}_c = \rho\, v\beta\, w^2 \quad \text{(parallel plate)}$$

combining equations (8) to (12) the expression for the mass flow rate can be obtained for both cases. The final results are

$$\frac{\dot{m}_c}{B} = \left[\frac{\pi L}{128} \frac{ND}{BL_1} D^3 \frac{g\beta \rho_1 F'}{\nu C_f(1+\gamma_p)} \left\{I_a - U_L(t_1 - t_a)\right\} \left(\frac{1}{2}\sin\theta + H\right) \right]^{\frac{1}{2}}$$

(parallel tube)

and

$$\frac{\dot{m}_c}{B} = \left[\frac{1}{12} W \frac{g \beta \rho_1 F'}{\nu C_f (1+r_p)} \left\{ I_a - U_L (t_1 - t_2) \right\} \left(\frac{1}{2} \sin \theta + H \right) \right]^{\frac{1}{2}} \quad (13)$$

(parallel plate)

3. THERMOSYPHON HOT WATER SYSTEM

A thermosyphon system is schematically shown in Fig.3. The essential

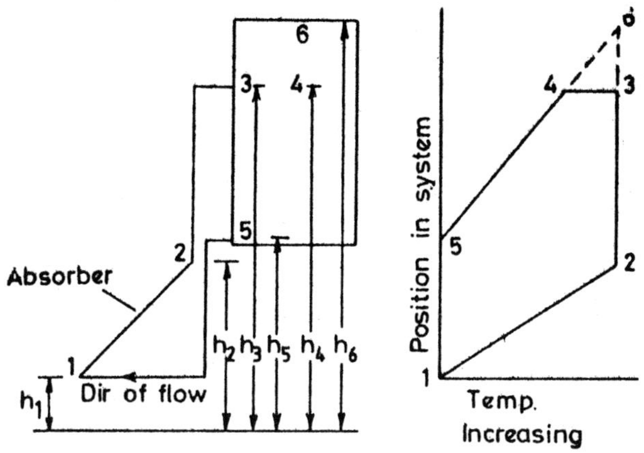

FIG.3 - SCHEMATIC OF THERMOSYPHON WATER HEATING SYSTEM

components are: a solar energy flat plate collector, storage tank, and connecting pipes or tubes. If the storage tank is kept under the pressure of city water supply, the system is termed as pressurrised, otherwise it is called a non-pressurised system. These systems are shown in Fig.4. It may be noted that the collector of the pressurised system has spiral fluid channel. This is done to create greater flow resistance, thereby reducing the flow rate of the liquid to achieve higher temperatures of collection. However, parallel tube collector can also be employed in a pressurised system.

The performance of the thermosyphon system depends upon the size and capacity of the storage tank, the thermal capacity of the collector and the connecting pipes including water and on the pattern of hot water use. Analytical expressions for the tank temperature as a function of collector parameters and solar insolation are available in the literature. It is found that the mean tank temperature is about the same as the mean collector temperature under steady conditions (Ref.1,2,3). These expressions also give the drop in the tank temperature during the night when solar insolation is zero.

All theories calculate pressure drops across the pipes etc. presuming them to be long and straight. There are, however, a number of joints and elbows which add to the pressure drops considerably. Moreover, due to degassing of water in the collector, air pockets can sometimes be

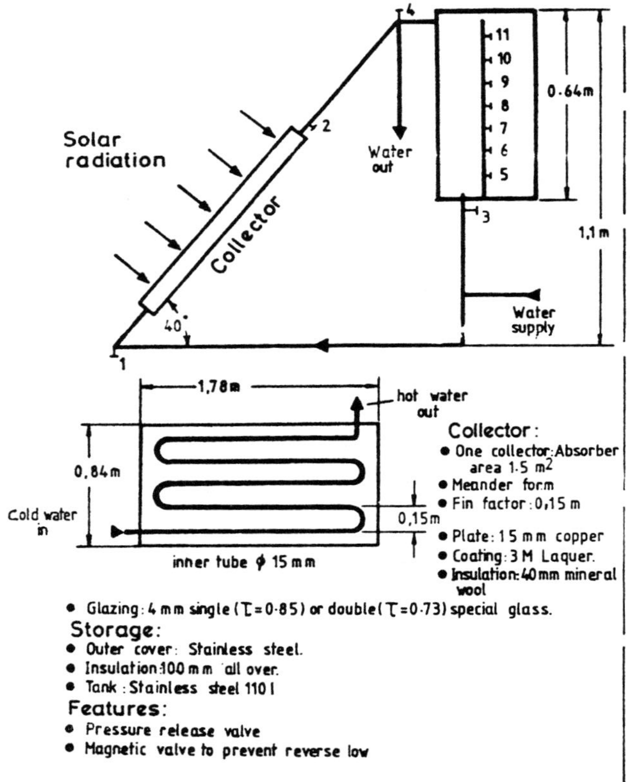

FIG.4(a) - SCHEMATIC OF THE PRESSURIZED DOMESTIC HOT WATER SYSTEM I

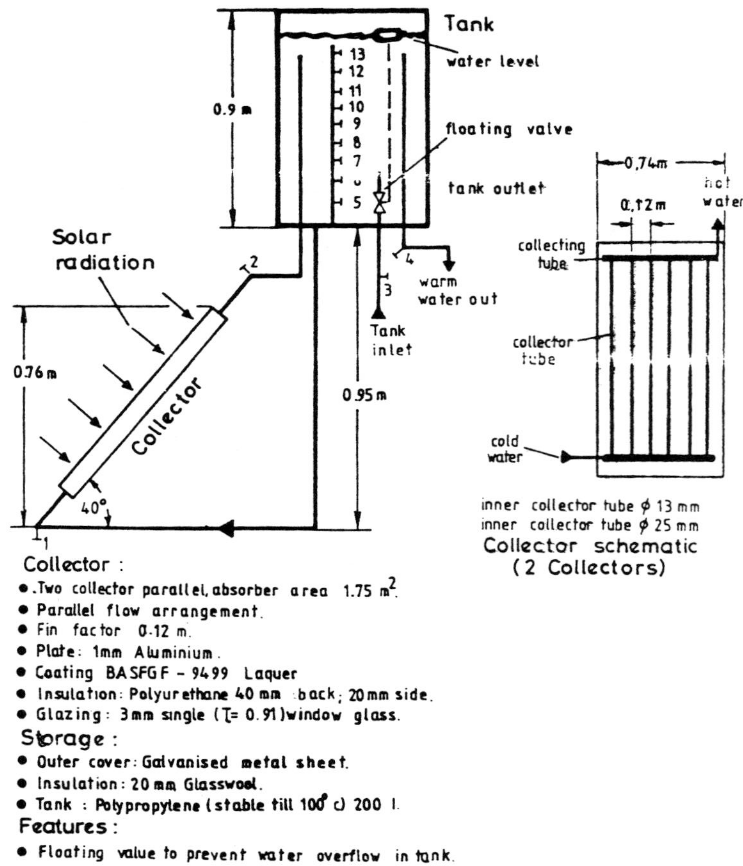

FIG.4(b) - SCHEMATIC OF THE NON-PRESSURIZED DOMESTIC HOT WATER SYSTEM II

created which tend to accumulate at these joints and completely obstruct the thermosyphoning process. In the design of the collector system, therefore, sharp bends should be avoided and smooth low curvature bends should be used.

It is natural that hot water will be withdrawn from the tank for use. In most systems, the hot water used from the storage tank is replaced by cold water resulting in lowering of tank temperature and disturbance in the temperature stratification of the tank. There are a number of methods available to estimate temperature profile of the stratified tank and the overall efficiency of the thermosyphon hot water system for simple pattern of hot water use.

4. REVERSE FLOW

Reverse flow can occur in thermosyphon systems during the periods of no sunshine, especially when the tank is located below the collector. Reverse flow, if present, can lead to considerable losses if not properly checked.

To understand the phenomenon of reverse flow, let us consider Fig.5 in which the solar water heater is looked upon as a U tube. During the day time the head of water in the tank plus that of the water in the return pipe constitutes the falling column of water, whilst the head of water in the absorber and in the flow pipe constitutes the rising column.

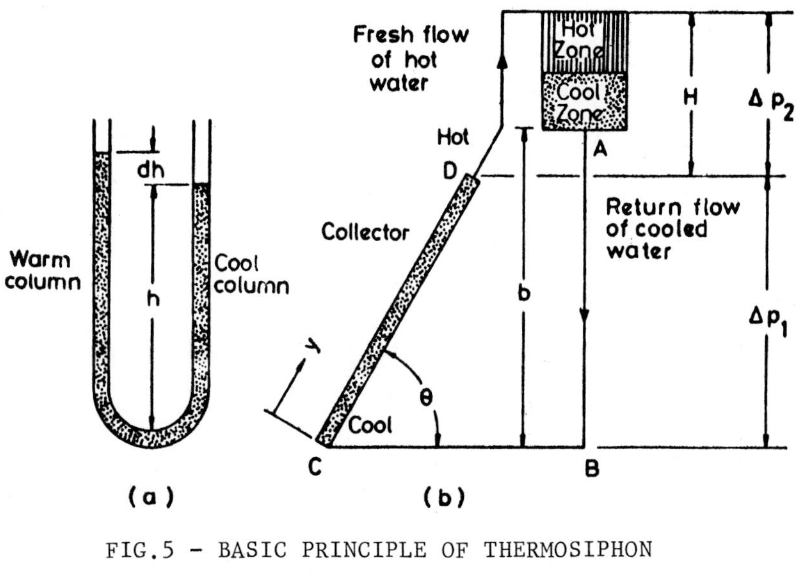

FIG.5 - BASIC PRINCIPLE OF THERMOSIPHON

The temperature in the return and in the flow pipe varies with height and it is thus useful to introduce the concept of mean density and mean temperature. The mean temperature, for example, is defined as

$$t_m = \frac{1}{h}\int_0^1 f(h)\,dh \tag{14}$$

where $f(h)$ is the function describing the temperature at any given height.

Let ρ_1 and t_1 be the mean density and mean temperature of water in the right leg of the tube representing head of water in the falling column and ρ_2, t_2 be the corresponding values for the rising column. t_1 is normally less than t_2 and hence $\rho_1 > \rho_2$. Under these conditions, the normal thermosyphonic action takes place and heat is removed from the collector to the tank. If, however, $t_2 < t_1$, i.e. $\rho_2 > \rho_1$, then the circulation is likely to occur in the reverse direction, resulting in flow of hot water from the tank to the collector where it loses heat and returns to the tank. For the reverse flow to occur either the solar insolation is zero or heat may be lost rapidly from the collector or both. On a cold night, the heat losses from the collector could easily result in such a situation. The reverse flow can be checked if $t_1 = t_2$.

By careful design and installation, reverse flow can be completely avoided, say for example, by placing the tank as high as possible, above the collector. The minimum height of the bottom of the tank from the top of the collector should be 20 cm. It has been suggested that reverse flow could be stopped by putting a one way valve in the fluid circuit. This is not very reliable because, the valve does not function satisfactorily for low pressure difference and for longer duration also.

Measurements on reverse flow in collectors have recently been reported by Uhlemann and Bansal (Ref.4). The results for pressurised as well as non-pressurised systems represented in Fig.4 are shown in Figs. 6, 7 and 8 where the time variation of $\Delta t_c = (t_2 - t_1)$ and average tank temperature (Δt_m) over 30 minutes interval are plotted. System I is equipped with a valve and time switch which interrupts the water flow from the collector to the tank or vice-versa. Figure 6 represents the case when the valve is kept open overnight. Due to insolation Δt_c and Δt_m vary during the sunshine hours (08.00-19.00 h). Around sunset at 19.15 h, however, Δt_c is seen to rise again to a rather high value in the night due to the fact that a back flow from the hot side of the tanks through the collector starts, resulting in high thermal losses. It is observed that Δt_c rises up to $30^\circ C$ as a result of this reverse flow. The corresponding heat losses in the tank are indicated by a negative value of Δt_m of about 0.5-$0.7^\circ C$ per 30 minutes, resulting in a drop of about $12^\circ C$ in the temperature of the tank overnight. In the morning from 7.00 h onwards, the back flow gradually stops and the collector starts heating the tank again from about 09.00 h due to increase in the solar radiations incident on the collector.

The rate at which the reverse flow occurs, i.e. $\dot{m}_R$ can be calculated by calculating the energy balance equation for collector and the tank, i.e.

$$MC_w \frac{dt_m}{dt} = -(\dot{Q}_{LC} + \dot{Q}_{LT}) \tag{15}$$

where $\dot{Q}_{LC}$ and $\dot{Q}_{LT}$ are the rates at which the heat is lost from the collector and the tank respectively. $\dot{Q}_{LC}$ can be expressed as

$$\dot{Q}_{LC} = \dot{m}_R C_w \Delta t_c \tag{16}$$

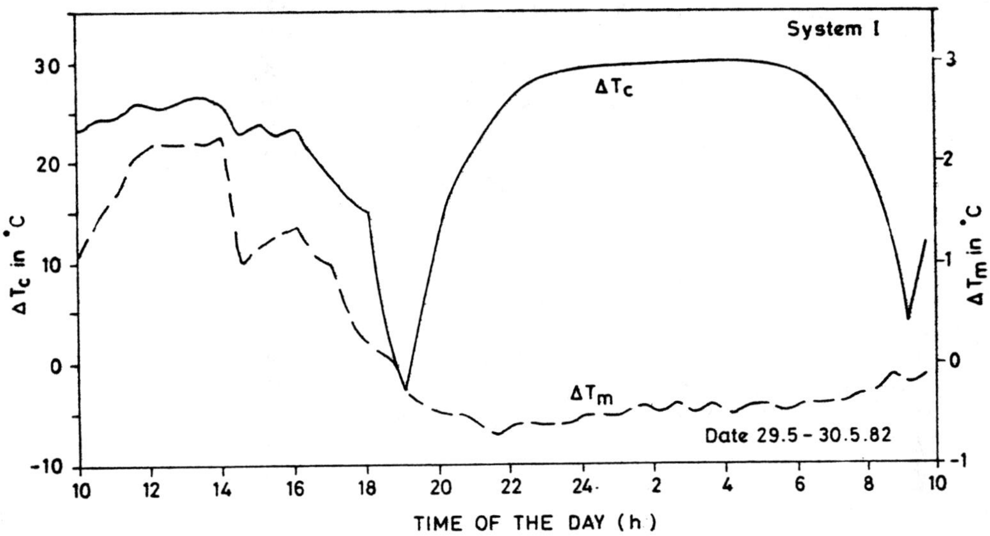

FIG.6 - MEASURED VARIATION OF ΔT_c AND ΔT_m (1/2 HOUR AVERAGES) IN SYSTEM I WITH THE VALVE KEPT OPEN OVERNIGHT

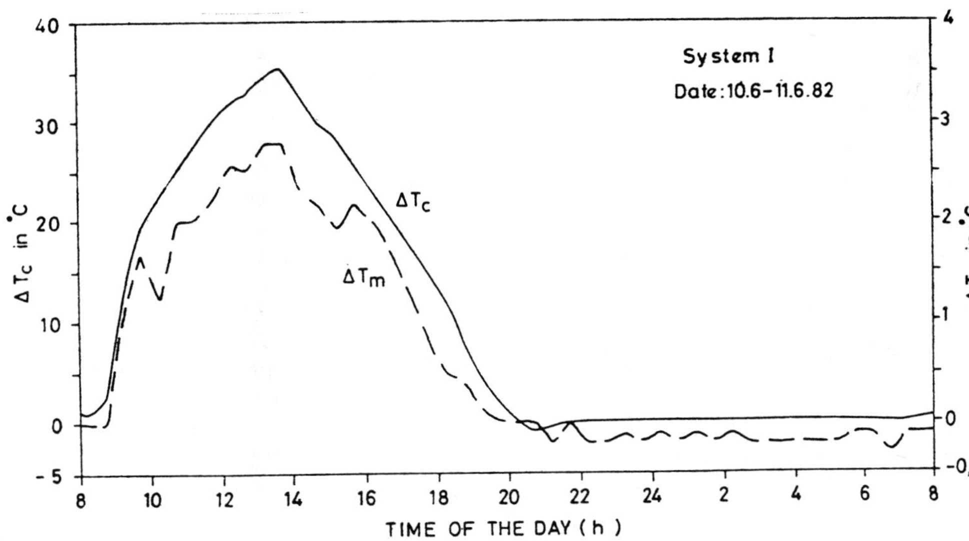

FIG.7 - MEASURED VARIATION OF ΔT_c AND ΔT_m (1/2 HOUR AVERAGE) IN SYSTEM I WITH THE VALVE CLOSED OVERNIGHT

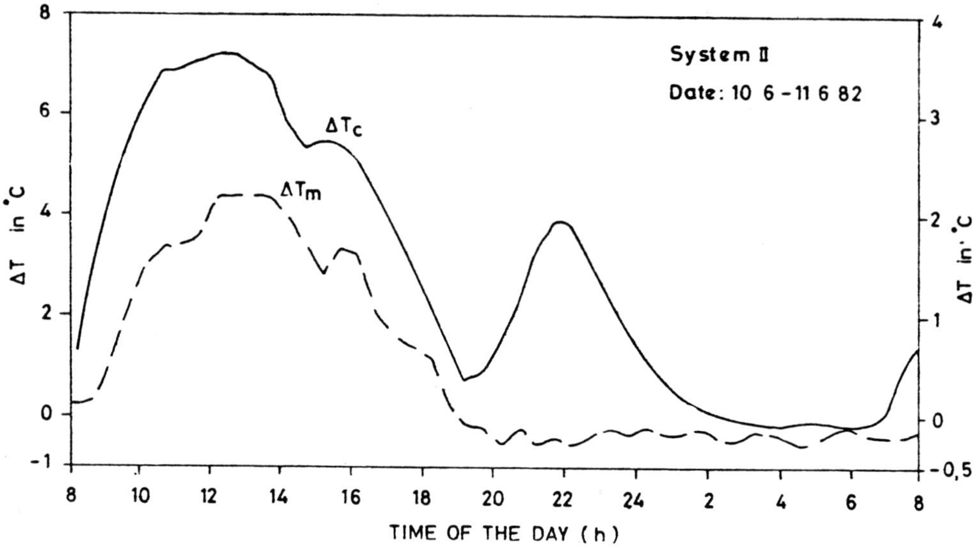

FIG.8 - MEASURED VARIATION OF ΔT_c AND ΔT_m (1/2 HOUR AVERAGES) IN SYSTEM II

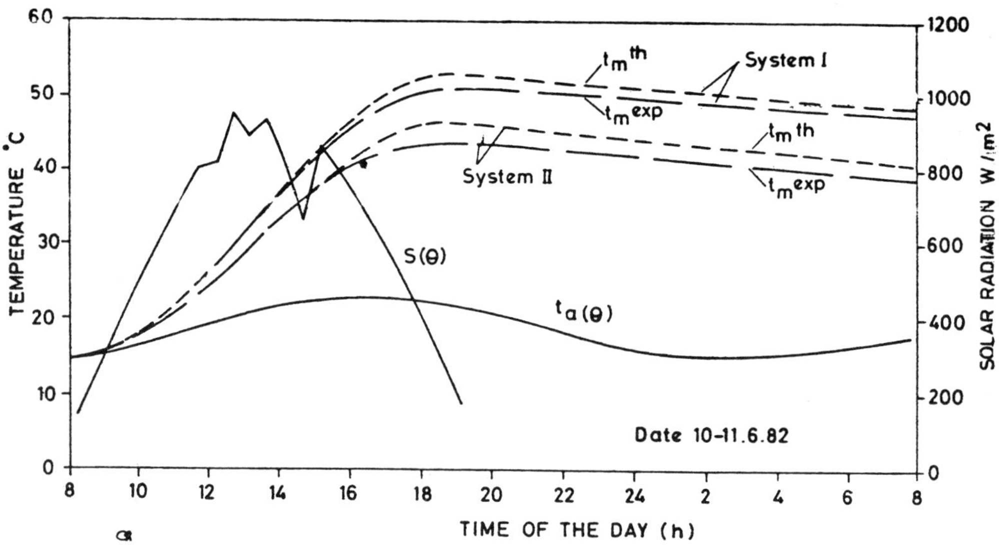

FIG.9 - THEORETICAL (t_mth) AND EXPERIMENTAL ($t_m$exp) VARIATION OF THE AVERAGE TANK TEMPERATURE DUE TO THE OBSERVED VARIATION OF SOLAR RADIATION S(θ) AND AMBIENT TEMPERATURE T_a(θ) ON A SELECTED DAY FOR BOTH THE THERMO-SYPHON SYSTEMS

where

$$\dot{Q}_{LT} = U_t A_t (t_m - t_a) \tag{17}$$

Substituting equations (16) and (17) in equation (15), the rate of reverse flow is given by

$$\dot{m}_R = -\frac{H\,\Delta t_m}{\Delta t \Delta t_c} + \frac{U_t A_t (t_m - t_a)}{C_w \Delta t_c} \tag{18}$$

Using equation (18) and using the measured values of various temperatures shown in Figs.6 and 8, the calculated values of reverse flow rates are as follows:

Reverse flow rates in Systems I and II at different hours of the day

Time	System I	System II
20.00 h	2.4 g/s	
22.00 h	1.5 g/s	0.34 g/s
04.00 h	0.65 g/s	-
08.00 h	0.45 g/s	-

Figure 7 shows the case where the valve of system I is closed overnight thus stopping the reverse flow. Hence Δt_c remains zero and the decrease in the average tank temperature, overnight, is only 3.7°C ($\Delta t_m \approx 0.2$°C per 30 minutes interval).

For the system II, the tank is located above the collector and the effect of reverse flow is very marginal. It is noticed that after sunset Δt_c increase to about 4°C and then decreases gradually to zero. The tank temperature falls by about 5°C.

5. PERFORMANCE OF THERMOSYPHON SYSTEMS

5.1 Open Loop Systems

Uhlemann and Bansal (Ref.4) recently made measurements on two commercially available thermosyphon systems of domestic size. The description and other parameters of the system are given in Tables 1-3. The time variation of the average system temperature as measured experimentally and calculated theoretically are shown in Fig.9. It is observed that in the non-pressurised system, the maximum value of the average tank (temperature reached in the non-pressurised system (System II) is 44°C, starting from an initial temperature of 15°C of fresh water in the tank at 8.00 h.

The maximum temperature reached in the tank of the pressurised system is about 51°C. The useful energy, corresponding to these temperatures, obtained from both the systems and the associated thermal efficiencies, are given in Table 3. The lower efficiency of the pressurised system may be due to higher temperatures recorded in this system resulting in more heat losses than in the non-pressurised system. Table 3 also contains the magnitude of the fall in the average tank temperature during night time and hence the night losses. During night time, the

TABLE 1 - Technical Data of Pressurised and Non-pressurised Solar Water Heating Systems of Figure 4

PROPERTIES	SYSTEM I SKETCHED IN FIG.4(a)	SYSTEM II SKETCHED IN FIG.4(b)
System Type	Pressurised	Non-pressurised
No. of collectors with absorber area	One, 1.78 m long and 0.84 m wide with an absorber area of 1.5 m^2	Two, each 0.74 m long and 1.19 m wide with a total absorbing area of 1,75 m^2
Collector Tubes	One 9.5 m long tube arranged in meander form, internal diameter of the tube 15 mm	Six copper tubes of 13 mm internal diameter connected in parallel to each other along the length of each collector
Fin factor	0.15 m	0.125 m
Glazing	Single or double glazing of 4 m thick glass	Single glazing of 3 mm thick ordinary glass
Collector Plate	Copper, 1.5 mm, coated with 3 M laquer	Aluminium, 1 mm thick, coated with BASF GF 33-9499 laquer
Casing	1.9 x 0.95 m x 0.13 m frame of 2mm light metal insulated with 4.0 mm mineral wool	1.25 x 0.8 x 0.11 mm frame of galvanised iron for each collector insulated with 4 cm thick polymethane at the bottom and 2.0 cm thick polymethane at sides
Water Storage Tank	Stainless steel tank consisting of two coaxial cylinders outside cylinder measurements are 0.9 m length and 0.6m diameter. Inside capacity 110 1 and the hollow space between the cylinders is filled with glass wool of 10 cm thickness	Double wall cylindrical tank outside material is galvanised iron of 1.0 m length and 0.65 m diameter. Inside cylinder is of polypropylene (stable till 100°C) of 200 1 capacity. The hollow space between the coaxial cylinder is filled with 2 cm thick glass wool
Special features	Overpressure valve, magnetic valve to prevent reverse flow	Floating valve to prevent the overflow of water from the tank

TABLE 2-Physical and Geometrical Parameters of Collector and Tube System

Parameter	System I (Pressurised)	System II (Non-pressurised)
Total collector area A_c	1.5 m^2	1.75 m^2
Transmission of collector cover glass		
Single	0.85	0.91
Double	0.73	
Absorption of collector plate	0.94	0.93
Collector heat loss coefficient U_L		
Single glass	4.5 W/m^2K	5.3 W/m^2K
Double glass	3.9 W/m^2K	-
Collector plate efficiency factor F'		
Single glass	0.7	0.82
Double glass	0.77	-
Collector efficiency η_c (averaged over the day)	48%	58%
Loss coefficient of the tank U_t	1 W/m^2K	1.9 W/m^2K
Medium tank height H	0.56 m	0.89 m
Inner diameter of the tubing d	15.10^{-3}m	$d_1 = 13.10^{-3}$m (d_1) (tubes) $d_2 = 25.10^{-3}$m (d_2) (collector)
Geometrical tube length:		6 x 1.18 m (d_1)(parallel)
Inside collector	9.65 m	= 2.69 m (d_2)
To and from tank	3.70 m	5.41 m (d_2)
Total geometrical tube length l	13.35 m	8.1 m (d_2)
Coefficient of resistance	2.85 (= 16 curves)	6.8 (6 knees with d) 45.5 (12 branches with d_1)
Mass flow	5 g/s	35 g/s (2 collector)
Flow velocity V	$2.85.10^{-2}$m/s	$2,2.10^{-2}$m/s inside collector (d_1) $3,6.10^{-2}$m/s outer tubing with d_1
Reynolds number R_e	540	360 (collector, d_1) 1130 (tubes, d_2)
Coefficient of friction	0.12	0.18 (collector, d_1) 0.057(tubes, d_2)

TABLE 3 - Main Thermal Characteristics of the Pressurised and Non-pressurised Natural Circulation Water Heating Systems.

Physical Quantity	SYSTEM I (Pressurised system)	SYSTEM II (Non-pressurised system)
Total incident energy per unit area of the collector surface	6.75 kWh/m^2	6.75 kWh/m^2
Total collector area	1.5 m^2	1.75 m^2
Total incident energy	10.1 kWh	11.8 kWh
Initial water temperature in the tank	15°C	15°C
Maximum average temperature of water in the tank at 19.00h	51°C	44°C
Tank capacity	110 1	200 1
Total useful energy	4.6 kWh	6.7 kWh
Average thermal efficiency of the system over 24 h	41%	47%
Average thermal efficiency of the system over daytime (8-20 h)	45%	57%
Fall in the average water temperature of the tank during night	3,7°C	5°C
Average heat loss coefficient during night (19-8 h)	1 W/m^2K	1.9 W/m^2K
Total heat loss overnight	0.46 kWh	1.16 kWh
Percentage of heat loss overnight	10%	17%

percentage of heat losses from the non-pressurised system is more than the heat losses in the pressurised system (in the case of no reverse flow), which has a better insulated tank.

Fig.10 shows the variation of the water temperature with withdrawal of water from the tank. The systems as described above were first filled with fresh water. No water was withdrawn on the first day. On the next two days, the following quantities of water were withdrawn at different times:

	System I	System II
7.30 h	33 l	60 l
12.30 h	16.5 l	30 l
19.00 h	44 l	80 l

As the average tank temperature after the last withdrawal in the evening is about the same as before the withdrawal in the morning, the total withdrawn energy from the systems over the day is nearly the energy taken up by the systems, i.e. the consumption profile corresponds to the capacity of the systems. The total energy withdrawn from the systems over the day was 3.8 kWh for system I and 4.45 kWh for system II, corresponding to nearly 40% efficiency for both the systems.

Recently, Norton and Probert (Ref.5) reported the measured performances of a few natural circulation solar energy water heaters. The study was made for a variety of ambient conditions and vertical heights of the storage tank above the collector. The absorbers used in the fabrication of the system were made of single glazed aluminium flat plate collectors using metallic black paint; their flow channels were of 0.025m internal diameter. The collectors were back insulated with a 0.055 m layer of polystyrene. The total harnessing area of one collector was 1.4 m^2 connected to a 0.1125 m^3 capacity water storage tank. Each of the collectors had fourteen riser channels. They were inclined at an angle of 60° to the horizontal. The plastic upriser and downcomer, each of 0.038 m internal diameter, were thermally insulated with foamed plastic. The surface area of the tank was 1.65 m^2, height 1.6 m and it was insulated using 0.02 m uniformly thick layer of glass fibre insulation.

Typical patterns of experimental data obtained by Norton and Probert (Ref.5) are shown in Figs.11 and 12 for two different heights of the tank. The results correspond to a very clear day with diurnal global insulation received on the collector plane being 6 kWh/m^2. The measurements were performed without withdrawing any water. It was observed that the pattern of temperature variation of the plate and hot water in the tank remains nearly the same in both the cases and both the units have similar diurnal heat gain efficiencies defined as $\eta_s = MC_w \; t_m / I_{total} A$, where Δt_m is the rise in the mean temperature of the hot water in the store during the period, i.e. approximately 24 h. However, for low insulation level a higher height of the tank (resulting in a multipass system) was more effective. Such a behaviour can also be achieved with a same level unit, if the fluid flow resistance is decreased and the total vertical height of the system is sufficient to allow a relatively high rate of thermosyphonic mass flow to occur. In a same level system

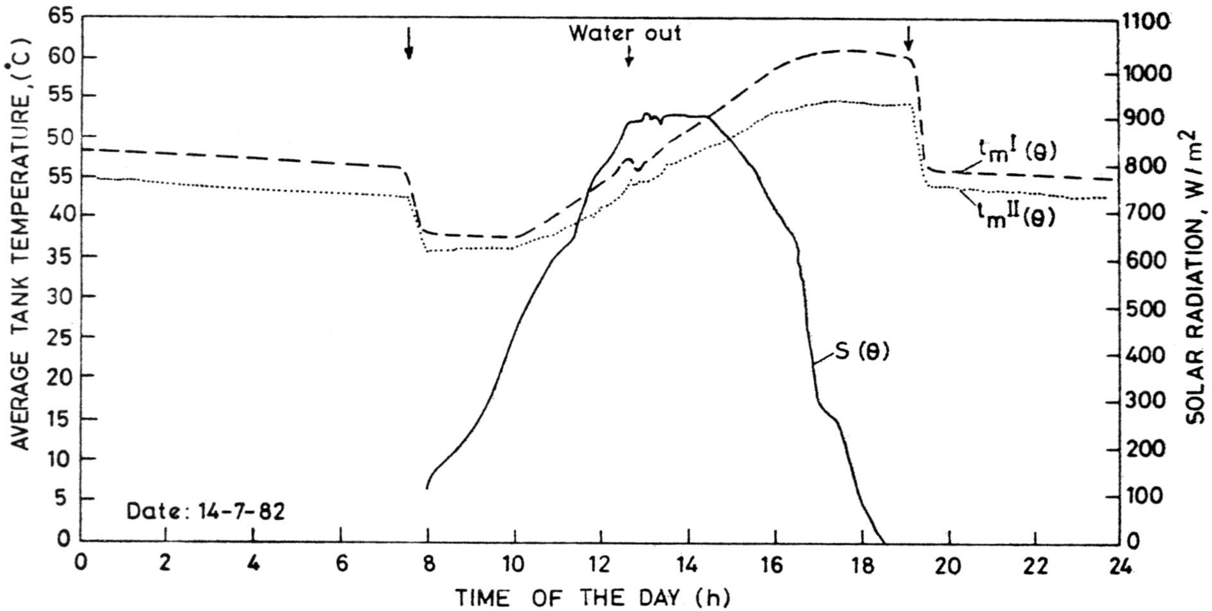

FIG.10 - VARIATION OF SOLAR RADIATION S(θ) AND AVERAGE TANK TEMPERATURE $t_m{}^{I}(\theta)$ IN SYSTEM I AND $t_m{}^{II}(\theta)$ IN SYSTEM II, IF WATER IS WITHDRAWN FROM THE TANK.

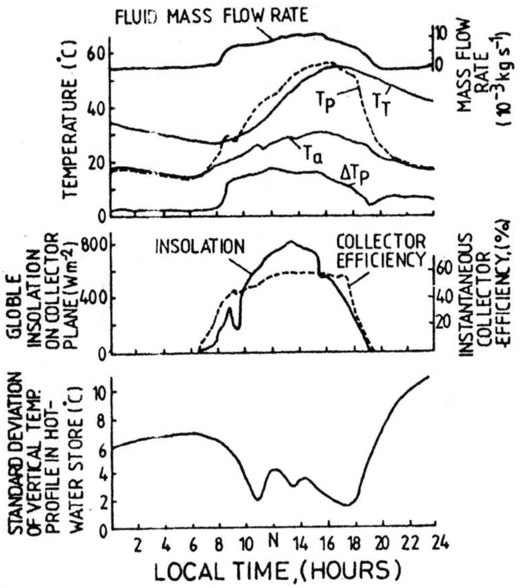

FIG.11 - DIURNAL TIME HISTORIES OF SYSTEM TEMPERATURES, MASS FLOW RATE, STANDARD DEVIATION OF VERTICAL TEMPERATURE PROFILE OF THE HOT WATER IN THE STORE AND THE INSTANTANEOUS FLAT PLATE COLLECTOR EFFICIENCY FOR AN N.C.S.E.W.H. WITH H_{TOT}/H_T=2.25. N.C.S.E.W.H.=NATURAL CONVECTION OF SOLAR ENERGY WATER HEATING SYSTEM. (REF.5)

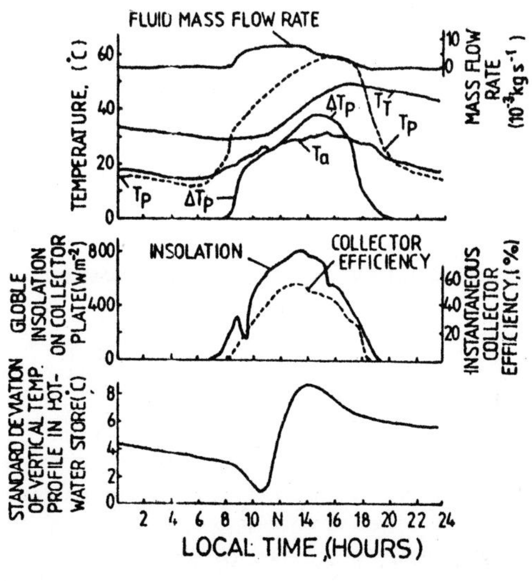

FIG.12 - AS FOR FIG.11 BUT WITH H_{TOT}/H_T=1.0 AND A ONE WAY CHECK VALVE INCORPORATED IN THE FLOW CIRCUIT. (REF.5)

(collector top and tank's bottom in the same plane), it is essential to incorporate a non-return valve with a low pressure differential (Norton and Probert, Ref.6) to prevent reverse circulation during night time. From various other studies like that of Close (Ref.1), Shitzer et al. (Ref.7) and Young and Bergquam (Ref.8), one may arrive at a conclusion that nearly 1 m^2 collector area is required for about 70 l of hot water storage tank capacity. The tank capacity should in turn be 1.5 to 2.5 times the daily hot water demand; the exact sizing depend on the climate. The system described in Fig.4 and the performance of which is discussed above are therefore really underdesigned. One should however exercise a caution that the actual system size should be determined for each specific location, and for typical users profile. The use of the described mathematical model in that context will play a significant role in optimizing the size of the collector area corresponding to the desired storage capacity and the typical hot water consumption.

5.1.1 Design variations. There are many design variations of solar water heating systems with thermosyphonic flow. As far as the number of manufacturers and design sophistication is concerned, Japanese are probably the world leaders. One such firm, known as Matsushita, has developed "Keeping Warm" units shown in Figs.13 to 16 (Ref.6 and 9). The basic difference between the designs discussed earlier and this design is the incorporation of two collectors inclined at different angles; one inclination collecting maximum energy in summer and the other in winter. In these designs, a special feature is the prevention of reverse circulation by providing inlet from the tank to the collector at the top of a limb (Figs.15 and 16) of the horizontal store. The limb of the tank prevents reverse circulation by being at a mean temperature above that of the collector at night. With this arrangement one has to however make sure that the tank limb does not attain too high a temperature during the insolation periods. If this happens, then the thermosyphon flow will be smaller, resulting in higher outlet temperatures and the reduced efficiencies. By feeding the cold water from the limb, the limb is kept cool. This arrangement may not be suitable for a system with a vertical tank as it would require the maintenance at a low temperature during the day of a more thermally stratified vertical body of water. Reverse circulation to the lower absorber of "Keeping Warm" type thermosyphon water heater is prevented solely by the height difference between the absorber and the store (Ref.6).

Some units incorporating a plane mirror reflecting the direct insolation on to the absorber and thus enhancing the performance have also been designed (Ref.10).

To overcome the conditions of water freezing in these compact units, one usually drains the systems manually, if subzero ambient temperatures are anticipated. One can also design the collector so that flow channel can expand to safely accommodate freezing water (Ref.9,11,12). If such low temperatures are likely to be experienced, attention should also be given to the design of the system for accommodating differential contractions of metallic structures and glass components.

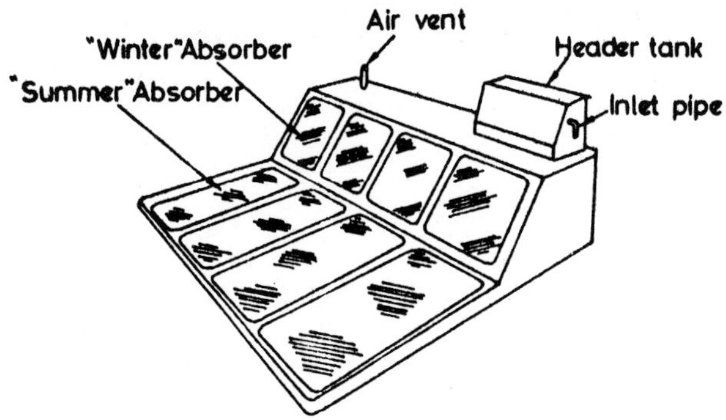

FIG.13 – EXTERNAL APPEARANCE OF THE MATSUSHITA "KEEPING WARM" TYPE COMPACT THERMOSYPHON SOLAR WATER HEATING SYSTEM.

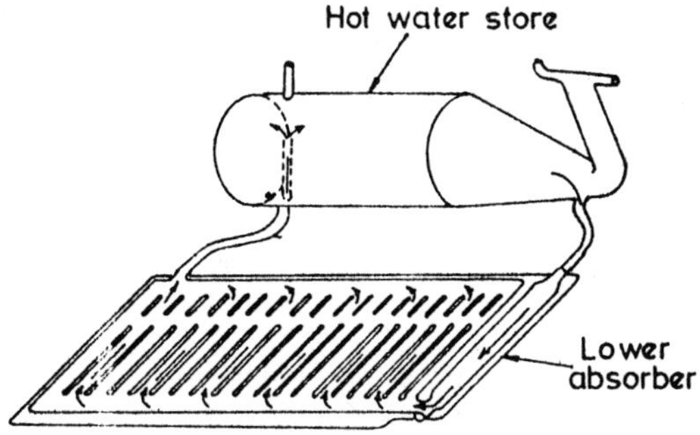

FIG.14 – THERMOSYPHON FLOW THROUGH THE LOWER ABSORBER AND THE STORAGE TANK OF THE "KEEPING WARM" TYPE SOLAR WATER HEATER.

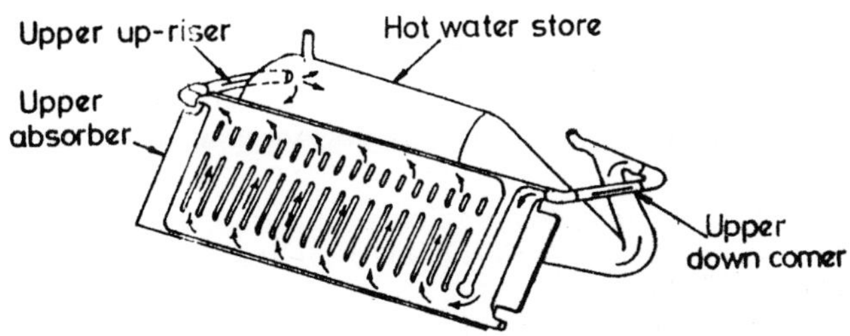

FIG.15 – THERMOSYPHON FLOW THROUGH THE UPPER ABSORBER AND STORAGE TANK OF THE "KEEPING WARM" TYPE COMPACT SOLAR WATER HEATER. THE ARROWS INDICATE THE WATER FLOW DIRECTIONS DURING THE SOLAR ENERGY GAIN PERIOD.

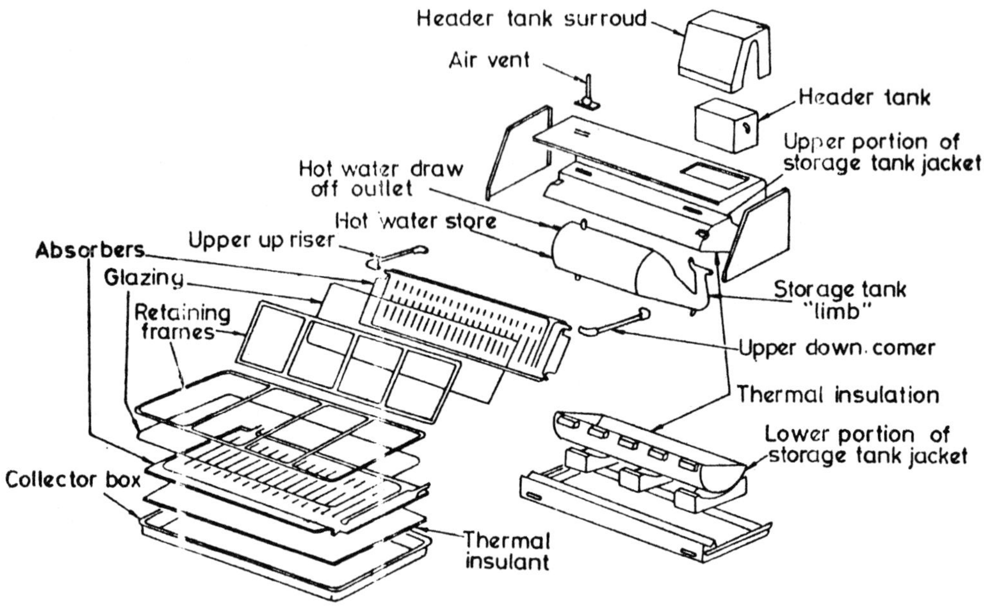

FIG.16 - EXPLODED VIEW SHOWING THE COMPONENTS OF THE MATSUSHITA "KEEPING WARM" TYPE COMPACT SOLAR WATER HEATER.

5.2 Closed Loop Systems

Actual performance data on closed loop thermosyphon water heating is really scarce. Mertol et al. (Ref.13), however, developed an analytical model for liquid solar thermosyphon systems with heat exchangers and studied the performance of such systems. One of the prime requirements in closed loop systems is the choice of heat transfer fluid in the collector loop. The heat transfer fluid should be nontoxic, i.e. it must have a Gosselin toxicity rating of 1, which means that the likely lethal dose for a human being is greater than 15 g/Kg body weight. In addition to being nontoxic the fluid must be chemically stable over the operating temperature range of the system and also be non-freezing at very low ambient temperatures. While p-glycol meets the toxicity and antifreeze requirements (Ref.14), it has a tendency to decompose into acids when exposed to high temperatures and oxygen. Some recent industry tests, however, indicate that p-glycol containing proper inhibitors can operate at temperatures up to 175°C for extended periods without significant degradation (Ref.15).

The toxicity hazard of p-glycol in thermosyphon systems can be minimized by diluting and using a limited amount of liquid in the collector loop. The liquid in addition has to be diluted before use. A simple calculation assuming 0.0075 m^3 (2 gal) of p-glycol and 0.15 m^3 (40 gal) storage tank shows that homogeneous dilution in the storage tank of the collector fluid will produce a dose of less than 0.1 gm/Kg body weight, which is well below the lethal dose of fluids with a toxicity rating of 1. P-glycol may therefore be safely used for solar water heating systems employing a single wall heat exchanger (Ref.16). Ethylene glycol, being readily and cheaply available, is a frequent choice as the heat transfer fluid. Ethylene glycol is, however, toxic and necessitates the use of double walled heat exchanger. This, however, results in still lower efficiencies. Usually p-glycol solutions containing necessary inhibitors are found to be a good choice for use.

Mertol et al. (Ref.13) developed an analytical model for the closed loop thermosyphon systems by writing continuity, momentum and energy equations for each of the components of the system. The equations were solved numerically for a system with two collectors (of 1.95 m^2 area each) connected to 300 litres tanks. Each of the collectors has $\tau\alpha = 0.86$ and $F'U_L = 4.2\ W/m^2\,^oC$. The results of a computer programme, developed using finite difference method for solving the equations and the trapezoidal rule to calculate the integrals, over a number of days show that at least three days are required to attain essentially a steady state performance of a system which includes the water withdrawal. Another important result of the study is that the effective heat transfer coefficient from heat exchanger to the fluid in the tank is not much affected by the fluid flow rate in the collector circuit.

An indication of the effect of heat exchanger on system's performance can be obtained from Figs.17(a)-17(b) as a function of the parameter r defined as

$$r = \frac{HTR}{1+HTR}$$

where $HTR = (UA)_{he}/(UA_c)_c$

$(UA)_{he}$ and $(UA_c)_c$ being the total heat exchanger and collector's heat loss coefficients respectively and defining a cumulative efficiency η_{cu} as

$$\eta_{cu} = \frac{\text{Total energy transferred to the storage tank}}{\text{Total energy incident on the system}}$$

The maximum cumulative efficiency, $\eta_{cu,max}$ thus determined is then used to set the asymptotic limit for Fig.17(b), which shows plot of normalised system's efficiency (third day) vs HTR for the cases with and without draw. The results of the figure suggest that a good performance of the system could be achieved for a thermosyphon with single heat exchanger composed of straight tubes and p-glycol in the collector loop.

Regarding the effect of tank stratification, the results of the model show that a condition of maximum tank stratification (about 11°C temperature) results in a system performance about 10 per cent better than for the condition of a fully mixed tank.

A number of important conclusions may be drawn from the study of Mertol et al. (Ref.13) on this domestic size (0.302 m^3 tank and 3.9 m^2 of collector area) closed loop thermosyphon water heating system:

- the performance of a closed loop heat exchanger system could be 90 per cent of that expected from a standard thermosyphon without a heat exchanger;

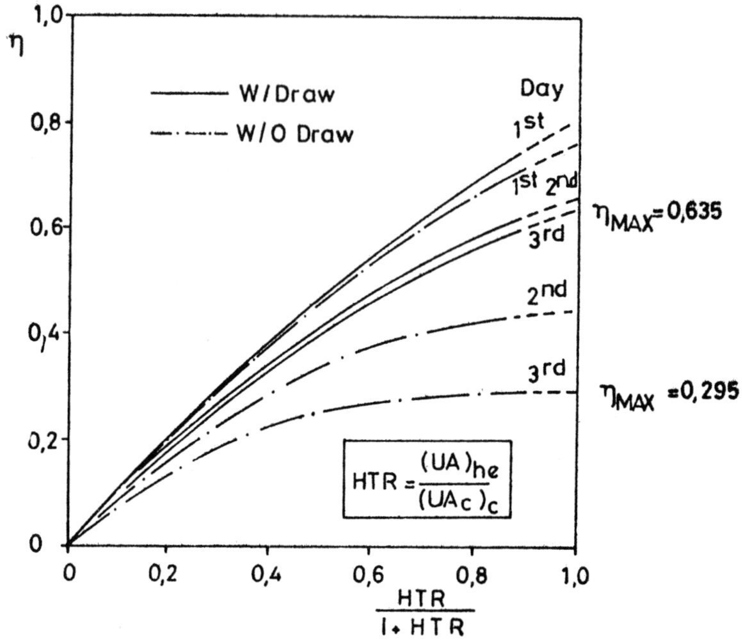

FIG.17(a) - EFFECT OF HEAT EXCHANGER ON SYSTEM PERFORMANCE

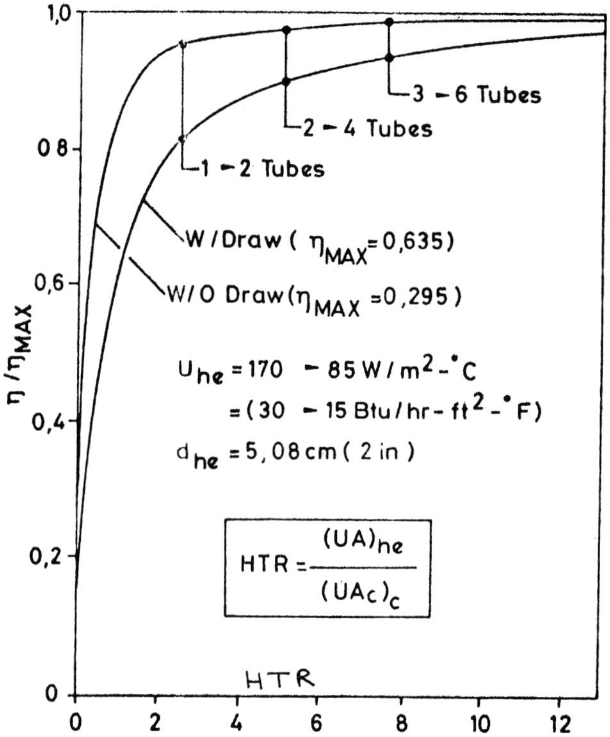

FIG.17(b) - EFFECT OF HTR ON SYSTEM EFFICIENCY

- the system's performance is relatively insensitive to the tank stratification. Detailed tank models may, therefore, be not necessary;
- varying tank elevation relative to the collector showed no significant effect on the 24 h performance until the bottom of the tank was below the top of the collector. The daytime performance, however, remains unaffected by tank elevation. For lower tank elevation, reverse flow occurs during night resulting in substantial thermal losses from the tank;
- different collector tube cross-sections though affect the flow rate (30% when increasing tube diameter from 6 mm to 12 mm), the daily average performance nearly remains the same;
- reduction in the connecting pipe size results in a considerable reduction of mass flow rate (87 per cent from 25 mm to 6 mm), the day's thermal performance reduces by 16 per cent.

A system has been developed in which water containing a corrosion inhibitor flows in a closed circuit, comprising a solar collector and a shell and tube heat exchanger. Experimental observations on this system have shown that the influence of the heat exchanger on the overall system performance was decreased by reducing the ratio of the collector area to the heat exchanger surface area.

Another novel thermosyphon water heating systems involve black heat transfer fluid (Ref.17) flowing in the gap between collector plate and the glazing.

6. TEST METHODS FOR THERMOSYPHON SYSTEMS

In contrast to the very well adopted test methods for the solar collectors (Ref.18, 19, 20), the standard test procedures for complete systems are still in the evolving stage. There have been draft proposals but a final version is still to be adopted. ASHRAE (Ref.21) has proposed a standard test procedure for domestic water heating systems of 0.45 m^3 (120 gal) storage capacity or less and for demands of the order of 0.38m^3 (100 gal) of hot water per day or less.

The requirements for instrumentation and other measurements are the same as prescribed in the ASHRAE Standard 93 (Ref.20). The test procedure recommended by ASHRAE applies to three categories of solar domestic hot water systems, namely (i) solar only systems, (ii) solar preheater systems, and (iii) solar plus supplemental systems. The test can be done by assembling the complete system in the laboratory and irradiating the collector array by use of a solar simulator. The test procedure of ASHRAE do not take into account the freezing conditions.

6.1 Installation Requirements

Test, on the system, is to be performed with the system components installed in accordance with the manufacturer's published installation instructions. The collectors are normally mounted remote from the storage, the test shall be performed with the total connecting pipe length between the storage tank and the collectors equal to 15 m (50 ft),

7.5 m (25 ft) in both the supply and between lines. The connecting piping and ducting shall be insulated in accordance with the manufacturer's instructions for installation. The collectors have to be mounted at the tilt angle specified in an associated rating standard.

6.2 Forcing Parameters

The incident solar radiation conditions to be used in the test should be specified for the test solar day as per example given in Table 4.

TABLE 4 - Example of Test Solar Day Conditions

Solar Time (h)	Ambient Temperature (°C)	Irradiance (W/m^2)	Incident Angle (deg)
8.00	20	233	62.0
9.00	20	539	48.4
10.00	20	747	35.5
11.00	20	873	24.8
12.00	20	917	20.0
13.00	20	873	24.8
14.00	20	747	35.5
15.00	20	539	48.4
16.00	20	233	62.0

Similarly the ambient air temperature surrounding the collector array has to be specified. The average ambient air temperature at the storage tank and components during the test should be controlled to a value specified in an associated rating standard ± 2°C on a continuous 24 hours basis. An example is given in Table 5. The water supply temperature to the system has to be controlled to a value t_{main} within ±1°C(± 1.8 F). Water has to be withdrawn as per specifications given in Table 5.

TABLE 5 - Example of Test Conditions for Solar Hot Water Systems

t_{main}	: 20°C
t_{set}	: 55°C
Volume flow rate during hot water draw	: 0.2 l/s
Duration of hot water draw	: 7 min
Time of hot water draw	: 800h, 1200h, 1600h
Ambient air temperature around solar hot water system components during the test	: 20°C
Tilt angle of collectors	: 45°
Time at the beginning of the first and subsequent test days	: 1600 h

7. MODIFIED THERMOSYPHON WATER HEATING SYSTEM

In thermosyphon systems described above, the circulation of the collector liquid starts at small temperature difference between hot and cold column. Therefore, the temperature of water contained in the storage tank rises slowly and rather uniformly over the whole volume. Only after a few hours of insolation, higher temperatures will be achieved.

In order to make the water available at the specified temperature, Boettcher et al. (Ref.22) suggested a modification of the normal natural convection water heating systems. In this modification, the collector liquid enters the storage tank only after reaching the specified temperature even if the whole system is cold. This was achieved by sparing the liquid of the collector and the storage side by an overflow area in the highest part of the connecting tube (Fig.18). As soon as insolation starts the liquid of the collector is heated and expands until it has reached the overflow level and starts flowing into the storage tank. The present distance between the level of the cold liquid and the overflow level Δh determines the temperature difference. Δt of the cold and hot liquid-columns in the moment of starting overflow. This happens the earlier the smaller the volume of liquid contained per m^2 of collector is. For this reason high vacuum tube collectors are specially suited for such systems independent whether they have direct cooling (Fig.18) or a central heat pipe (Fig.19).

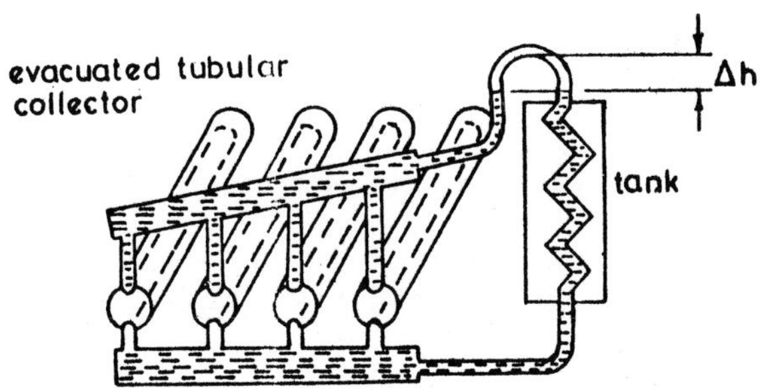

FIG.18 - PRINCIPLE DESIGN OF A CONTROLLED THERMOSYPHON FLOW SYSTEM

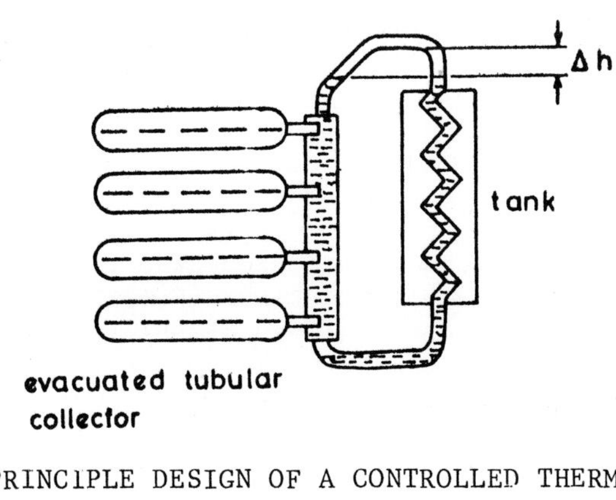

FIG.19 - PRINCIPLE DESIGN OF A CONTROLLED THERMOSYPHON FLOW SYSTEM

Contrary to the normal thermosyphon systems, the temperature of the collector liquid entering the storage tank, in this case, does not depend strongly on the insolation rate, but it is nearly constant. Instead, the quantity of collector fluid entering the storage tank under standard conditions is approximately proportional to the insolation rate.

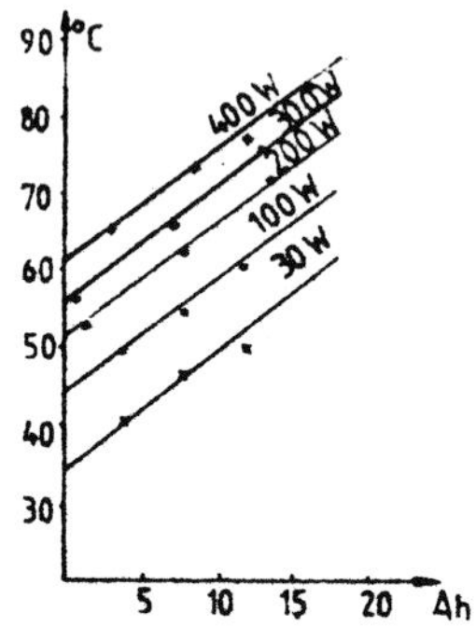

FIG.20 - CORRELATION OF Δh AND COLLECTOR-OUTLET TEMPERATURE FOR DIFFERENT HEAT-RATES

The results of a simulated experiment using electrical power to heat the collector tubes are given in Fig.20. An electric power between 30 W to 400 W per tube corresponding to solar insolation of 250 to 1000 W/m^2 and being reflected by mirror boosters of concentration 3 was used for the experiments. The main results of the simulated experiments were:

1. In case of high flow resistance in the system, as it resulted with several collectors assembled in series, boiling starts within the tube for higher insolation rates. The resulting two phase flow produces high circulation speed. In this case, the temperature of the fluid entering into the storage tank from the collector does not depend upon the magnitude of Δh but simply on the boiling temperature of the liquid at a given pressure.

2. For a given thermal power per simulated collector tube, the resulting ΔT between the hot column and the cold column and hence the collector outlet were approximately proportional to the present Δh. For both types of systems (Figs. 18 and 19), the resulting ΔT as well as the fluid temperature at the collector outlet increased slightly with growing thermal power applied per simulated collector tube.

3. For sudden changes of the electric power, heating the tube, periodic oscillations of the temperature of the liquid at the collector outlet occurred especially when the collectors were mounted in parallel. The oscillations, however, did not have any substantial influence on the system's performance.

Based on the experience of the above simulated experiments, outdoor experiments were performed by Boettcher et al. (Ref.22) on a system consisting of 6 vacuum tube collectors (1.13 m^2 absorber area) connected in parallel and possessing non-tracking mirror boosters with a concentration of 3. The collectors are connected to a 225 litre indirectly heated storage tank (Fig.21); the heat exchanger has a specifically high exchange area in the upper part of the tank. The performance of the system for continuous withdrawl of water at the rate of 10 l/h is shown in Fig.22. It is seen that the temperature of the collector fluid (water) rises steeply in the morning hours, reaches a plateau between 12.00 h and 15.30 h (German summer time) and falls back with decreasing insolation. The temperature of water in the upper region of the storage tank also rises quickly and reaches a flat maximum between 15.00 h and 17.00 h.

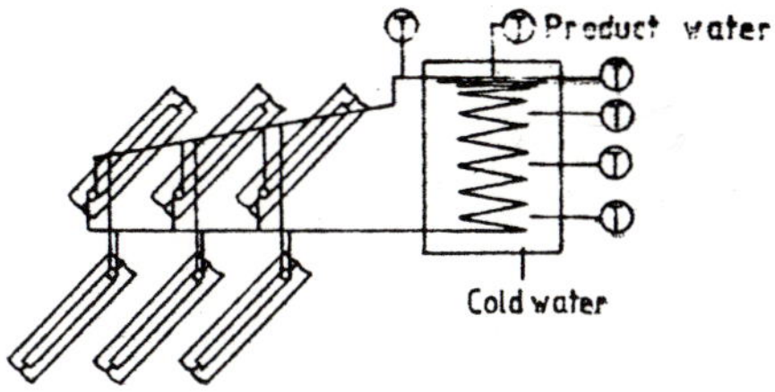

FIG.21 - SCHEMATIC DESIGN OF THE EXPERIMENTAL UNIT

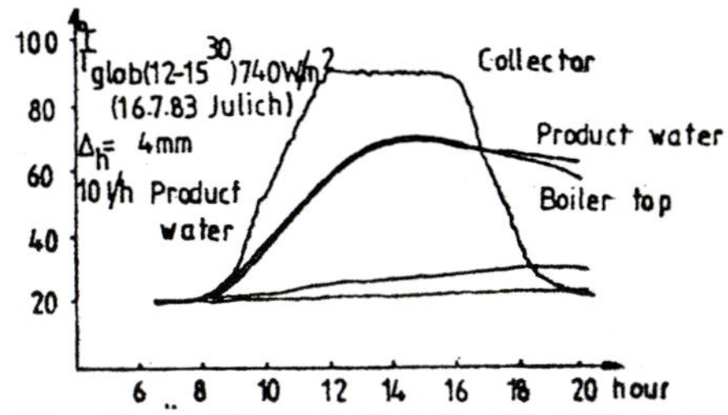

FIG.22 - TEMPERATURE IN THE PRIMARI AND SECONDARY CIRCUIT (SCHEMATIC)

If one increases Δh to 12 mm, the temperature at the outlet of the collectors reaches values of 115-120°C, thus yielding higher temperatures in the tank.

REFERENCES

1. D.J. Close, (1962), 'The performance of solar water heaters with natural circulation', Solar Energy, 6, 33-40.
2. C.L. Gupta and H.P. Garg, (1968), 'System design in solar water heaters with natural circulation', Solar Energy, 12, 163-182.
3. K.S. Ong, (1976), 'An improved computer programme for the thermal performance of solar water heater', Solar Energy, 18, 183-191.
4. R. Uhlemann and N.K. Bansal, (1985), 'A study of pressurised and non-pressurised solar water heating systems with thermosyphonic flow', Solar Energy (In Press).
5. B. Norton and S.D. Probert, (1983), 'Measured performance of natural circulation solar energy water heater design', Applied Energy, 16(1), 1-26.
6. B. Norton and S.D. Probert, (1983), 'Recent advances in natural circulation solar energy water heater designs', Applied Energy, 15(1), 15-42.
7. A. Shitzer, D. Kalmanoviz, Y. Zuirin and G. Grossman, (1979), 'Experiments with a flat-plate solar water heating system in thermosyphonic flow', Solar Energy, 22, 27-35.
8. M.F. Young and J.B. Bergquam, (1981), 'Performance characteristics of a thermosyphon solar domestic hot water system', J. Solar Energy Engineering, 103, 193.
9. H. Iwata, A. Morie and T. Masuda, (1980), 'Development of a thermosyphon solar water heater', Technical Report of the Matsushita Electric Works Ltd. No.20, pp 43-49.

10. H. Andoh, (1981), 'Natural Circulation Solar Water Heaters', National Technical Report, 27(3), 36-41.
11. S. Buckley, (1976), 'Thermal diode solar panels: Passive and modular' Proc. Passive Solar Heating and Cooking Conf., Albuquerque, New Mexico, USA, pp. 293-299.
12. J.M. Bradley, (1977), 'The development of a freeze tolerant solar water heater using cross linked polythylene as a material of a construction', Report COO/2959/8, Division of Solar Energy, ERDA, Washington D.C., USA.
13. A. Mertol, W. Place and T. Webster, (1981), 'Detailed loop model (DLM) analysis of liquid solar thermosyphons with heat exchangers', Solar Energy, 27, 367-386.
14. J.Q. Searcy, (1978), 'Hazardeous properties and environmental effects of materials used in solar heating and cooling (SHAC) technologies', Interim Handbook, Report DOE EV-0028, Department of Energy, Washington D.C., USA.
15. Houston, (1978), 'Solar heat transport fluids for solar energy collection system', Report No.DOE/NASA CR-150569, Houston Chemical Company, Division of PPG Industries, Inc., USA.
16. ANSI, (1979), American National Standard, Solar Heat Exchangers, ASME/SESI.
17. Y. Meas, J. Quintana, A. Samano and A. Fernandes, (1981), 'Black fluids, A new way to collect solar energy', Proc. Annual Meeting of Am. Sec. of ISES, Philadelphia, USA, 4(1), 231-234.
18. J.E. Hill, E.R. Stread, G.E. Kelly, J.C. Geist and T. Kusuda, (1976), 'Development of proposed standards for testing solar collectors and thermal storage devices, NBS Technical Note 899.
19. BSE, (1978), 'Guidelines and directions for determining the usability of solar collectors', Bundesuerband Solar Energie.
20. ASHRAE, (1977), 'Methods of testing to determine the thermal performance of solar collectors', ASHRAE Standard 93-97, American Society of Heating, Refrigeration and Airconditioning Engineers, Inc., Atlanta, Georgia, USA.
21. ASHRAE, (1981), 'Methods of testing to determine the thermal performance of solar domestic water heating systems', ANSI/ASHRAE-95-81, American Society of Heating, Refrigeration and Airconditioning Engineers, Inc., Atlanta, Georgia, USA.
22. A. Boettcher, H. Heybutzki and W. Krug, (1982), 'Self-regulating solar water heaters with very short start up phase', Solar World Congress, Perth, Australia.

HOW TO GET THE MOST SOLAR HEATED WATER FOR THE LEAST COST IN INDIA

Walt Chappell
President, Solar Company, USA, and
Former Communications Branch Chief
Solar Energy Research Institute, USA

1. INTRODUCTION

I offer the following recommendations to the Indian Solar Industry as methods of assuring optimum solar performance and quality. These design concepts are for Indian solar hot water installations and are meant to reduce long term maintenance costs for the manufacturers and contractors while increasing overall performance for the user.

Each of these areas of the solar hot water system are essential and need to be incorporated in the design before the bid is tendered. Quality materials installed using appropriate procedures will then assure that the system performs as designed. Close supervision during installation and commissioning is vital to minimize costly errors which will have to be corrected later to get optimum solar heat.

2. ELIMINATE HEAT EXCHANGERS

Wherever possible, remove the heat exchangers from the collector loop and plumb the supply and return pipes directly to the storage tanks (Fig.1). The performance will improve by 10 to 30%. The only time heat exchangers may be required in India is to take heat out of water to transfer to air, or when a toxic heat exchange fluid is required to keep the collectors from freezing. Neither case applies to most Indian hot water installation. Hard water scaling or corrosion should be minimal at most sites.

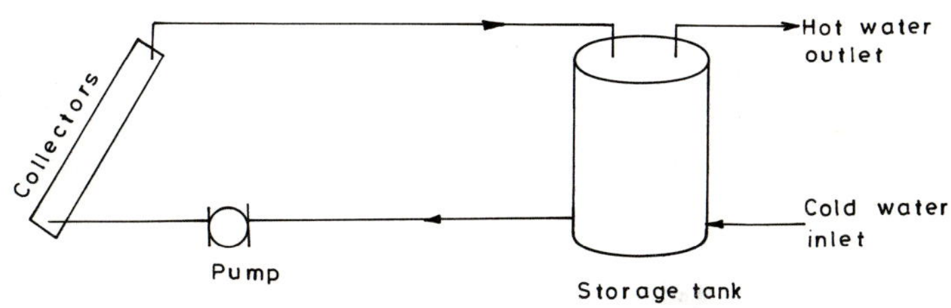

FIG.1 - SOLAR WATER HEATER WITHOUT HEAT EXCHANGER

H. P. Garg (ed.), Solar Water Heating Systems, 327–335.

In cases where heat exchangers are necessary to prevent freezing or to reduce the effects of "hard water", make sure that adequate heat exchange surface area is provided for the most effective heat transfer. In large collector systems, very large heat exchangers are necessary. To increase the heat transfer surface area, two or more exchangers should be plumbed in parallel. This will reduce static pressure and help maintain optimum flow rates.

3. USE PRESSURIZED TANKS

Seal the storage tanks so they can be pressurized and plumbed in a series cascade. That way, the user is always drawing the hottest water from the top of the second tank. The collectors will also work more efficiently since the coldest water at the bottom of the first tank will be circulated to provide more rapid heat exchange. The plumbing schematic as shown in Fig.2 should be used. Notice the placement of the by-pass gate valves to isolate the solar system from the domestic supply. This arrangement will allow repair work on the solar loop without interrupting the user's supply. Isolating the system is also helpful to demonstrate system performance.

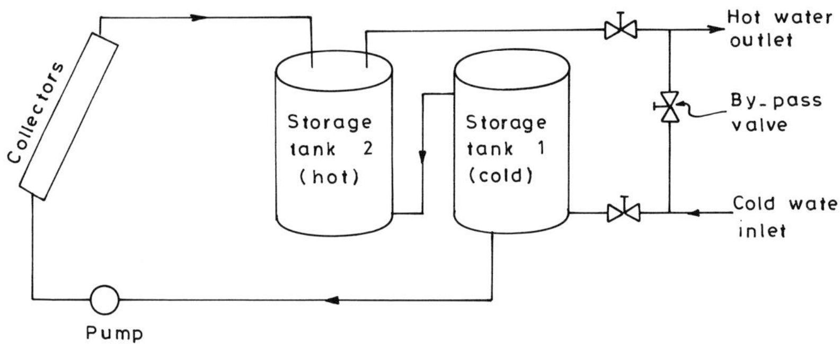

FIG.2 - TWO TANK SOLAR WATER HEATER SYSTEM

Not only will the collectors perform better, but there will also be less maintenance due to less oxygen in the tanks to corrode float valves and the inside of the tanks. Use a pressure temperature relief valve on the top of the tank set at 100 psi. If the collector loop is isolated from the tank by a heat exchanger, then the heat transfer fluid should be pressurized to 25 psi.

The storage tank should also be mounted inside the building out of the weather. This procedure will reduce heat loss at night, eliminate the chance that rain water will ruin the insulation material, plus reduce the expensive labor and material cost of adding aluminium sheeting over the tank. The tank should be located as close to the ground as possible with a vent pipe extended from the pressure temperature relief valve to a water floor drain or outside the building.

4. USE ADEQUATE STORAGE

It should be the primary goal to design and instal solar hot water systems which provide the least cost per million BTU's. The only way to achieve that goal is to constantly remember that the sun only gives a fixed amount of heat any given day. Each component in the system should be designed to transfer that energy and save it for the use it was intended.

One key way of meeting this goal is to provide adequate storage. This means one gallon of storage per square foot of collector if very hot water is needed. Use one and one half gallons per square foot of collector if medium heat is to be stored. For most efficient collector performance for a large quantity of hot water, use two gallons of water per square foot of collector.

5. PLUMB COLLECTORS IN COUNTER-FLOW

Use a counter-flow plumbing schematic to balance the pressure drop across the total array. This technique will prevent short circuiting water through the front rows of the collectors while providing only minimal flow to the majority of the panels further from the plumb.

In addition, various valves should be drawn into the plumbing schematic provided to the installers prior to starting the job. For the collector loop, a one-way check valve should be installed. The pressure-temperature relief valve (PTRV) on the storage tank should be shown. Likewise, if a heat exchanger isolates the collectors from the storage, then a second pressure-temperature relief valve is needed in the collector loop to release pressure if the expansion tank is too small.

For large arrays of over 10 collectors, balancing valves are needed to assure equal flow rates through each array of collectors in the system. Drain valves should be installed at the top opposite end of each array to verify that equal flow has been achieved. Likewise, an automatic air purging valve should be placed at the highest point of the system to bleed off air which is trapped in the plumbing. Isolation gate valves should be used at various points to allow repairs or to stop water leaks. At least three of these gate valves should be installed as a way for the customer to by-pass the solar system as shown in section 3.

6. PLUMB NO MORE THAN (5) COLLECTORS IN SERIES

Plumb the collectors in two (2) parallel sets of four (4) in series. This procedure will reduce friction loss, and maintain optimum flow rates without overworking the pumps. More importantly, any more than five (5) collectors in series causes the additional panels in the series array to add negligible heat to the water passing through. As a result, the customer who buys 9 panels in series only gets heat from 5. The last 4 collectors lose more heat than they gain.

This problem is very acute on larger systems where 100 panels have been purchased but in the series arrays of 10 collectors only 50 are able to operate efficiently. The other 50 lose most of their heat to the outside air.

The schematic (Fig.3) shows how the counter-flow, series/parallel plumbing is done for a (24) collector systems in (3) arrays of (2) parallel sets of (4) collectors in series.

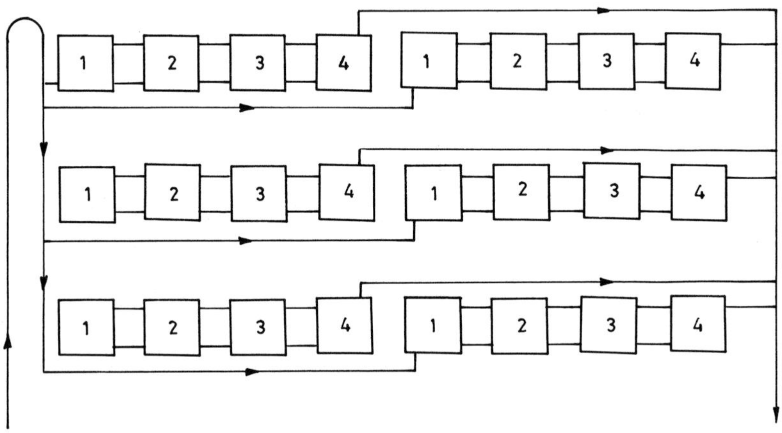

FIG.3 - RECOMMENDED WAY OF CONNECTING BANK OF COLLECTORS

7. SIZE PIPES AND PUMP TO MATCH OPTIMUM COLLECTOR FLOW RATE

Quality pays: This is especially true in solar hot water systems which are installed to last 10 to 20 years in very hot sun, with high winds and dust storms to test their strength. Quality designs, materials, manufacturing and installation all pay high profits in satisfied customers and low maintenance costs.

One key place for quality is to properly size the pipes and pump to supply the appropriate flow rate through each and every collector in the total system. For example:

If each collector is most efficient at 150 liters per hour (l.p.h.), then a 100 collector system plumbed in (10) arrays of (2) parallel sets of (5) collectors in series would require:

(150 l.p.h. per each 5 collectors in series) x
(20 parallel sets) = 3,000 l.p.h. for the pipes and pump to provide for the total system flow rate.

Next determine the pipe size based on the pressure drop across the total system including all series/parallel collectors, elbows, pipe runs of different diameters and heat exchanger if required. Then check the curves of different pumps to find the size which matches the total system flow rate required at the pressure drop in the plumbing.

Also insist that your engineers use precast elbows in their calculations and have the installers use these fittings at the job site rather than cut a pipe and weld it at right angles. Abrupt, welded joints greatly restrict the flow and throw off the pump and pipe sizing calculations substantially.

8. ALWAYS USE A DIFFERENTIAL CONTROLLER

To get maximum daily output from an active system, a differential controller should be used to compare the temperature between the bottom of the storage tank and the collector outlet (Fig.4).

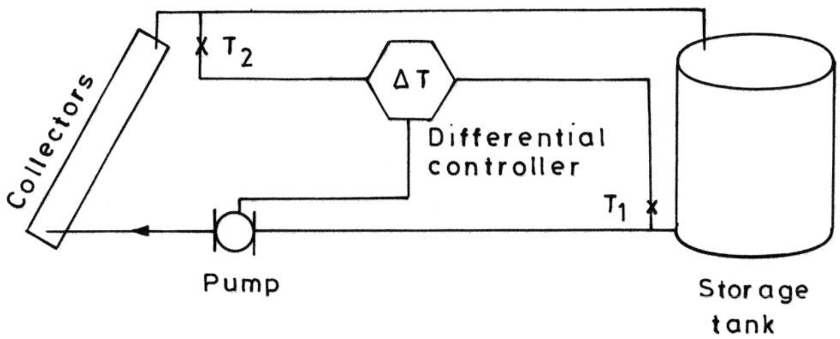

FIG.4 - POSITIONING OF DIFFERENTIAL CONTROLLER

Using fixed setting control logic turns the system on much later in the day and shuts the collectors off too soon in the afternoon. As a result, these installations are only pumping 3 hours per day. If the customer is using hot water during the day and the storage is sized for one square foot of collector per one to two gallons of water, the pump should stay on while adding heat to the storage for at least 4 to 6 hours per day.

The differential controllers with different ratings are now available in many countries including India. For large capacity pumps, relay which is operated by the controller should be of proper capacity.

One of the biggest problems which the solar industry creates for itself is to build up false expectations for its customers. An example is to assume that solar energy is free - therefore the systems should be free or lower priced than can be profitably installed.

Anyone can build a collector which stagnates or performs inefficiently to produce high temperatures at the lowest bid price.

However, high collector outlet temperatures or high storage tank temperatures are no sign of a good system. In fact, if the storage is undersized or the collectors are inefficient, these high temperatures may be a sure sign of a poorly designed or installed system.

Customers should not be led to expect 60°C temperatures all the time, nor should the systems wait until 60°C before turning on the pump. Depending on the quantity and frequency of the use of the water, the storage temperatures will vary greatly.

At no time should an efficient solar hot water system take 30°C water and on a single pass produce 60°C output temperature. The key variables are the flow rate and total system efficiency over the period of a day - not how hot the collector output gets on one pass or the tank temperatures at any given time.

In fact, an efficient, well designed and balanced flat plate collector system which matches the customers use pattern may never reach 60°C and be performing better than expected. Only by isolating the system with the gate valves and checking the BTU's stored versus the BTU's supplied by the sun during a 2 hour period can the total system performance be determined. It is then the system which provided the most BTU's (not temperature) for the least cost which is the best choice.

9. PROJECT SYSTEM WITH ADEQUATELY SIZED EXPANSION TANK

For a large hot water system, special consideration needs to be given to an expansion tank and the pressure relief valves. The daily temperature swings will be significant - especially if the collectors should stagnate. The volume of water in the collector loop expanding due to temperature rise from 30 degrees C up to the collector's stagnation limit (± 200°C) plus 20% margin should be provided. Without proper allowance for fluid expansion, particularly in large arrays under stagnation conditions (caused by such things as power failure, or controller or pump problems), the pressure relief valves become fountains, joints burst and costly repairs result.

Diaphragm type expansion tanks can be purchased on the world market to maintain a constant pressure at different temperatures. An expansion tank similar to the ones used with boiler heating systems can be substituted if adequate air space is allowed at the top of the tank.

In any case, make sure a pressure relief valve is installed to protect the collectors and plumbing. Taking collectors apart to rebraze risers to headers can be a very expensive repair job which could have been prevented with an appropriately sized expansion tank and a Rs.120 pressure-temperature relief valve (PTRV).

10. USE SILICONE CAULK FOR SEALING COLLECTOR JOINTS AND GLAZING

To prevent water leakage into collectors, either at the joints or under the glazing rubber gaskets, use a silicone caulking compound such as Dow Corning No.795. Approximately one tube per collector is needed to dispense the sealant. Purchase caulk guns to apply the silicone in a thin bead.

For approximately Rs.60 (US$5) per tube when purchased wholesale in 12 or 24 tube cases, this caulking will last for years to seal out water and dust. It stays soft to expand with the hot metal and glass and is not deteriorated by direct sunlight or high collector temperatures.

11. COLLECTORS MOUNTED AT LATITUDE FOR YEAR ROUND HOT WATER

To get optimum performance from your installations for year round hot water heating, the collectors should be mounted at the latitude of the site. Since the sun's arc is seasonal, the best fixed angle for year round use is equal to the site latitude. The only time latitude plus 15 degrees would apply is when the primary use of the heat from the collectors is during the winter months when the sun is lowest in the sky.

In fact, many of the most important design decisions in any solar energy system are site specific. Before bidding on a tender notice, the installing contractor must visit the site and engineer the equipment to fit the constrains of the site. Only then can the cost of installation be determined and an adequate profit margin be added to cover expenses and stay in business to back the warranty. The necessary wiring, plumbing, mounting, structural supports, etc., can be expensive costs to a proper installation.

Only by going to the site where the sytem is to be installed can these costs be calculated, after the site variables are taken into account. For large systems which require detailed engineering before a bid can be given, the customer who wants the system should be willing to pay the engineering expenses to design the system and calculate the installed costs. At that point, the customer can make an informed decision whether the solar price will save enough heat to be justified.

12. ALLOW FOR ADEQUATE EXPANSION IN THE PLUMBING

Pipe connections and collector brazed joints can be greatly stressed as the sun heats up the system. Aluminium, steel, copper and glass all expand at different rates. Find the expansion coefficient of each material and allow enough room for this movement without causing glass to break against the metal frame or pipe to bend. For long runs of pipe in the collector headers, a "U" shaped plumbing layout can be used to allow pipes to expand without breaking.

It is also important to separate dissimilar metals to prevent galvanic corrosion. Putting teflon tape on the threaded joints between G.I. pipe and copper headers will retard this corrosion. Also, attaching aluminium battens to the steel collector frame should be done with galvanized steel screws which are separated from the aluminium with a rubber washer.

13. USE PROPER INSULATION

Mother Nature plays no favorites. If the equipment is not assembled with the proper materials or the system designed with the proper sizes or flow-rates, the contractor has the job of re-doing the work. It is the responsibility of the Government and Academic Institutions to supply accurate information to help the manufacturers and contractors make the correct decisions. They may all want to share the expenses of repairing systems which are not functioning efficiently, since the manufacturers, contractors, Government and Academic Institutions are all part of the same team.

In a young industry like solar, many variables may not be known, as to how various materials will respond to the conditions in a solar collector or to the direct rays of the sun. Much of this information is known in other countries however.

It is in everyone's interest to share their experience openly, so that problems are identified early and misconceptions not repeated. The process should not be to blame people for mistakes but to share the learning process and cost of fixing systems which have failed or are performing poorly.

One place where this practice of cooperation is essential is in the choice of glazing gaskets. Most of the products on the market are designed for building glazing systems where the temperatures are not high. EPDM or silicone gaskets will tolerate this high stagnation temperature and the sun's ultraviolet radiation.

However, if these two rubbers are not available, then samples of the rubber gaskets supplied for the glazing should first be tested in an oven at various temperatures up to the stagnation level plus 20%. these tests will help verify the manufacturer's claims as to whether the proposed gaskets can stand the heat without off-gassing or cracking. Either breakdown can create major maintenance problems in the future.

Since the Government of India contractors carry a ten year warranty, each gasket test should be done for at least one 24 hour period at three graduated temperature levels up to 240 degrees C in a sealed container. After each test, check for a thin film on the container walls and cracks in the rubber.

If the gaskets will not stand the high temperatures, the Dow Corning No.795 may also be used in place of a glazing gasket by applying caulk directly to the metal frame, since the silicone rubber will handle temperatures up to 450 degrees F. Plan on one tube extra per collector if the caulk is substituted for the gasket. Water leakage and off-gassing is the largest cause of collector performance loss.

14. LEVEL COLLECTOR ATTACHMENT TO ROOF

Finally, to provide a quality installation which will last, a level metal frame needs to be attached to the terrace. This frame will allow the collectors to be mounted in a straight row without blowing away, breaking glass or putting stress on header joints which cannot be aligned properly without a level base. Installing and maintaining threaaded or soldered pipe connections so they don't leak over time is not easy. The installation labor time to mount the collectors can be cut significantly and the call-backs to fix leaky joints can be reduced substantially if the collector foundation is level and securely attached to the terrace. Shopping the metal fabricators should turn up a suitable angle iron or "U" channel for the purpose.

Also, on a flat terrace, installers should use roofing tar on a daily basis to seal any cracks created by equipment or tools dropped on the hot roofing material. Preventing water leaks into the building below is very important to good customer relations.

It has been an honor to share this solar hot water design information with the Indian solar industry. Over the last 15 years, many facts have been learned about how to make these systems provide the most heat for the least cost. Through conferences, workshops and on-the-job training, the Indian solar industry will be able to transfer this state-of-the art knowledge without each manufacturer having to reinvent the wheel.

NOVEL DESIGN CONCEPTS IN SOLAR WATER HEATING

Atam Kumar
Allied Metal and Engineering Works
29 Okhla Industrial Estate
New Delhi 110 020
India

1. INTRODUCTION

We in Allied Metal and Engineering Works have been working on Solar Water Heating Systems since 1975, in association with Bharat Heavy Electricals Limited (BHEL) till 1978 and independently from 1979 onwards. We have had the opportunity of learning from the failures and successes of the system designed, manufactured and commissioned under the guidance of BHEL Solar Systems group, and the performance of the system designed and commissioned by us. We have come to the following conclusions:

(i) All the systems must be designed considering that human failures in maintenance and operation can occur and even if the system fails to perform due to these human failures, it should not get damaged.

(ii) It is unlikely that continuous effective control on water composition can be maintained and system design should cater to the quality of water as existing.

(iii) Emphasis on all system designs should be on meeting the specific needs of the particular end use rather than maximising efficiency at the cost of end use.

(iv) Mixing of hot water and cold water causes lowering of temperatures achieved during the daytime in the tank. The temperature of hot water in the tank goes on lowering with the ingress of cold water in hot water storage tank till it reaches lower non-usable temperature. The mixing should be avoided, if possible.

(v) The controls for the systems should be made least dependent on manual adjustments.

(vi) System layouts have got to be made considering the area availability and orientation and sizes of available area but symmetry in collector layout results in better flow equalisation and better operating conditions than use of instrumentation and control mechanisms for such purposes.

(vii) The system design should ensure lower system thermal capacities to ensure better outputs.

H. P. Garg (ed.), Solar Water Heating Systems, 337–345.

2. LARGE SYSTEM DESIGN

Based on the limitations enumerated above, we have evolved a schematic for the large capacity systems. A large size Solar Water Heating System is schematically shown in Fig.1.

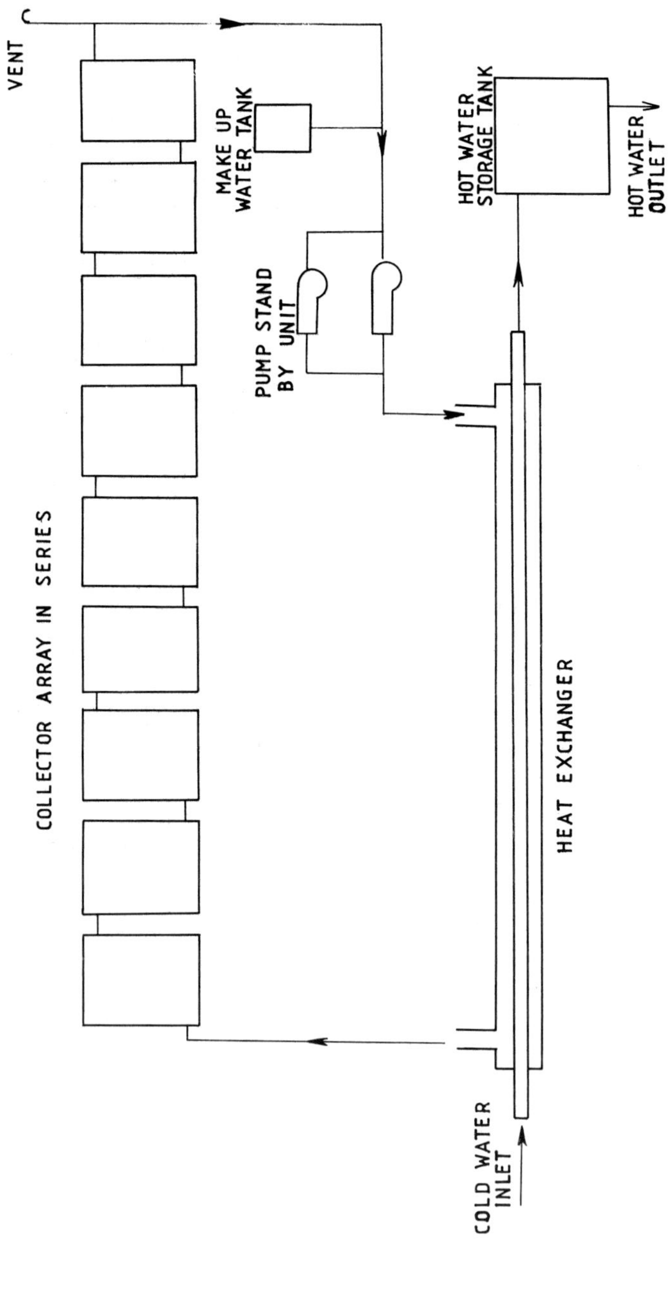

FIG.1 - LAYOUT-SOLAR WATER HEATING SYSTEM WITH HEAT EXCHANGER

The basic features of this arrangement are:

(a) Arrangement of solar flat plate collectors in series and parallel combination.

(b) Use of a heat exchanger and a closed loop flow in the collectors.

(c) A make-up water tank installed at the highest point in the system.

(d) A differential temperature control for the closed loop flows.

2.1 Operation of the System

In the system detailed in Fig.1, as the sun rays reach the collector absorber surface, the absorber temperature starts going up heating the water in it. When the hot water temperature at the outlet reaches a preset value compared to water temperature in pipes, the differential temperature controller operates and the circulation of water in the collector loop starts. With this circulation the temperature of water starts rising in the collector loop. The temperature of water in the secondary circuit of the heat exchanger also starts rising up. For the control of hot water flow, two modes of controls can be used:

(a) Hot water starts flowing to the storage tank as soon as the temperature at outlet point from the heat exchanger reaches a preset value.

(b) After a certain interval of time (after starting primary circuit flow) set to heat up the stored water in the primary circuit and heat exchanger the water starts flowing into the storage tank. We have so far utilised the first mode of water flow, but the second mode is proposed for use by us in some new systems. The advantage in the second mode will be to heat water even when the sun intensity is low.

As soon as the temperature of water near the outlet of the collector becomes equal to the water in the pipe of heat exchanger, the differential controller switches off the pump.

2.2 System Layout

There can be several layouts of the system but based on our practical experience, we have decided the following arrangements:

(i) Parallel arrays of collector connected in parallel by pipe headers and pipe header connected in series to form a series arrangement as per Fig.2.

(ii) Parallel arrays of collectors connected in series.

(iii) Series array of collectors connected in parallel (Fig.3).

Of these arrangements, the last arrangement has the least amount of water hold up capacity and consequently the smallest thermal capacity. But depending upon a specific site condition any one arrangement may turn out to be most beneficial.

2.3 Heat Exchanger (Fig.4)

The use of heat exchanger in the system has been thought of to avoid the dependence of system life and performance on quality of water. In earlier systems, a large reservoir containing demineralised water was

FIG.2 - COLLECTORS (25) IN PARALLEL

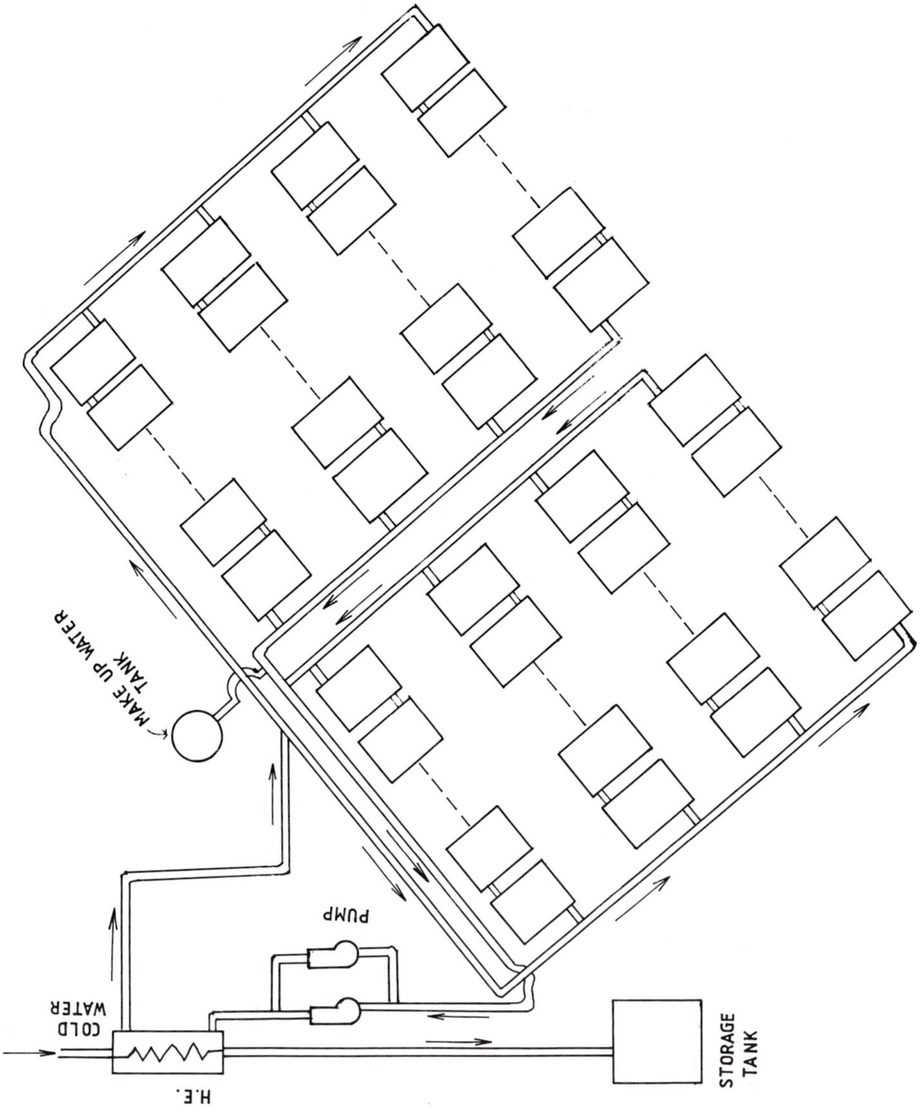

FIG.3 - SYSTEM LAYOUT SERIES ARRAYS IN PARALLEL

used in which a coiled heat exchanger was dipped. The water in this reservoir is heated up during day. Any water to be drawn is passed through a tube coil dipped in the reservoir and is heated up before reaching the use point. This arrangement is found to have the following limitations:

(a) Large quantity of demineralised water is required for starting the system.

(b) The temperature of large reservoir is to be raised before effective utilisation of water can take place.

(c) The heat exchanger working on stagnant water on one side had

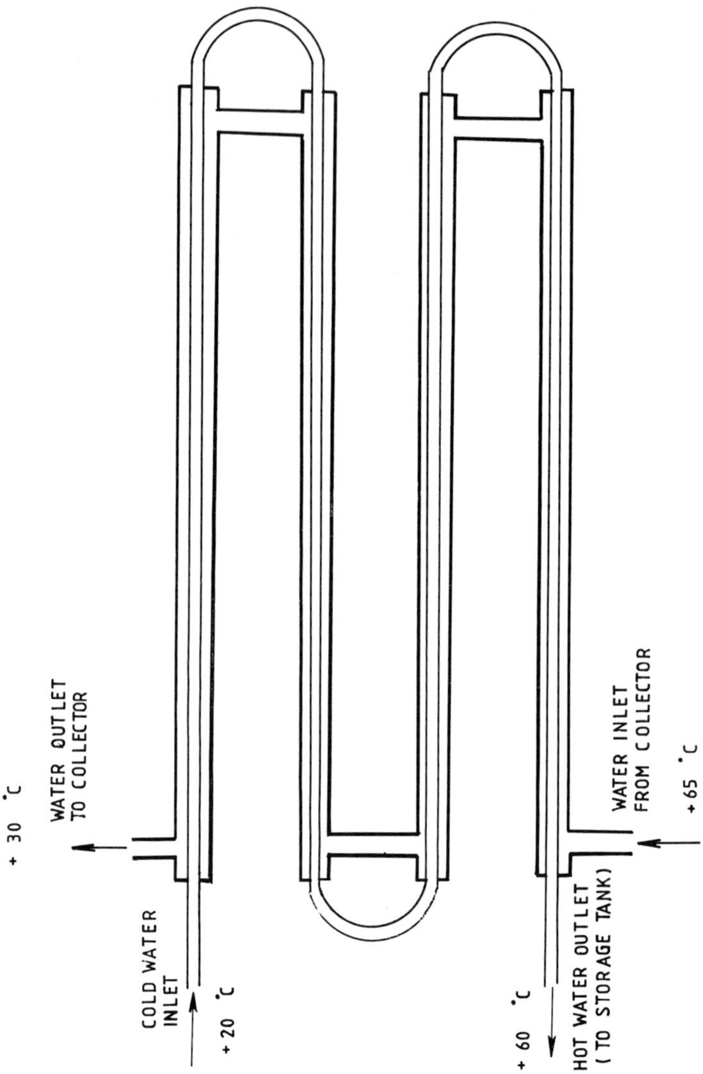

FIG.4 - SCHEMATIC OF WATER TO WATER HEAT EXCHANGER

poor heat transfer coefficients, which meant the temperature difference in the reservoir water and water drawn is large. This has resulted in poor utilisation efficiency of the system.

In the arrangement used here, the flow of water through the primary and secondary side of the heat exchanger is maintained more or less continuously throughout the sunny period with the result that the heat transfer coefficients are high and lower temperature differences are maintained. Only the hot water is sent to the hot water reservoir and is drawn from there as and when needed, thus effective utilisation can start immediately after the heating has started.

The arrangement of the heat exchanger is in counter current flow condition so that the maximum temperature achieved in the primary flow and secondary flow differ by a few degrees only.

2.4 Make-up Water Storage Tank and Pumps

Make-up water storage tank is installed at the collector outlet and the pumps suction side. This tank is vented and serves the following functions:

(i) The expansion of the water with heating is absorbed by this storage tank.

(ii) Positive suction head is available to the pump as level of water in the tank is maintained.

This arrangement, however, results in pump being subjected to the highest temperature of water in the system and these have to be capable of sustaining such a temperature.

In this arrangement the major pressure drop takes up in the heat exchanger and collectors are subjected to low pressure.

3. DESIGN OF SMALLER SYSTEM

We have evolved two arrangements working on pressurised water supply available for domestic use. Both these types of systems have advantages and disadvantages.

3.1 High Temperature Thermosyphon System (Fig.5)

In this design, the mixing of cold water with hot water is avoided and sufficient high water temperature can be obtained. Also better collector efficiencies at lower costs are sought to be achieved by using good quality selective coating only in the top half of the collector. A small tank is incorporated in the system in which cold water is allowed to flow and get heated in the collector arrays and return after getting heated. The hot water flows to the hot water storage tank after the desired temperature has been achieved. There is no flow of water to the storage tank below the desired temperature.

3.2 System with Temperature Controlled Valves at the Collector Outlet (Fig.6)

Another system design proposed to be utilised is by using thermally controlled valves. In this mode no electricity will be required and system becomes completely independent of power supply provided a water head of 1.5 metre of water column is available at terrace floor level. The hot water storage tank in this kind of system can be kept at lower level also within the building. The thermostatic valve using special bellows allows the hot water at a fixed temperature from collector to the hot water storage tank. Such thermostatic valves are now available in India.

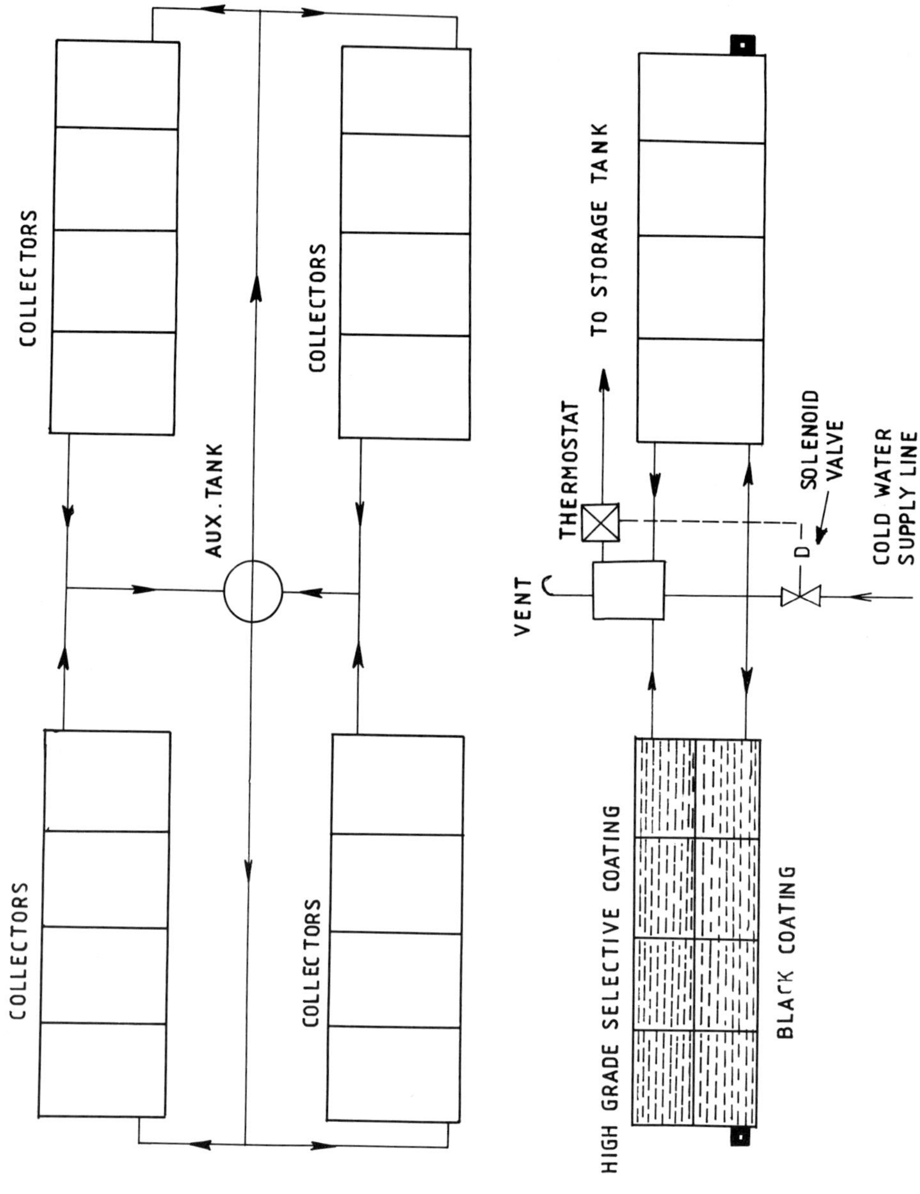

FIG.5 - HIGH TEMPERATURE THERMOSYPHON SYSTEM

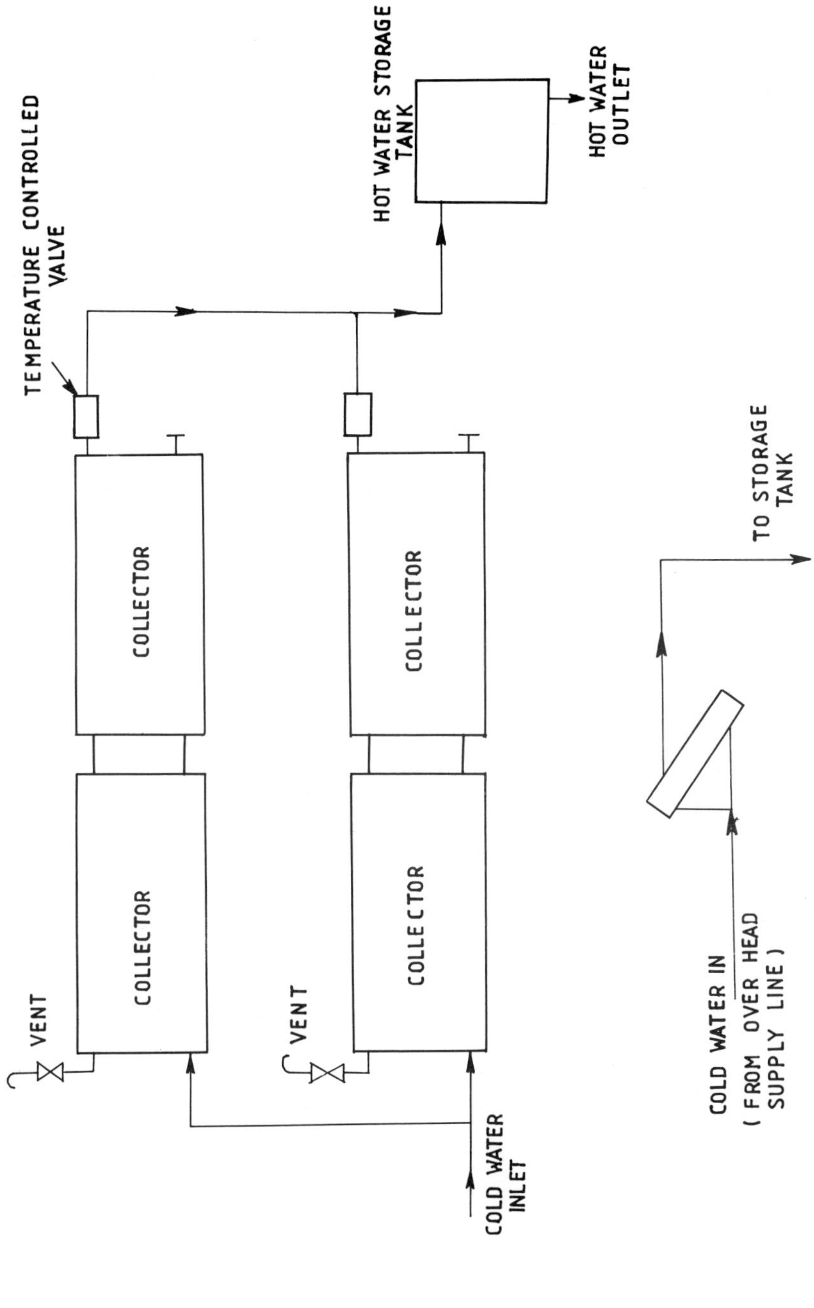

FIG.6 - SYSTEM LAYOUT WITH TEMPERATURE CONTROLLED VALVE WITHOUT PUMPSET AND HEAT EXCHANGER

SOLAR WATER HEATING SYSTEM IN A TEXTILE INDUSTRY - A CASE STUDY

K.S. Rao
Gujarat Energy Development Agency
B.N. Chambers
Vadodara 390 005
India

1. INTRODUCTION

Textile industry is one of the largest energy consuming industries in India with an annual consumption of 4.5 million tonnes of coal. Of the total energy consumed in a textile mill, about 90% is in the form of thermal energy (steam) and the rest is in the form of electricity. The fuel used for steam generation is mostly coal except in Bombay where fuel oil is used. The operating temperatures of various processes in the industry range from 50°C to 220°C. Of the total thermal energy input 80 to 85 per cent is used for processing and drying. (Table 1 gives the estimates of specific hot water consumption in a textile industry for various processes and the desired temperatures.) Solar energy, which is pollution free and inexhaustible, can easily provide the hot water at the desired temperatures and the technology is already available for such solar collectors. In a textile mill solar energy can be used in various forms e.g. as hot water, hot air and low pressure steam. Also the textile mills have large roof area, generally south slopping, which can be used to install the solar collectors for harnessing the solar energy.

In view of high potential of solar systems application in textile industry, the Department of Non-Conventional Energy Sources/Department of Science and Technology, Government of India, has sponsored several demonstration projects in different parts of the country. Table 2 shows the details of the demonstration systems. The first installation under this programme is at Jahangir Textile Mills of Ahmedabad of National Textile Corporation (Gujarat). The project was sanctioned in 1980 and the system was commissioned by May 1981. This paper presents the scope of solar energy in textile industry and a detailed description of the solar hot water systems at Jahangir Textile Mills along with performance data of the system for one year. The managerial and operational problems encountered in installing and operating the system are also discussed.

H. P. Garg (ed.), Solar Water Heating Systems, 347–360.

TABLE 1 - Estimates of Specific Hot Water Consumption

Process	Hot Water Temperature °C*	Hot Water Consumption litres/Kg of cloth
Boiler feed water	40 - 90	Variable
Rope washing	50 - 60	10 - 17
Kier Boil		
(a) Open	50 - 90	3 - 8
(b) Pressure	50 - 90	5 - 10
Continuous Bleaching		
(a) Saturator	50 - 70	1 - 2
(b) Rope washing	50 - 60	10 - 15
(c) Open-width washing	50 - 60	10 - 15
Cloth Mercerising	50 - 80	25
Yarn Mercerising	70 -	15
Jiggers		
(a) Desize/Bleach	50 - 80	5 - 8
(b) Dyeing	40 - 90	18 - 20
Winch		
(a) Desize/Bleach	50 - 80	45 - 60
(b) Dyeing	40 - 90	45 - 85
Yarn Dyeing Machine		
(a) Boil	50 - 90	10
(b) Dyeing	40 - 90	20 - 25
Beam Dyeing Machine		
(a) Scovring	50 - 80	30 - 40
(b) Dyeing	50 - 90	35

* Shows the temperature at which hot water can be used, does not show process temperature.

TABLE 2 - Solar Hot Water System Installed in Various Textile Mills

Sr. No.	System location	Capacity of the system	Outlet temperature	System description		Year of instal-lation	Sponsorer	Remarks
1.	Jupiter Mills Ahmedabad	1500 lpd	80°C	Total No. of flat plate collectors: No. of single glazed black paint collectors: No. of double glazed black paint collectors Collector size: Capacity of storage tank:	24 Nos 12 " 12 " 1m x 2m 1000 lits	1980	Govt. of Gujarat	
2.	Jahangir Textile Mills, Ahmedabad	6000 lpd	80°C	Total Nos. of flat plate collectors: No. of single glazed black paint collectors: No. of double glazed black paint collectors: No. of single glazed selectively coated acrylic glazed collectors: Capacity of storage tank:	92 Nos 48 " 24 " 20 " 6000 lits	1981	CASE/DST	Sloping asbestos roof
3.	Western India Mills, Bombay	6000 lpd	80°C	- do -		1984	DNES	Sloping roof

TABLE 2 (cont'd)

4.	Muir Mills, Kanpur	6000 lpd	$80^{o}C$	Total No. of collectors: No. of single glazed black paint collectors: No. of double glazed black paint collectors: No. of single glazed selectively coated collectors: Capacity of storage tank:	84 Nos 40 " 24 " 20 " 6000 lits	1984	DNES	Flat terrace
5.	Bhal Nalkantha Khadi Gramodyog, Mandal, Ranpur	5000 lpd	$70^{o}C$	Total flat plate collectors: Selectively coated size of collectors:	64 Nos 2mx1m	To be installed by May 1985	DNES & GEDA	Flat terrace
6.	Indore Malwa Mills	12000 lpd	$85^{o}C$	Total No. of flat plate collectors: Size of collectors	192 Nos $1.9m^2$	1984	Madhya Pradesh Urja Vikas Nigam	-
7.	Avanthi Processors	15000 lpd	$70^{o}C$	Total No. of flat plate collectors: Size of collectors:	225 Nos $1.9m^2$	Yet to be installed	"	-
8.	Tamil Nadu Coop. Textile Mills Ltd., Erode	48000 lpd	$70^{o}C$	Total No. of flat plate collectors: Size of collectors:	560 Nos $1.9m^2$	1984	Planning & Development Department, Tamil Nadu	-

CASE - Commission on Additional Sources of Energy
DST - Department of Science and Technology
DNES - Department of Non-conventional Energy Sources

2. JAHANGIR TEXTILE MILLS

Jahangir Textile Mills has an installed capacity of 39,732 spindles and 788 looms of which 740 are plain and the rest are automatic. The mill has a processing capacity of 85,000 metres cloth per day, the cloth having 90% cotton and 10% synthetic. It is a composite textile mill and has most of the capabilities of a modern textile mill. It is housed with Research and Development Unit of NTC (Gujarat). The energy budget of the mill is shown in Table 3. Of the total energy consumption, 90% is consumed in the form of thermal energy and 10% is in the form of electrical energy. The mill has a roof area of 20,640 m^2 of which 17,080 m^2 is south facing roof which can be used for installation of solar systems. From Table 2 it is evident that steam requirement for different purposes are used in the form of water (28.5%), steam (57%) and hot air (14.5%).

The hot water in the textile industry needs temperatures from 60°C to 80°C. An ordinary flat plate collector with or without selective coating, single glazed/double glazed will be able to deliver the energy at the above temperatures.

3. DETAILS OF SOLAR HOT WATER SYSTEM AT JEHANGIR TEXTILE MILLS

The system consists of 92 collectors and an insulated storage tank. In order to keep the pressure drop through the system low, the collectors are placed in 4 parallel rows in the south facing asbestos roof for grey cloth and cotton waste department of Jehangir Textile Mills. Each row of collectors consists of 12 single glazed collectors with black absorber, 6 double glazed collectors with black absorber and 5 single glazed collector with selectively coated absorber and acrylic glazing. A rotameter is provided in each row for monitoring the flow rate. A single centrifugal pump circulates water through all the four rows and balancing valves are provided in the circuit to maintain uniform flow through the rows. The circulation of water through the system will be once through type, i.e. cold water enters the collector field and hot water is forced from collector field to the storage tank.

4. SYSTEM CONTROL

For automatic operation of the system a thermostatic control is provided for the pump and soleniod valves are provided in the suction line and at the end of the delivery hot water line. A thermostatic temperature control system with the sensor placed at the outlet of the collectors automatically switches on the pump simultaneously opens the solenoid valves, when the water temperature rises the set value. The pump circulates water through the collectors resulting in lowering of the water temperature. When water temperature drops below the set value, the pump automatically switches off and simultaneously the solenoid valves get closed. The whole cycle repeats again when water temperature rises.

TABLE 3 - Energy Budget of Jahangir Textile Mills.

Fuel consumption : 750 Tonnes/month*
Electricity consumed : 10,88,960 KWh/month

Machine	Steam Demand %	Steam Consumption Tonnes/ Month	Ultimate media of heating
Bleaching & Finishing Department			
Singeing Bath	0.99	37.1	Water
Pressure Kiev Scouring	13.40	502.5	Water
Open Kier Scouring	0.51	19.1	Water
Open Kier Bleaching	3.75	140.6	Water
Preshrinking machine	6.79	254.6	Steam
Calenders	4.15	155.6	Steam
Cloth mercerising	1.53	57.4	Water
Cylinder drying	20.38	764.3	Steam
Printing Department			
Hot Flue Dryer	5.60	210.0	Hot air
Cylinder Dryer	2.64	99.0	Steam
Ager	7.11	266.6	Steam
Dyeing Department			
Yarn Dyeing	2.01	75.4	Water
Yarn Drying	0.87	32.6	Hot air
Jigger Dyeing	4.55	170.6	Water
Cylinder Drying	5.47	205.1	Steam
Flout Dryer	8.00	300.0	Hot air
Sizing Department			
Cylinder Dryers	10.52	394.5	Steam
Size Vats	1.81	67.9	Water
Total:	100.00	3750.0	

* Estimate for the period March 1980.

This simple and straightforward control system offers automatic operation and capability of adjusting the outlet temperature to any desired value.

In order to avoid wastage of water through overflow from hot water storage tank, an electronic level controller with probe is provided at the top of the storage tank. This controller switches off the power supply to the pump when the storage tank is full. The level controller also indicates the availability of hot water in the tank.

5. STORAGE TANK

The capacity of the storage tank is decided by the use pattern and the rate of system output. Normally in the textile mills the first shift starts around 8 a.m. and all the process equipment is loaded by 9.30 to 10.00 a.m. The hot water is required at this time and the solar system is not in a position to supply it so early in the morning. Once loaded, the equipment generally do not require hot water again till 4.00 p.m. This means that all the system output between 10.00 a.m. to 4.00 p.m., which is nearly the full output, will have to be stored in a tank and used in the second shift. Therefore the storage tank holds 6000 litres of hot water at 80°C. A 100 mm thick mineral woold insulation is provided to keep the overnight temperature drop at less than 5°C. In actual practice the hot water is not stored overnight but only a few hours in the afternoon. Therefore the temperature drop is in the order of 1 or 2°C. An insulated removable lid is provided with proper sealing arrangement to prevent evaporation losses. The level probe is fixed in the lid to provide adjustability of the level.

6. PIPING AND PIPING INSULATION

The piping between the collectors and rows is of 26 mm diameter and hot water pipelines are insulated by 50 mm thick mineral wool insulation. The insulation is covered with aluminium cladding for weather protection. The pipe line between the collector field and the storage tank is of 40 mm diameter.

7. PERFORMANCE OF THE SYSTEM

The system was commissioned in May 1981 and the performance data was collected until August 1982, shown in Table 4. The average thermal output is 6.54 million Kcals/month. In computing the efficiency the radiation data from Indian Meteorological Department (IMD) for Ahmedabad was used instead of actual incident solar radiation at the factory site. Few spot checks at the factory site revealed that radiation level is 10% less compared to the meteorological data reported by IMD, Ahmedabad. The annual average efficiency from the performance is 27% as compared to the 30% of the designed efficiency of the system. During the monsoon months the set temperature is lowered to 60°C. The solar heated water is used for second shift operation in the jiggers.

TABLE 4 - Performance of Solar Hot Water System

Month and Year	Total quantity of hot water/ month in Kilo-litres	Average outlet temperature °C	Number of days the system in operation	Heat output Million Kcal/month	Efficiency
August 1981	56.0	80	30	2.81	16.4
October 1981	155.9	75	26	6.56	26.8
December 1981	190.0	79	31	9.92	33.0
March 1982	221.0	81	31	10.50	33.3
May 1982	155.0	80	28	6.40	25.9
June 1982	139.3	79	28	6.10	31.2
August 1982	105.3	63	28	3.50	22.6

8. ECONOMICS OF THE SOLAR HOT WATER SYSTEM

The total installed cost of the solar hot water system works out to be Rs.3,25,000 and the annual thermal energy output of the system is 94 million Kcal. The economic evaluation of the system based on the actual system performance has been calculated for several options such as (i) different interest rates, (ii) different state and central subsidies, (iii) the type of fuel saved, e.g. coal and fuel oil.

An annual depreciation is taken as 60% for the first year and 40% for the second year.

The payback period of the system ranges between less than one to nine years. It is shown in Tables 5 and 6.

9. PROBLEMS IN RETROFITTING SOLAR HOT WATER SYSTEMS IN TEXTILE INDUSTRIES

(i) Textiles are one of the earliest industries and these mills are located in the crowded areas where the space availability is a problem. Open ground space is not available in most of the existing mills and the solar collectors have to be installed in a shadow free south facing roof near the process where the bulk of the hot water is being utilized. Since these roof structures are not meant for taking any significant loads it may become necessary to strengthen the roof to take up the collectors' load. If the roof is a slopping structure with corrugated asbestos sheets as roof top, it is necessary to build a separate structure along with cat walls for placing the collectors at proper inclination.

(ii) The hot water storage tank has to be installed in the process house near the utilization points. The roof for installing solar collectors and the storage tank should be so selected that the length of insulated hot water lines are minimised and the additional pipelines do not intervene with the operations of the process house. It should be kept in mind that the pipelines contribute significant thermal losses and hence the lengths of pipelines from storage tanks to the usage points should be minimised.

(iii) In choosing the site for collector installations, care should be taken to see the dominant wind directions which should be such that the deposition of dust, ash etc. on the collectors from the factory chimney and other dust sources are minimum.

(iv) The roof chosen for installing the collectors should be easily accessible so that installation and monitoring of the system become easy.

10. TECHNICAL PROBLEMS FACED IN OPERATING THE SOLAR HOT WATER SYSTEMS

(i) Deterioration of Acrylic glazings.

The system originally installed has about 20 collectors with selectively coated absorbers with acrylic glazing which has transmissivity better

TABLE 5 - Economic Analysis for Solar Hot Water System in Private Sector.

System capacity : 6000 lpd at 80°C
Alternative : Fuel oil
Total investment on solar system : Rs.3,25,000 (1 US$ = Rs.12)

Year	Cash flow Rs.	Interest Rate 12%	Depreciation 20%	Saving in tax on	Net energy saving in Rs.	Gross Saving (5+6)Rs	Net Savings
(1)	(2)	(3)	(4)	(5)	(6)	(7)	(8)
0	-2,17,750	-	-	-	-	-	-
1	-1,31,555	26,130	1,30,650	65,385	47,000	1,12,325	86,195
2	- 56,792	15,787	87,100	43,550	47,000	90,550	74,763
3	- 16,607	6,815	-	-	47,000	47,000	40,185
4	+ 28,401	1,992	-	-	47,000	47,000	45,008
5	= 75,401	-	-	-	47,000	47,000	47,000

<u>Assumptions</u>: Subsidy given by DNES : 33%
Calorific value of fuel oil : 10500 Kcal/Kg
Cost of fuel oil : 3000/Ton
Income tax rate : 50%

TABLE 6 - Economics of Solar Hot Water System

Sr. No.	Fuel Saved	Rate of Interest	Cooperative Sector Payback period in years: With 87.5% subsidy		Cooperative Sector: With 75% subsidy		Public Sector Corporation Payback period in years: With 65% subsidy		Public Sector Corporation: With 50% subsidy		Private Sector 33% subsidy Payback period in years
			Profit making sector	Non-profit making sector	Profit making sector	Non-profit making sector	Profit making sector	Non-profit making sector	Profit making sector	Non-profit making sector	
1.	Coal	12%	<1	1-2	2-3	4	3-4	7-8	4-5	-	8
2.	Fuel oil	12%	<1	<1	1-2	2-3	2-3	3-4	2-3	4-5	3-4

Assumptions:

Calorific value of coal	:	5000 Kcal/Kg
Calorific value of fuel oil	:	10,500 Kcal/Kg
Cost of coal	:	Rs.700 per ton
Cost of fuel oil	:	Rs.3000 per ton
Income Tax rate	:	50%
Depreciation for first year	:	60%
Depreciation for second year	:	40%

than 0.93. During the initial operation period it was observed that the glazings sagged in the middle. Wooden pegs were fixed to prevent the sagging. However, during the next six weeks of operation hairline cracks developed at the edges and over a period of six months these cracks spread to the centre. The acrylic glazings also turned yellowish in colour showing the U.V. degradation. The contractors replaced all the acrylic sheets with glass at their cost.

(ii) Flow Distribution in Collector Arrays.

It was reported during the second month of operation that the system was doing exceedingly well and delivering more than 10,000 lpd compared to the designed capacity of 6000 lpd. When we visited the system it was observed that the uniform flow distribution was disturbed. The control system monitors the outlet temperature of one of the rows and switches on and off the pump. Because of non-uniform flow distribution some rows delivered water at much lower temperatures because the flow was higher in these rows compared to the flow in the control row. One has to be careful that the monitored temperature is proper. Sometimes air pockets develop near the temperature sensors and it is advisable occasionally to purge the air through the vents provided in the collectors.

(iii) Inner Glass Breakages in Double Glazed Collectors.

In the beginning it was reported that the frequent breakage of inner glass in the double glazed collectors occurred. The problem was carefully analysed and several solutions were tried. The use of non-wavy glass for inner glazing solved the problem.

(iv) Operation During Monsoon Months.

Since the system is designed for delivering hot water at 80°C, the performance of the system is rather poor during monsoon months. Since hot water at 60°C has also utility, the set point of control system is lowered to 60°C so that some useful energy can be extracted. Actually, for process industries where hot water is needed continuously it is always advisable to eliminate control system and have water flowing once through the collectors. Such systems can deliver nearly 30 per cent of more than the thermal output compared to on-off control systems and costs less. The water flow rate for such systems has to be selected such that it can deliver the desired temperature only at noon time.

(v) Seepage of Water.

It is observed in few collectors that the sealing of the glazing is not proper and small amount of rain water seeped into the collector. This water could be observed as tiny droplets dripping on the inner surface of the glazing. The water in the insulation causes significant decrease in the efficiency. The manufacturers were asked to seal these collectors properly.

(vi) Dust on the Collectors

Since the mills are located in the industrial areas, the air carries significant amount of dust particles. The dust particles were not only deposited on the glazing but also reduced the incident solar radiation by nearly 10 per cent compared to the meteorological data reported by IMD Ahmedabad. However, the dust on the collectors have to be cleaned once in a week. Some dust creaped into the inner glass absorber over a period of couple of years of operation. The collector assemblies were opened and the glass/absorbers were cleaned.

(vii) Leakage of Unions.
The collector headers are joined by GI unions. Few of the unions started leaking after three years of use. The unions have been replaced easily with new ones.

11. MANAGERIAL PROBLEMS INVOLVED IN OPERATING THE SYSTEM

(i) The Attitude of Management and Workers.
When the project was undertaken by Central Salt and Marine Chemicals Research Institute, Bhavnagar, and Ahmedabad Textile Industry's Research Association, Ahmedabad, in collaboration with National Textile Corporation (Gujarat), the Managing Director and General Manager of the unit were extremely helpful and they were able to enthuse their technical personnel to cooperate with the project personnel in implementing the solar water heating system. In the Phase II hot air system was also installed at Jahangir Textile Mills. But later there was a change in top management of the mill and the attitude of the mill personnel has changed. In their view the hot water delivered by the solar system is less than 5 per cent of their need and the energy contribution is less than 0.5 per cent and they could not afford to put manpower to look after the systems and collect data. Therefore, it is important to have the size of demonstration unit such that its contribution of energy saving is visibly felt by the management.

(ii) The Financial Crisis in the Textile Industry.
The Textile Industry in general is passing through financial crisis, any additional investment in terms of equipment or in terms of personnel is difficult to obtain as there are many pressing demands for funds in terms of modernisation, which directly affect the productivity. The two private industries showed great interest in installing solar water heating system of 50,000 lpd capacity in their textile mills. They have visited the Jahangir Textile Mills and were fully satisfied by the performance and economics of the systems. However, when we have asked them to deposit their share of money, they requested us to consider their application for the year 1985-86.

(iii) The textile mills' process houses can be classified as large, medium and small according to their scale of operation. In the case of existing large and medium mills the space may become a critical factor for installing solar hot water system which can contribute significantly to their energy requirements. However, the small process houses use crude heating devices and simple sun drying. They do bleaching, printing and dyeing for which they need hot water comparatively small quantities which can be supplied by solar systems. Thrust should be given in popularising solar systems in these small process houses. There are hundreds of such units in Ahmedabad.

12. CONCLUSIONS

(i) Solar water heating systems have good potential in providing process hot water in textile industry. The small process houses seem to be better suited than the existing large and medium sized mills.

(ii) The pay back period of solar hot water systems is quite attractive ranging from less than a year to nine years depending upon the subsidy, interest rate, fuel saved, etc.

INDUSTRIAL EXPERIENCES WITH SOLAR WATER HEATERS

M.S. Ramaprasad and Sudhir Mohan
Energy Division
Jyoti Limited
Vadodara 391 410
India

1. INTRODUCTION

Energy Division of Jyoti Limited, a pioneering organisation in the area of Renewable Sources of Energy, started its activity on solar energy sometime in 1976. In the initial stages, i.e. from 1976-78, greater emphasis was given on basic research and on development of a suitable design of a flat plate collector system which could be effectively used to meet hot water demand in industries as also to power Vapour Absorption Refrigeration Systems (VARS) for cold storages in rural areas. In order to assess the potential for solar hot water systems in industries, a survey was conducted in selected textile, chemical, pharmaceutical, dairy and engineering industries. The survey revealed that there existed a tremendous potential for solar water heating systems in industries in general and process industries in particular (Table 1).

TABLE 1 - Energy Requirements of Various Industries in the Form of Hot Water/Hot Air

Industry	% of total thermal energy consumption	
	Below 100°C	Below 140°C
1. Coffee	100	100
2. Liquid milk processing	100	100
3. Tea	90	100
4. Tobacco	90	98
5. Cotton Textile	74	90
6. Rayon	40	79
7. Soap	25	25

H. P. Garg (ed.), Solar Water Heating Systems, 361–369.

After completing the survey, a concerted effort was made for commercialising the product on a large scale. Beginning from the year 1979, we have been able to market quite a good number of systems all over the country for a variety of applications. The major areas of application have been industrial canteens, boiler feed water heating, hot water for process requirements, for maintaining temperature in L.S.H.S. tanks and solar powered cold storage units. An attempt is made in this paper, to share our experiences gathered over these installations, in respect of materials of construction, manufacturing processes, control systems, support structures, site related problems, customer related problems and last but not the least, problems relating to capital availability and cost-effectiveness.

2. INSTALLATION OF SYSTEM

We have been installing solar water heating systems of different capacities for a variety of applications all over the country since 1979. In the initial stages, we used to face a lot of resistance from prospective clients for installing solar water heating systems. The main reason for this was that people were not aware or were apprehensive towards their viability. Of course, another major obstacle was capital investment from the client's side. In spite of this, we somehow managed to convince some of the industrial clients who had an inclination to try out thisnewly emerging technology. One of our first large capacity systems was installed at the Works Canteen of Hoechst Pharmaceuticals, Bombay. Being the first large capacity industrial system, we did face a number of problems mostly pertaining to controls, while installing the system. The client was quite cooperative and took a keen interest in not only installing the system but also in using it very effectively. An engineer was subsequently assigned the task of looking after the maintenance of the system and keeping a day to day record regarding performance of the system. After a year, they were so very impressed with the performance of this system that they decided to go in for a larger capacity system of 15,000 litres/day at 70°C (the capacity of the first system was 5,000 litres/day at 60°C). This system of 15,000 litres capacity was installed for supplying feed water to the boiler. The system was quite simple in nature with a simple fixed temperature controller. (Detailed systems specifications are given in Table 2, while a photograph of one of the systems is shown in Fig.1.) An alarm system was also installed inside the boiler house so that the boiler attendant could have an indication whether hot water was available from the solar system or not.

The absorber panels in this system were made of galvanized iron tubes and CRC sheets. The system is working quite satisfactorily even today, i.e. after six years. This amply proves that if proper manufacturing process is employed, G.I. absorbers are quite efficient as far as solar water heating systems are concerned. Therefore, we are of the firm opinion that there should not be any hesitation in using G.I. absorbers in solar collectors (Ref.1). We do not have any reservations against using alternative materials such as copper or copper aluminium in the manufacture of solar collectors but our only concern is that large scale consumption of copper in making solar collectors may aggravate the existing crisis as far as availability of copper in the country is concerned. As far as copper aluminium is concerned, there should not be any difficulty provided good metallurgical bonding is assured between the copper tube and the aluminium fin.

3. MATERIAL PROBLEMS

As the water used in the system was not soft, engineers of M/s. Hoechst were curious to know the extent of scaling in the absorbers. After using the system for a period of two years, they opened out one of the collectors and observed the scale formation by actually cutting the absorber in the lateral direction. They observed a negligible amount of scale formation and even today, the system is operating without any problem. We also carried out a study at this stage to find out effect of scaling, if it is there at all, on performance of solar collectors. The study showed that nominal scaling may not affect the performance drastically (Ref.2).

In order to reduce still further the possibility of scaling inside the absorber tubes, we also developed our own scale prevention device patented under the name of 'Descaltron'. This particular equipment utilised a magnetic field which ionises the dissolved solids so that they do not stick to the tube and are flushed along with the water and deposited in the storage tank. By periodic draining of the storage tank, it is possible to precipitate all dissolved solids. We have installed such units along with some of the solar systems and they are working quite effectively. Therefore, it is feasible to use this unit as an alternative to heat exchangers because it avoids additional pump for heat exchanger loop and also additional piping.

To begin with, we were using ordinary glass as glazing material. Although from the performance point of view there are no problems, it does create a lot of problems during handling of collectors. There is a limitation of the glass which can be easily handled. In order to obviate these difficulties, we thought of replacing the same with UV stabilised acrylic sheets. One of the local manufacturers came forward to supply these sheets for manufacturing solar collectors. After using these sheets in some of our systems, we found that they were not suitable because they used to turn slightly opaque after some time and also developed hairline cracks probbably due to temperature cycling. Of course, the transmittivity of acrylic and even some other plastic materials was much better than glass (please see Table 3 (Ref.3)). At this stage, we contacted several manufacturers of glass who were manufacturing toughened variety and after a lot of persuasion from our side, we could convince one of the leading manufacturers in the country to supply us with toughened glass for our collectors in the size of 1m x 2m. Even today, this company is the largest manufacturer in the country who can supply good quality toghened glass for manufacturing solar collectors. Lately, this has also posed a serious problem because the glass is not available in the required quantity. Therefore, efforts

TABLE 2 – Specifications of Solar Water Heating Systems at Hoechst Pharmaceuticals, Bombay

A. Systems for 5000 lites/day at 60°C

No. of collectors	:	42
Nominal collector area	:	1m x 2m
Material of absorber	:	galvanised iron
Material of box	:	CRC sheet
Glazing	:	Two-piece toughened glass
Type of control	:	FTC
Tank capacity	:	5000 litres

B. System for 15,000 litres/day at 70°C

No. of collectors	:	118
Nominal collector area	:	1m x 2m
Material of absorber	:	galvanised iron
Material of box	:	CRC sheet
Glazing	:	Two-piece toughened glass
Type of control	:	FTC
Tank capacity	:	15,000 litres

FIG.1 – PHOTOGRAPH OF COLLECTOR PANELS OF SOLAR WATER HEATING SYSTEM OF HOECHST PHARMACEUTICALS, BOMBAY

TABLE 3 - Measured Values of Transmittivity for Solar Radiation

No.	Material	Thickness in mm	Sensor Reading Direct	Sensor Reading With Sample	Transmittivity %
1.	Acrylic	2	800	750	93.7
2.	Toughened Glass				
	i) Hindustan	3	800	690	86
	Safety (HSG)	4	800	670	83.7
	ii) Sardar Vallabh (SVG)	4	800	670	83.7
3.	FRP				
	i) Sunlit	1	800	735	92
	ii) Morlit	1	800	530	66
	iii) Miki	1	800	715	89
	iv) Filon	1	800	720	90

are being made to import solar glass from other countries so that the implementation of various projects in the country is not affected. However, due to the existing duty structure, it is not possible to import this glass because this will lead to escalation in the collector prices. Therefore, all efforts should be made to develop more number of parties in the country who can supply good quality toughened glass to all the manufacturers. Only then, will the country's programme in this area be achieved without any difficulty.

There are a number of insulating materials which can be used in the manufacture of solar collectors. The mainly used insulating materials are mineral wool, glass wool, polyurethene foam etc. Lately, there have been some reservations against use of mineral wool in solar energy applications. However, from our experience, if proper grade of mineral wool (Spintex 300) is used, there should not be any problem. Collectors having mineral wool as insulation are in active operation for more than 6-7 years. This by itself proves that this material can be used without any difficulty.

The various components used in the solar system such as pumps, solenoid valves, overflow controls, temperature controller, water meter, rotameter etc. play a vital role in the overall performance of the system. Therefore, proper selection of these components is very essential in the larger interests of solar energy applications. From our experiences, we have observed that some of the above mentioned components available in the country are not up to the mark. Our efforts in identifying appropriate components have not yielded worthwhile results. In some instances, it has so happened that a large capacity system installed against a huge capital investment may not be effectively put to use because of the failure of one single component costing a few hundred rupees. Therefore, efforts should be made in developing reliable components which go into the solar energy systems. If it is absolutely necessary, some of these components may be imported.

4. OTHER PROBLEMS

Selection of site plays a very important role in effective utilisation of solar energy systems. A lot of parameters have to be considered before finally selecting the site, such as shadow free area, approximate cold water supply and hot water use points, availability of electric power, etc., etc. In this direction it is absolutely necessary for both the implementing agencies and for the manufacturers to be well conversant with the parameters for site selection and an all round analysis made by both the parties before finally selecting the site. There are some instances where a site has been selected which is totally unsuitable for installing a system. This problem can be very easily avoided if the agencies decide to collect all the relevant information regarding the site before placing the orders on the manufacturers.

Another important factor which greatly influences the performance of the system is proper flow distribution of water within the system. A lot of R&D work has been carried out in this regard and we have found that it is quite feasible to have 15-20 numbers of collectors in series in any given row. It is feasible to have multiple number of rows having 15-20 collectors while installing a large capacity system. It is also desirable to have individual rotameters in every row in order to adjust the flow through the rows. It is also desirable to have the location of the pump and the storage tank diagonally opposite with reference to the collector array. It is also desirable to have the inlet and outlet connections coming from the top in a loop fashion so that the collector array is not drained at any given time.

Active involvement of the ultimate user plays a very important role in successful implementation of solar energy programmes in the country. From our experiences, we can definitely say that our clients in private industrial sector, who have purchased solar energy systems even without subsidy or tax benefits are effectively utilising the systems. They are maintaining these systems effectively without passing any burden to the manufacturers except during the guarantee period. There are some instances where the legitimate users have taken an indifferent attitude as far as proper maintenance of the solar system is concerned. Therefore, one has to seriously consider installing some of the major solar water heating systems for private industries even if it is to be subsidised to the extent of 75-100%. By installing these ystems in private industries, we can definitely expect to gain rich experience in proper utilisation of solar systems as we can afford to demand proper feedback from these industries.

Before concluding, I wish to share some of our experiences in installing some of the major solar water heating systems in the country. Here I would specifically mention about the difficulties we faced in installing the system at Erode Textiles. The system was required to be installed on an inclined asbestos roof and the supporting structure was in a very poor condition. Before actually installing the system, we were required to consult a structural expert for strengthening the existing structure so that it could take the load of the solar energy system. After installing the system with great difficulty, we were not in a position to use this system effectively because of non-availability of cold water and electrical power. On some of the days, the system was out of operation because they could not use the hot water already available in the storage tank. After a lot of discussions with the concerned authorities, we were finally able to convince them to provide all the above mentioned requirements and now, finally the system has been put into operation without any difficulty. The collectors are shown in Fig.2.

FIG.2 - PHOTOGRAPH OF COLLECTOR PANELS OF SOLAR WATER HEATING SYSTEM OF ERODE TEXTILES, TAMIL NADU

The Energy and Economy Department of TISCO, Jamshedpur, of their own approached us to find out the feasibility of using a solar hot water system for maintaining the temperature of low sulphur heavy stock (LSHS) storage tanks. To begin with, we were not confident whether the system could be used for such an application as before this such a system had not been tried anywhere. However, in order to explore a new avenue for solar energy applications, we agreed to supply a system. The system installed in early 1982, supplies, on an average, 2 lacs kilocalories of thermal energy/day at a temperature of 80°C so that the oil can be maintained at a temperature of 70°C. We have used the existing heat exchangers within the storage tank to transfer the thermal energy available from the solar system to the tank. The system has been operating without any problem for more than three years. Based on this experience, we ventured to install some more systems.

On similar lines and as on date, we have been successful in installing another system at TI Cycles, Madras. Table 4 gives detailed technical specifications working for the TISCO system. The system is shown in photo Fig.3.

TABLE 4 - Specifications of Solar LSHS Heating System at TISCO, Jamshedpur

LSHS Tank capacity	:	69,000 litres
Temperature required	:	80°C
No. of Collectors	:	60
Monsoon back-up	:	Steam
Steam savings	:	500 Kg/day
Oil savings	:	50 litres/day
Monetary savings	:	Rs.150/day
Direct payback period	:	4-5 years

FIG.3 - SOLAR WATER HEATING SYSTEM OF TISCO, JAMSHEDPUR

We also have the unique distinction of installing the largest capacity solar water heating system in the country at Bhopal Dairy. While installing this system we faced serious problems in locating the collectors at different levels which posed serious flow distribution problems. After several trials, we could manage balanced flow and now the system is working quite effectively. The system is shown in Fig.4.

FIG.4 - SOLAR WATER HEATING SYSTEM (30000 LITRES/DAY) OF BHOPAL DAIRY, BHOPAL

REFERENCES

1. "Corrosion of G.I. pipes and tanks in hot water", Internal Report, Energy Division, Jyoti Limited, 1984.
2. J.M. Reddy and N.R. Yardi, (1980), "Effect of scale on the performance of Jyoti Flat Plate Collector", Internal Report, Energy Division, Jyoti Limited, 1980.
3. "Transmittivity of different glazing materials for solar collectors", Internal Report, Energy Division, Jyoti Limited, 1984.

SOLAR WATER HEATING - DESIGN METHODS

C.L. Gupta
Solar Energy Unit
Sri Aurobindo Ashram
Pondicherry 605 002
India

1. INTRODUCTION

Solar water heating is the best proven and most cost effective thermal use of solar energy. During the last two years, nearly half a million square metre of collector area have been installed in India in systems with capacity varying from 100 l/day to 50,000 l/day. However, the design methods for choosing the reference month, for fixing up tilt/ orientation multiplier or incident angle modifier, for taking into account the radiation variability around mean monthly values, for relating the instantaneous test efficiency curve to collector sizing and for fixing the criteria for choice of system configuration have largely been ignored in actual practice. This paper seeks to present an overview and possible approaches to application practice in context of the systems adopted in developing countries in general but with particular reference to India.

2. TYPES OF SOLAR WATER HEATING SYSTEMS

Commonly used systems in India, China and other developing countries can be broadly classified into three categories namely Stand Alone Solar (SAS), Auxiliary Assisted Solar (AAS) and Solar Preheater (SPH). Prominent areas of use and current design features are summarised below and system configurations are shown in Figs.1, 2 and 3.

(i) STAND ALONE SOLAR (SAS)

- Low capacity domestic use.
- Temperature and volume specified.
- Lowest radiation month is the design month - Energy dumping is there in non-design months.
- Minor back up heating (if any) for excess demand/rainy days.

Thermosypon system for domestic use is a primary example of this type (Fig.1).

H. P. Garg (ed.), Solar Water Heating Systems, 371–381.

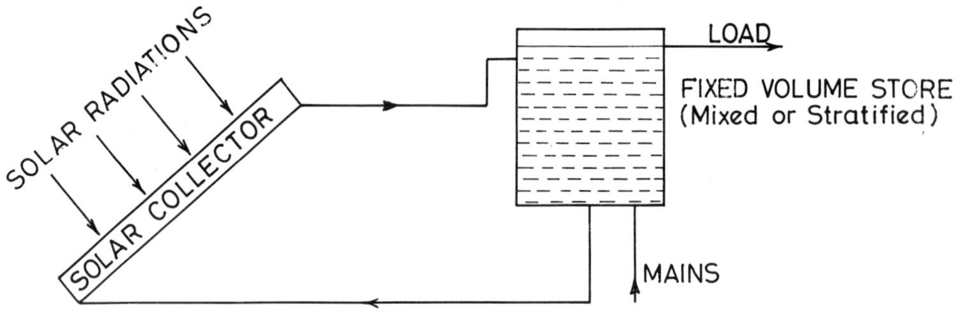

FIG.1 - THERMOSYPHON SYSTEM (SAS)

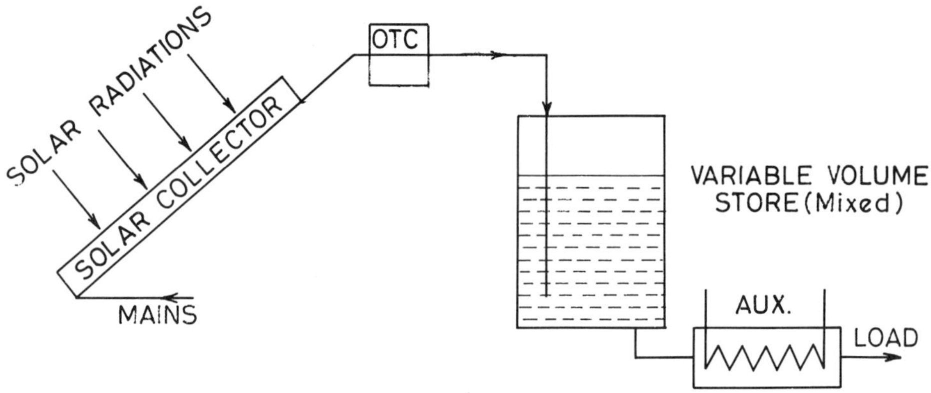

FIG.2 - OPEN LOOP (OTC) SYSTEM (AAS) (BATCH OPERATION)

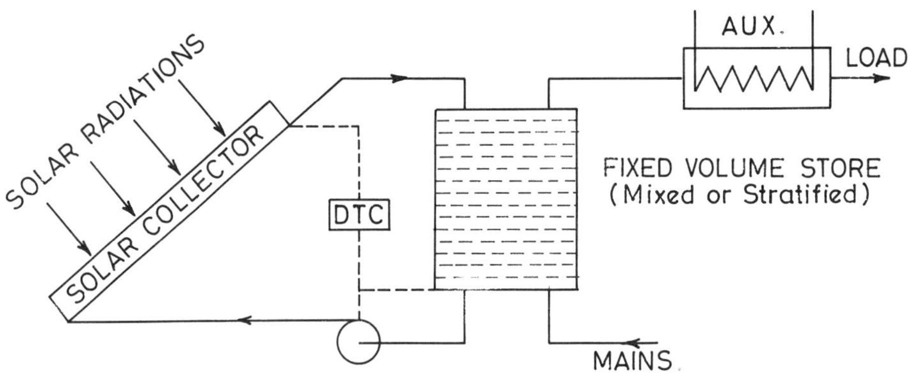

FIG.3 - CLOSED LOOP (DTC) SYSTEM (SPH)

(ii) AUXILIARY ASSISTED SOLAR (AAS)

- Large capacity public buildings.
- Temperature and volume specified.
- Lowest radiation month is the design month on mean basis - mass dumping is there in non-design months.
- Back up for topping temperature due to monthly/seasonal fluctuations.

Open loop output temperature controlled system is a principal configuration used in this category (Fig.2).

(iii) SOLAR PREHEATERS (SPH)

- Large capacity industrial systems.
- Energy replacement fraction is the principal criterion within some cost constraints.
- Annual design on mean radiation basis.

The closed loop differential temperature controlled system is most commonly used in this category (Fig.3).

3. DESIGN STATUS AND DATA BASE

(1) Most methods in literature are for SPH system, which is least used in India.

(2) Temperature specified systems (SAS and AAS) have excess energy in non-design months and have less energy on some days in the design month as currently designed. Higher temperatures than design value or larger volumes than specified are currently used as symbols of success. But they waste or rather burn capital money.

Minimum information needed for design is:

Collectors - Efficiency curve (ASHRAE Test)
- Time constant
- Incident angle modifier coefficient
- Installed cost Rs/m^2

Climate - Monthly mean G and $\overline{K}_T$, ambient t_{max} and t_{min}.

Load - Draw off pattern for choice of store/control strategy required.

4. COLLECTOR CHOICE

The first step is to use an appropriate type of solar collector for a given climate and temperature of hot water required with respect to product test data and installed cost quotation. Collector characterization (Fig.4) in broad classes can be done as per British Standard Practice (Ref.1). Collector choice is made as shown in Fig.5, following the procedure suggested by Howell et al. (Ref.2). The value of $\overline{X}$ ordinate is fixed as per current design procedure by taking $\overline{I}_{Gt}$ as eighty per cent of noon time peak value incident on the tilted surface.

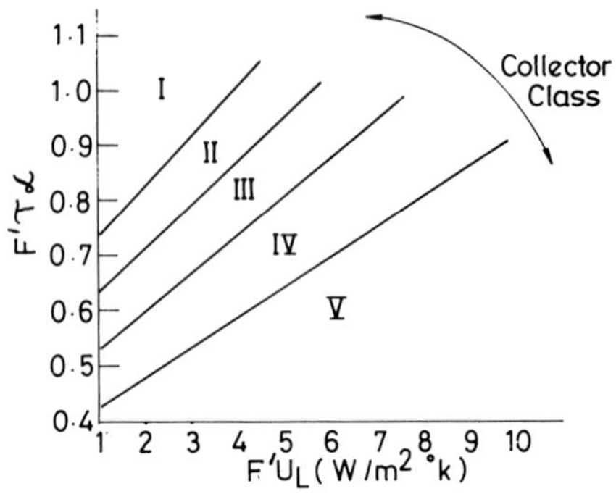

FIG.4 - COLLECTOR CHARACTERIZATION

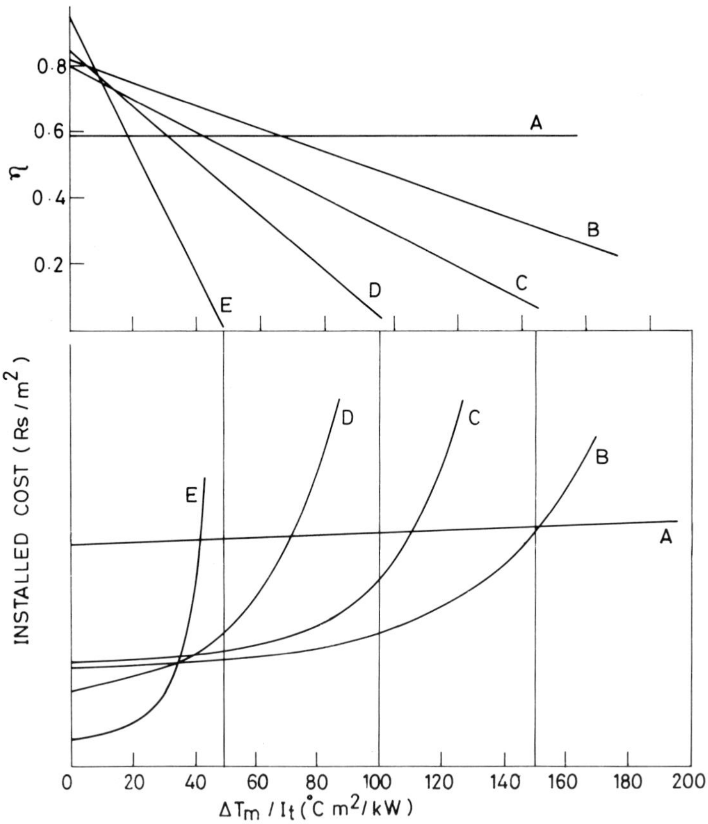

FIG.5 - COLLECTOR CHOICE FOR LOAD QUALITY AND CLIMATE

A = Evacuated Tube Collector (ETC); B = Single Glass Selective (SGS);
C = Double Glass Non-Selective (DGNS); D = Single Glass Non-Selective(SGNS)
E = Unglazed Non-Selective (UGNS); 1$(US) = Rs.12.00.

5. COLLECTOR SIZING

The Hottel-Whillier-Bliss (H-W-B) equation for instantaneous efficiency is:

$$q_u = F'\overline{\tau\alpha}(I_{Gt} - I_c) \tag{1}$$

where

$$I_c = \frac{U_L}{\overline{\tau\alpha}}(t_m - \overline{t}_a) \tag{2}$$

t_m = mean collector temperature

$\overline{\tau\alpha} = 0.95\ (\tau\alpha)_{normal}$

for the month of any hour

$$\sum_N q_u = F'\ \overline{\tau\alpha} \sum_N \left[I_{Gt} - I_c\right]^+ \tag{3}$$

for full month

$$\sum_{hrs} \sum_N q_u = Q_u = F'\overline{\tau\alpha} \sum_W \sum_N \left[I_{Gt} - I_c\right]^{++}$$

$$F'\overline{\tau\alpha}\ \overline{G}\ \overline{R}_\beta\ N\ \overline{\Phi} \tag{4}$$

where

$$\overline{\Phi} = \frac{\frac{1}{N} \sum_W \sum_N \left[I_{Gt} - I_c\right]^{++}}{\overline{G}\ \overline{R}_\beta} \tag{5}$$

= daily utilizability

Thus Daily Collector Efficiency ($\overline{\eta}_\Phi$) is given by

$$\overline{\eta}_\Phi = F'\ \overline{\tau\alpha}\ \overline{\Phi} \tag{6}$$

So,

$$(\overline{q}_u)_{\overline{\Phi}} = F'\overline{\tau\alpha}\ \overline{\Phi}\ \overline{G}\ \overline{R}_\beta \tag{7}$$

and collector area $A_{c\Phi} = \dfrac{\text{load}}{(\overline{q}_u)_\Phi}$ for any given month (8)

$\overline{\Phi}$ can be obtained graphically from plots of I_c and $\overline{K_T}^*$ as shown in Fig.6, following Evans et al. (Ref.3) and derived from radiation statistical correlation based on primary Indian measured data compiled by A. Mani (Ref.4). Alternatively Φ curves can be used from Duffie and Beckman (Ref.5). Night time store losses and line losses are added to the load for ease of calculation.

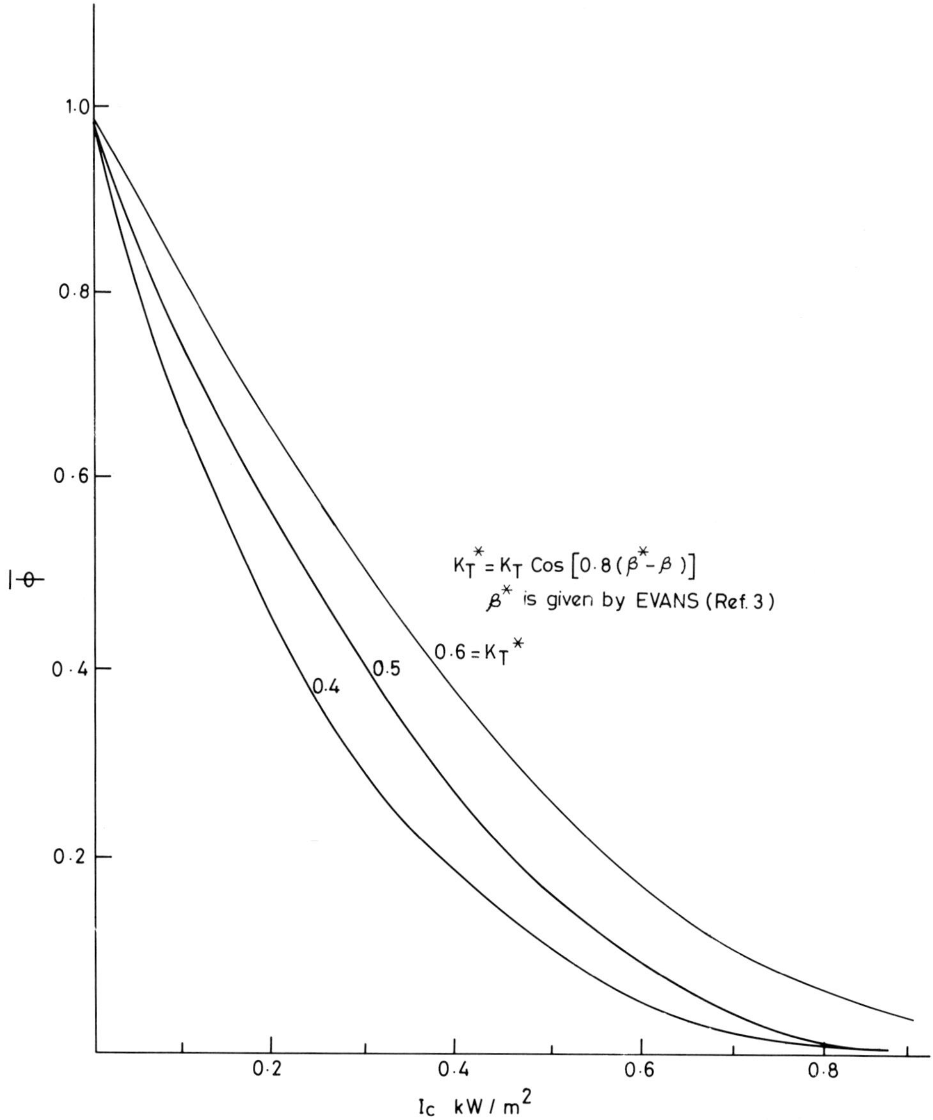

FIG.6 - LONG TERM PERFORMANCE PARAMETER ($\bar{\phi}$) EFFECT OF VARIABILITY OF G AROUND $\bar{G}$.

6. VARIABILITY FACTOR

Thus taking into account radiation variability over the month:

$$A_{\phi} = A_{mean}\left[\frac{I_{Gt}-I_c}{\bar{\phi}\;\bar{I}_{Gt}}\right] \tag{9}$$

where

A_{mean} = area sized on the mean monthly basis using

$\bar{I}_{Gt}$ = $(0.8\ I_{peak}\ \bar{R}_\beta)$ as daily representative solar radiation intensity

$$I_c = \frac{U_L}{\tau\alpha}(t_m - \bar{t}_a) \qquad (10)$$

$$\frac{\bar{I}_{Gt} - I_c}{\bar{\phi}\ \bar{I}_{Gt}} = \text{monthly variability factor}$$

and $\bar{\phi}$ is determined graphically for any location or month or collector tilt by a simple set of curves provided I_c is known.

and $\bar{K}_T = \frac{\bar{G}}{ETR}$ for the given month and location

= Clearness Index

ETR = Extraterrestrial Radiation Intensity.

7. PERFORMANCE INDEX

$$\text{Annual heat collected/m}^2 = \sum_{J=1}^{J=12} (F'\overline{\tau\alpha}\ \bar{\phi}\ \bar{G}\ \bar{R}_\beta\ N)_J \qquad (11)$$

(for the relevant season month
for domestic use)

$$\text{Performance Index} = \frac{\text{Annual heat collected/m}^2}{\text{Installed cost/m}^2}$$

in KCal a/Rs or MJ a/Rs (12)

8. SYSTEM DESIGN METHODS

So far we have sized collectors for the month (for maximum load and lowest radiation) and that is why the collector area sized is the maximum area(A_{max}) . But solar water heating system consists of collectors, piping, storage tank, control circuit for pumps and heat exchangers. Also A_{max} will involve lot of 'energy dumping' (temperature or mass) in better months.

So the choice of collector area (for maximum solar fraction or minimum capital cost/Joules annual) and choice of store volume in relation to A_c, load draw off, system configuration and control strategy are yet to be discussed for DTC, OTC and other systems.

During the last five years, considerable new thinking has emerged in these areas.

f charts (Ref.6) and $\bar{\phi}$-f charts (Ref.5) were primarily done for most popular DTC configurations, the primary emphasis being on collector area A_c. Effects of Mixing, Transients and Parasitic losses were not considered in relation to system efficiency as offered to collector efficiency.

Current thinking (Ref.7) is that low flow systems either with, (i) uniform flow and straified storage (Ref.8) with multiple pass, or (ii) variable flow (continuous) and variable volume with single pass, are much superior.

9. DESIGN PROCEDURE

9.1 Thermosyphon (SAS) Systems

For one day store and no daytime draw off:

(i) Choose design month in the winter (December or January). Fix orientation ($\lambda + 15^o$) and assemble data (λ is the latitude of plate).

(ii) Daily design load = Volume of water (litres) x 4.18 x ($t_{design} - t_{main} + \Delta t_n$),

where Δt_n = designed night time mean tank temperature drop ($< 8^oC$ ideally 5^oC)

(iii) Calculate I_c with respect to $t_m = \left(\frac{t_{design} + t_{main}}{2}\right)$

(iv) Calculate $\bar{\phi}$ $(\bar{q}_u)_\phi$ for unit collector area from Fig.6.

(v) Calculate collector area required (A_ϕ).

Daytime draw off:

Uptil 50% capacity excess can be assumed if water is drawn off before noon - use a low resistance system.

Allowance for rainy days (in winter):

(i) Provide for 1.5 days storage with 1 day's collector area - 75 l/m^2 is maximum. Keep the collector area as for one day's load with same t_m.

(ii) Provide auxiliary heater of rating that it can heat one l/minute uptil 50^oC from main temperature and use it only in night/early morning.

$$t_{mains} = (0.8\ t_{min} + 0.2\ t_{max})$$

where t_{min} and t_{max} is the daytime minimum and maximum temperature respectively.

9.2 OTC System with Variable Store (Low Flow Single Pass, AAS System)

For Continuous Variable Flow systems:

Design is exactly similar to Thermosyphon systems (t_m is same).

Batch Flow systems:

This is most commonly used large SWH system in India and is most poorly researched one. It is used year round for public buildings/industry.

Tentative design procedure -

(1) Fix collector tilt equal to latitude, and assemble data including collector time constant and incidence angle modifier.

(2) For calculation of I_{CD}, use $t_m = t_{Design}$

$$\text{i.e., } I_{CD} = \frac{U_L}{\tau\alpha}(t_D - \bar{t}_a)$$

This is valid as experimental observations show that cycle time is not very large and the collector heats up in a stagnating mode.

(3) For operating time, use I_{co}, where

$$I_{co} = \frac{U_L}{\tau\alpha}\left[(t_D - \bar{t}_a) + 0.632(t_D - t_{main})\right]$$

(4) Use a sinosoidal curve for $\overline{R}\,\overline{G}$ with

$$\text{Peak value} = \left(\frac{\pi}{2}\frac{\overline{R}\,\overline{G}}{\text{Day length (hrs)}}\right), \quad \text{where}$$

$$\text{Day length} = \frac{\cos^{-1}}{7.5}\left[-\tan\lambda\tan\delta\right]$$

Determine operating time with respect to I_{co}.

(5) Determine Collector effective Capacity/Module

$(MC)_e = F'\,U_L\,A_c$* (Time Constant), or

determine time constant from effective capacity as

$$(MC)_e \text{ of module} = (MC)_{water} + F'\left[(MC)_{plate} + (MC)_{insulation} + (MC)_{cover}\right]$$

$= 16$ KJ/^{o}K for collector with single glazing.

(6) Cycle Time T = $-$ (Time Constant) ln (X)

$$\text{where} \qquad X = \frac{I_{Gt}\tau\alpha - U_L(t_D - t_a)}{I_{Gt}\tau\alpha - U_L(t_{fi} - t_a)}$$

$$= \frac{I_{Gt} - I_{CD}}{I_{Gt} - I_{ci}}$$

where $I_{ci} = \frac{U_L}{\tau\alpha}(t_{ft} - t_a)$

I_{Gt} = Radiation falling on collector face at the time.

(7) Repeat calculations for each hour and use incidence angle modifier to get $(\tau\alpha)$ from $(\tau\alpha)_n$ get total number of cycles during the operating time.

(8) Water delivered at $t_D{}^oC$ during day corresponding to mean $\overline{G}$ of the month is given by

V = No. of cycles x A_c x (water in collectors/m^2)

Calculate A_c for given V or vice versa.

(9) Repeat calculations for each month and get f_{annual}.

9.3 DTC Systems with Well Mixed Store (High Flow Multiple Pass SPH System)

Ø-f chart method:

(i) Monthly calculations-fix collector tilt = latitude and assemble data

(ii) Calculate loads as in system 1 and multiply by number of days in the month to get monthly load in Joules. Choose trial collector area (m^2) = SWH Capacity/75(1)

Parameters	Range
$(\tau\alpha)_n$	$< 0.6, 0.9 >$
FRA_c	$< 5, 120m^2 >$
U_L	$< 2, 8\ W/m^2\,{}^oK >$
β	$< 30^o, 80^o >$
$\frac{\text{Store cap.}}{m^2}$	$< 37.5, 300\ 1 >$

(iii) Calculate $(\overline{\tau\alpha}\ F_R)$, $(F'_R$ if heat exchanger is there in collector loop,

if η vs $\Delta t_m/I_t$ curves are given by

$$FR\ \overline{\tau\alpha} = F'\tau\alpha \left[\frac{1-\exp(-NTU)}{NTU}\right]$$

where NTU $= (F'U_L/GC_p)$

where G can be taken as 50 l/m^2hr (pump design accordingly).

(iv) Calculate $I_{c,min}$ for T_{mains} and $\overline{\phi}_{max}$ for this I_c.

(v) Calculate $X = \frac{A_c F_R U_L (100)\,\Delta t}{\text{Load (Joules)}}$

where Δt is number of seconds in month and U_L is in $W/m^2\,{}^oK$

$$Y = K_3 \frac{A_c F_R (\overline{\tau\alpha})\ \overline{G}\ \overline{R}\ \ N}{\text{Load (Joules)}}$$

Here N is number of days in month and $\overline{G}$ is J/m^2.

where $K_3 = \left(\frac{11.6+1.18\ t_{Design}+3.86\ t_{mains}-2.32\ \bar{t}_a}{(100-t_a)}\right)$

Here all temperatures are in ^{o}C.

(vi) Use X and Y and chosen storage capacity and load pattern to get solar fraction, f, from charts or programs available in literature (Ref.6).

(vii) Calculate monthly solar fractions for each month and obtain an annual fraction or seasonal fraction.

(viii) Repeat calculations for various collector areas and storage volume per unit collector area ratio for economic choices with respect to given constraints as for life cycle cost.

Following points should be kept in mind while using the $\bar{\emptyset}$-f chart method:

$\bar{\emptyset}$-f chart method is valid only if -

* minimum temperature level of load (return temperature to load) is constant during the month;
* load over the day is uniform (12 hrs/day or 24 hrs/day are allowed), i.e. not valid for week-end non-uses etc. as in industry, schools, etc.
* for temperature independent performance systems (not valid for solar air conditioning or organic Ranking Cycle systems).

REFERENCES

1. B.S.-5918, (1980), 'Code of practice for solar heating system for domestic water heating', British Standard Institution, London.
2. J.R. Howell, R.B. Bannerot and G.C. Vliet, (1982), Solar Thermal Energy Systems - Analysis and Design, McGraw Hill, New York.
3. D.L. Evans, T.T. Rale and B.D. Wood, (1982), 'A new look at long term collector performance and utilizability', Solar Energy, 28, 13-24.
4. A. Mani, (1980), Handbook of Solar Radiation for India, Allied Publishers, New Delhi.
5. J.A. Duffie and W.A. Beckman, (1974), Solar Energy Thermal Processes, John Wiley & Sons, New York.
6. W.A. Beckman, S.A. Klein and J.A. Duffie, (1977), Solar Heating Design by the f-Chart Method, Wiley-Interscience, New York.
7. M.C. Pereira, (1984), 'Single pass strategy for solar water heating', Sun World, 8(2), 49-51.
8. L.F. Jesch and J.E. Braun, (1984), 'Variable volume storage and stratified storage for improved water heater performance', Solar Energy, 33(1), 83-88.

TESTING OF SOLAR COLLECTORS

A.K. Sharma* and M.S. Sodha**
*Centre for Energy Studies
**Physics Department
Indian Institute of Technology
Hauz Khas, New Delhi 110 016
India

1. INTRODUCTION

The parameters which determine the performance of a collector are the operating temperature, fluid flow rate, solar insolation, orientation, tilt, time of the day, wind conditions, outdoor temperatures, clearness of the sky etc. Due to large variations in these parameters, the performance of these systems used to be often expressed in an arbitrary fashion. With increasing number of solar collectors in the market, a need was felt to adopt a standard testing and rating procedure for them. These test procedures should be able to provide an equitable basis for competition amongst manufacturers and should also provide an essential criterion for design and selection of the equipment. The test procedure should be based on analytical considerations, which enable the performance over a wide range of parameters to be uniquely characterized by a curve, which could be obtained with few observations.

2. PERFORMANCE EQUATIONS

2.1 Efficiency

Under steady state conditions, the energy balance equation for flat plate collectors is (Ref.1-3):

$$\frac{\dot{Q}_u}{A_c} = I_t(\tau\alpha)_e - U_L(\bar{T}_p - T_a) \tag{1}$$

A diagramatic representation of various heat transfer parameters providing the basis for writing equation (1) is shown in Fig.1. In a typical flat plate collector the measurement of plate temperature is not convenient. Therefore, the tests are based on the following equations, which can be derived (Ref.4-10) from equation (1):

H. P. Garg (ed.), Solar Water Heating Systems, 383–397.

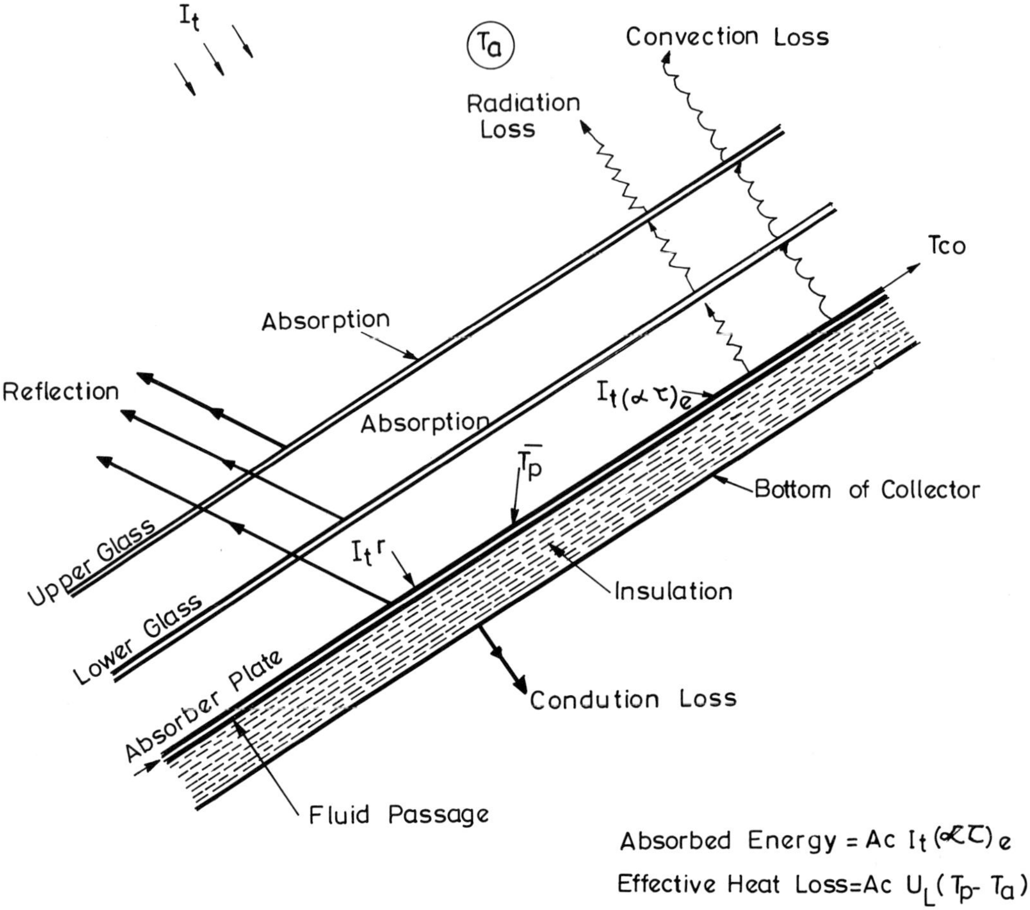

FIG.1 - HEAT LOSS FROM FLAT-PLATE COLLECTOR

$$\frac{\dot{Q}_u}{A_c} = F'\left[I_t(\tau\alpha)_e - U_L\left(\frac{T_{co}+T_{ci}}{2} - T_a\right)\right] \quad (2)$$

or

$$\frac{\dot{Q}_u}{A_c} = F_R\left[I_t(\tau\alpha)_e - U_L(T_{ci}-T_a)\right] \quad (3)$$

These equations are popularly known as Hottel-Whillier-Bliss equations; the collector plate efficiency factor F' and the heat removal efficiency factor F_R may be defined as:

$$F' = \frac{\text{Actual collected useful energy}}{\text{Collected useful energy if the collector plate were at the mean fluid temperature}} \quad (4)$$

$$F_R = \frac{\text{Actual collected useful energy}}{\text{Collected useful energy if the collector plate were at the inlet fluid temperature}} \quad (5)$$

$\dot{Q}_u$ is obtained from the equation

$$\dot{Q}_u = \dot{m}\, C_p(T_{co}-T_{ci}) \quad (6)$$

Using equations (1), (3) and (4), the collector efficiency may be expressed as

$$\eta = \frac{\dot{Q}_u}{A_c I_t} = (\tau\alpha)_e - U_L\left(\frac{\bar{T}_p - T_a}{I_t}\right) \quad (7)$$

$$\eta = F'\left[(\tau\alpha)_e - \frac{U_L}{I_t}\left(\frac{T_{co}+T_{ci}}{2} - T_a\right)\right] \quad (8)$$

and

$$\eta = F_R\left[(\tau\alpha)_e - U_L\left(\frac{T_{ci}+T_a}{I_t}\right)\right] \quad (9)$$

If one uses the collector outlet (T_{co}) as the base temperature then the efficiency equation is

$$\eta = F''\left[(\tau\alpha)_e - U_L\left(\frac{T_{co}-T_a}{I_t}\right)\right] \quad (10)$$

Equations (7) to (10) indicate that a plot of η versus $\frac{\Delta T}{I_t}$ is a straight line provided F' and U_L remains constant. The slope of the straight line is mainly determined by U_L and the η intercept is proportional to $(\tau\alpha)_e$. Parameters F',F" and F_R can be correlated with each other; the relevant conversion factors are given in Table I.

TABLE 1 - Conversion Factors for the Collector's Efficiency to Various Base Temperatures (Ref.2)

To obtain ↓ by / Multiply →	$F_R U_L$ or $F_R(\tau\alpha)_e$	$F'U_L$ or $F'(\tau\alpha)_e$	$F''U_L$ or $F''(\tau\alpha)_e$
$F_R U_L$ or $F_R(\tau\alpha)_e$	1	$\frac{1}{1+\frac{F'U_L}{2\dot{m}C_p}}$	$\frac{1}{1+\frac{F''U_L}{\dot{m}C_p}}$
$F'U_L$ or $F'(\tau\alpha)_e$	$\frac{1}{1-\frac{F_R U_L}{2\dot{m}C_p}}$	1	$\frac{1}{1+\frac{F''U_L}{2\dot{m}C_p}}$
$F''U_L$ or $F''(\tau\alpha)_e$	$\frac{1}{1-\frac{F_R U_L}{\dot{m}C_p}}$	$\frac{1}{1-\frac{F'U_L}{2\dot{m}C_p}}$	1

2.2 Time Constant

Wherever transient condition exists, the thermal performance of collector is given by (Ref.1 and 11):

$$\frac{C_a}{A_c}\frac{d\bar{T}_f}{dt} = F_R\left[I_t(\tau\alpha)_e - U_L(T_{ci}-T_a)\right] - \frac{\dot{m}C_p}{A_c}(T_{co}-T_{ci}) \tag{11}$$

In order to solve equation (11), the following simplifying assumptions are usually made:

(i) The outlet temperature of the transfer fluid is related to the average temperature by

$$\frac{d\bar{T}_f}{dt} = K\frac{dT_{co}}{dt} \tag{12}$$

where

$$K = \frac{\dot{m}C_p}{F'U_L A_c}\left(\frac{F'}{F_R} - 1\right)$$

(ii) During the transient condition, all the parameters, viz. I_t, $(\tau\alpha)_e$, U_L, T_a, $\dot{m}$, C_p and T_{ci} are constant.

The solution of equation (11) is then

$$\frac{F_R I_t(\tau\alpha)_e - F_R U_L(T_{ci}-T_a) - \frac{\dot{m}C_p}{A_c}(T_{co}-T_{ci})}{F_R I_t(\tau\alpha)_e - F_R U_L(T_{ci}-T_a) - \frac{\dot{m}C_p}{A_c}(T_{co,initial}-T_{ci})} = \exp(-t/\Upsilon_1) \tag{13}$$

where $\Upsilon_1 = KC_a/\dot{m}C_p$; the parameter Υ_1 is known as the collector's time constant.

2.3 Incident Angle Modifier

The effective transmittance-absorptance product, $(\tau\alpha)_e$, of a solar collector is described by the relation (Ref.12):

$$(\tau\alpha)_e = K_{\alpha\tau}\,(\tau\alpha)_{e,n} \tag{14}$$

where $K_{\alpha\tau}$ is known as incident angle modifier and is a function of the angle of incidence of the direct solar radiation.

3. TESTING PROCEDURES (Ref.13)

In order to test the solar collectors using water as the transfer fluid, the following testing procedures are suggested:

3.1 N.B.S. Testing Procedure

The N.B.S. testing procedure (Ref.14) requires a series of outdoor tests to be conducted on collectors to determine the near normal incidence efficiency over a range of temperatures. The test configuration is shown in Fig.2. The requirements for the size of collectors, its mounting, installation etc. are as follows:

(i) The size of the collector should be big enough to be representative of a module in solar energy system. If one module is to be tested, it should be so mounted and insulated as to minimise the back and edge losses to the extent that these are representative of actual losses in practice.

(ii) The collector should be installed in a location where reflected or re-radiated energy from surrounding buildings is not incident on the collector surface; further the collector should not be shaded during the test.

(iii) The orientation of the collector should be such that the incident angle is always $< 45^{\circ}$.

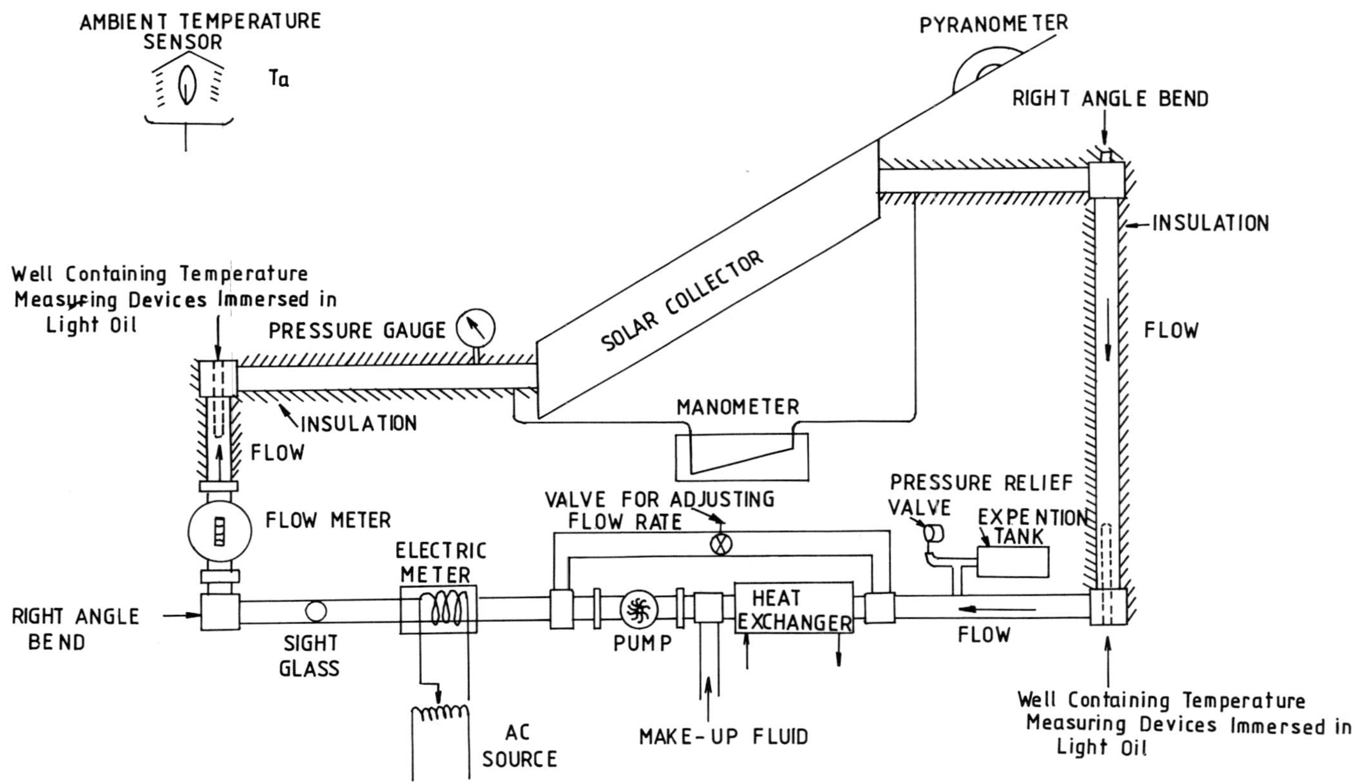

FIG.2 - NBS TESTING CONFIGURATION FOR THE SOLAR COLLECTOR WHEN THE TRANSFER FLUID IS LIQUID (REF.14)

(iv) The tests have to be performed on days and time when the 15 minute integrated average solar radiation values should be $\geqslant 630\ W/m^2$.

(v) The record of air velocity across the collector has to be maintained. The measurement has to be done at one meter distance from the collector surface.

(vi) Test should be performed at various constant inlet fluid temperatures, viz. 10°, 30°, 50° and 70^oC above the ambient temperature.

(vii) The diffuse part of the radiation has to be measured and reported.

(viii) There should be at least four data points for each value of the collector inlet fluid temperature (T_{ci}), two preceding solar noon and two following solar noon.

(ix) The measurement and the calculation of the efficiency has to be averaged over an interval of 15 minutes,

$$\eta = \frac{\dot{m}C_p \int_o^{15\ mts.} (T_{co} - T_{ci})\ dt}{A_c \int_o^{15\ mts.} I_t\ dt} \qquad (15)$$

(x) The theoretical pumping power of the transfer fluid mass to be specified and calculated by using the formula

$$P_{th} = \frac{\dot{m}\Delta p}{\rho}$$

Here Δp is the pressure drop across the collector.

(xi) The efficiency curve is based on equation (8). At least 16 data points are obtained for establishing the efficiency curve and using least square fit an equation for the curve is obtained. The slope of the best fit line and the intercept on efficiency axis yield $F'U_L$ and $F'(\tau\alpha)_e$ respectively.

3.2 ASHRAE Standard 93-77 Testing Procedure (Ref.15)

Figure 3 shows the test configuration. The ASHRAE standard 93-77 is similar to NBS procedure with the following distinctions:

(a) Efficiency equation (9) is utilized in place of equation (8) to get the collector parameters.

(b) Time constant test is performed.

(c) Incident angle modifier is determined.

The test procedure consists of following steps:

(i) Prior to test, the collector is exposed to solar radiation for three days under stagnant conditions with the incident solar radiation over the collector surface exceeding the value of $197\ W/m^2$.

(ii) The 'time constant test' is carried in the following way: The collector is exposed to solar radiation and after the existing fluid temperatures have stabilised, the collector is suddenly shielded from the sun. A record of the outlet fluid temperature is maintained on a

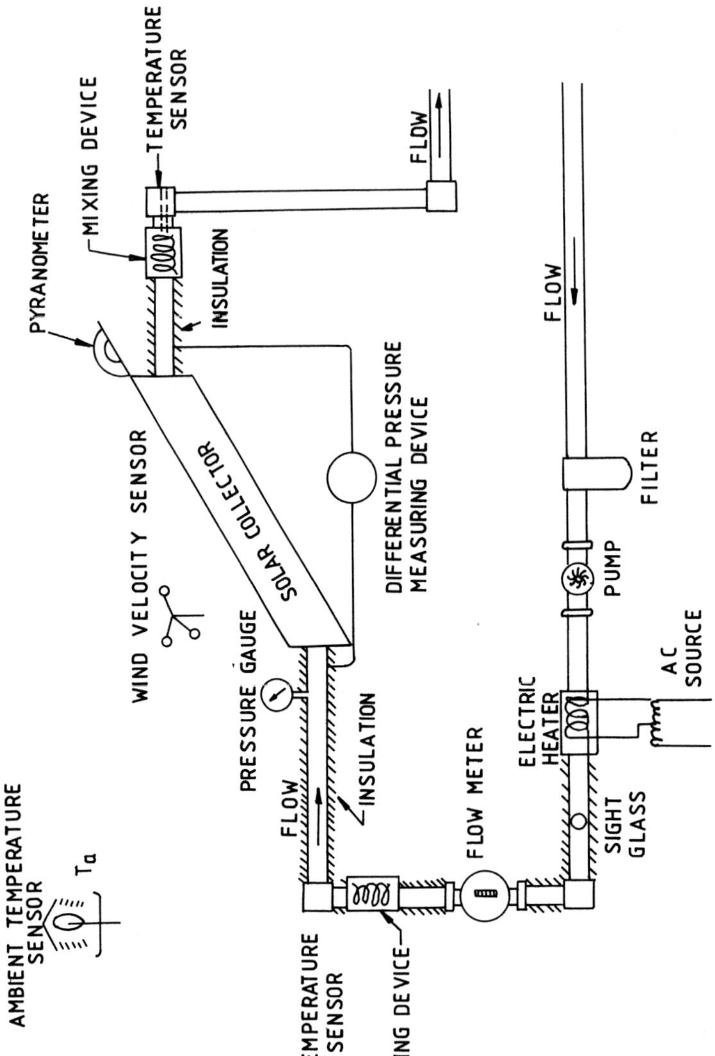

FIG.3 - OPEN LOOP TESTING CONFIGURATION FOR THE SOLAR COLLECTOR WHEN THE HEAT TRANSFER FLUID IS A LIQUID AND IT IS SUPPLIED CONTINUOUSLY (REF.16)

strip chart recorder. The inlet fluid temperature has to be measured with an accuracy of $\pm 1^{o}C$ and has to be above the ambient temperature. The radiation level should be maintained $\geqslant$ 790 W/m^2.

(iii) In the 'efficiency test' the average efficiency is integrated over a 5 minute interval ($\geqslant$ time constant) and is calculated over a range of temperatures. The flow rate is required to be steady and vary by $<$ 1% for the duration of the test. The heat transfer fluid should be of known specific heat and should not vary more than 0.5% over the temperature range of experiment.

(iv) At least 16 data points are required for the complete test. These data points must be taken symmetrically with respect to solar noon. The incident solar radiation must be quasi-steady during each test period.

(v) The 'incident angle modifier' test is performed after a series of tests. Here, first of all the solar efficiency curve is obtained for near solar noon efficiency and then the test is made for incident angles of 0°, 30°, 45° and 60°. By comparing these results with those obtained in establishing the efficiency curve at incident angle of $< 30^{o}$, values of $K_{\alpha\tau}$ can be computed as a function of incident angle up to 60°.

3.3 B.S.E. Testing Procedure (Ref.17)

The B.S.E. testing procedure was developed by German Solar Energy Society (Ref.18) and requires that the optical efficiency ($\eta_o = \dot{Q}_u/I_t A_c$, when Q_I the energy loss is zero) and thermal loss ($\dot{Q}_L$) of a collector were determined through a series of outdoor and indoor tests. The procedure is similar to N.B.S. testing procedure but differ in the determi nation of thermal loss coefficient of the collector which is determined under indoor laboratory conditions (no incident solar radiation). The two separately determined properties are then used to construct normal incidence efficiency curve as a function of various operating conditions. The need of developing this test procedure was to improve the reproducibility of test results and to decrease the time required for collector testing. To improve the reproducibility of test results the indoor testing is reduced to minimum and it is combined with the controlled indoor testing. In the process, the required testing time is also considerably reduced.

4. INSTRUMENTATION

4.1 Solar Radiation Measurement

A pyranometer is used to measure the total solar radiation incident on a solar collector. The pyranometer should have the following characteristics:

(i) It should either have built-in temperature compensation circuit or its temperature coefficient should be predetermined.

(ii) The time constant should be $<$ 5 sec.

(iii) The pyranometer is calibrated within six months of the collector test against the standard pyranometer.

(iv) It should be mounted on a surface parallel to the collector so that it does not cast a shadow on the collector.

(v) No reflected and re-radiated energy from the collector is incident on to the pyranometer.

4.2 Temperature Measurements

(i) The temperature measurements are made (Ref.19) in accordance with ASHRAE Standard 41-66.

(ii) Thermoelements or thermopiles or platinum resistance sensors, which are all suitably calibrated, are utilized for measuring the temperature at inlet/outlet of solar collector.

(iii) The thermopile is made by taking copper-constantan thermocouple wires from a single spool. The wire diameter should not be larger than 0.51 mm (Ref.20). The ends of thermopile are located near to the inlet/outlet of the solar collector.

4.3 Flow Rate Measurements

Liquid flow rates are usually measured with rotameters. These measuring instruments are designed in such a way that the maximum tolerance should not be greater than ±1%. Measurement error caused by air bubbles in the test section (used for measuring the flow rate) should be avoided. In general, air bubbles are less likely to form in high pressure sections and in cold sections. For this reason, the flowmeter is to be installed before the lower inlet side. During indoor tests, the lowest point after the collector outlet is the optimal side for the flowmeter. Depending on the construction of flowmeter, it could be advantageous to change the tilt angle (e.g. to a sloping position) in order to prevent the air pockets in the test section in which the flow rate is measured.

4.4 Pressure Drop Measurement Across the Collector

The pressure drop across the collector is measured using static pressure tap holes and a manometer. The edges of the holes should be free of burrs on the inside surface of pipe and should not be exceeding 1.6 mm (1/16') diameter (Ref.21). The thickness of the pipe wall should be 2.5 times the hole diameter.

4.5 Integrators and Recorders

(i) Strip chart recorders with an accuracy of ±0.5% of the temperature difference or measured voltage and of a time constant $\lesssim$ 1 sec are used.

(ii) Electronic integrators which have an accuracy of ±1% of the measured value are used.

4.6 Time and Mass Measurements

These are made to an accuracy of ±2% (Ref.22).

4.7 Wind Velocity Measurement

The wind velocity is measured with an instrument which can determine the integrated average wind velocity of 15 minutes interval during test period with an accuracy of ± 0.08 m/s (1.8 mph).

5. PRESENTATION OF RESULTS

From various measurements performed under each test, the test data has to be presented along with a table of test conditions (e.g. see Table 2). The N.B.S. and B.S.E. procedure demand a plot of the efficiency versus $(T_{co}+T_{ci})/2-T_a\ /I_t$.

A typical curve (Ref.23) is shown in Fig.4.

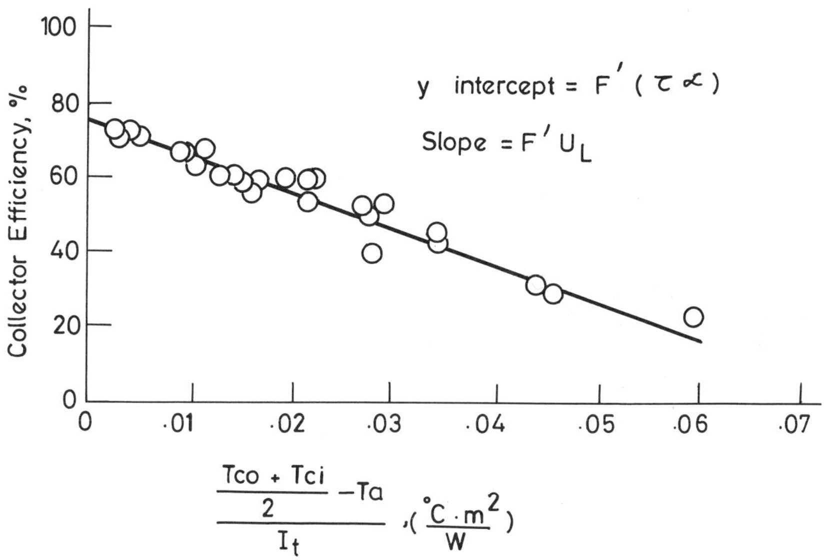

FIG.4 - EFFICIENCY CURVE FOR A FLAT-PLATE COLLECTOR USING WATER AS THE TRANSFER FLUID (REF. 23)

The ASHRAE method is based on equation (9). The time constant can be theoretically calculated using equation (13). There is, however, large uncertainty in the determination of effective heat capacity (C_a). The upper limit for this heat capacity is the sum of mass multiplied by specific heat for each component that makes the collector (absorber, glass, insulation, etc.), plus that of the heat transfer fluid required to fill the collector. If this value were used for Ca, the time constant might be over estimated. The time constant has therefore to be determined experimentally. A typical performance equation curve for a double glazed collector using ASHRAE procedure is shown in Fig.5.

TABLE 2 - Test Data to be Recorded (Ref.14)

S.No.	Item
1	Date
2	Observer(s)
3	Equipment name plate data
4	Collector tilt angle
5	Collector azimuth angle (as a function of time, if movable
6	Collector aperture area or frontal transparent area
7	Local standard time, at the beginning of collector warm-up and at the beginning and end of each 15 minute test period
8	Ambient air temperature, at the beginning and end of each 15 minute test period
9	$\Delta T = (T_{co}-T_{ci})$ across solar collector (either as a continuous function of time or as a 15 minute integrated quantity)
10	Inlet temperature, T_{ci} (as a continuous function of time)
11	Outlet temperature, T_{co} (as a continuous function of time)
12	Liquid flow rate
13	Gauge pressure at solar collector inlet
14	Pressure drop across solar collector
15	Height of the collector outlet above the collector inlet
16	Wind velocity near the collector surface or aperture (15 minute average)
17	I, the incident solar radiation onto the collector (as a continuous function of time and as a 15 minute integrated quantity, if required)
18	I_d, the diffuse component of the solar radiation onto the collector (at the beginning of the 15 minute period and after the completion of the 15 minute period

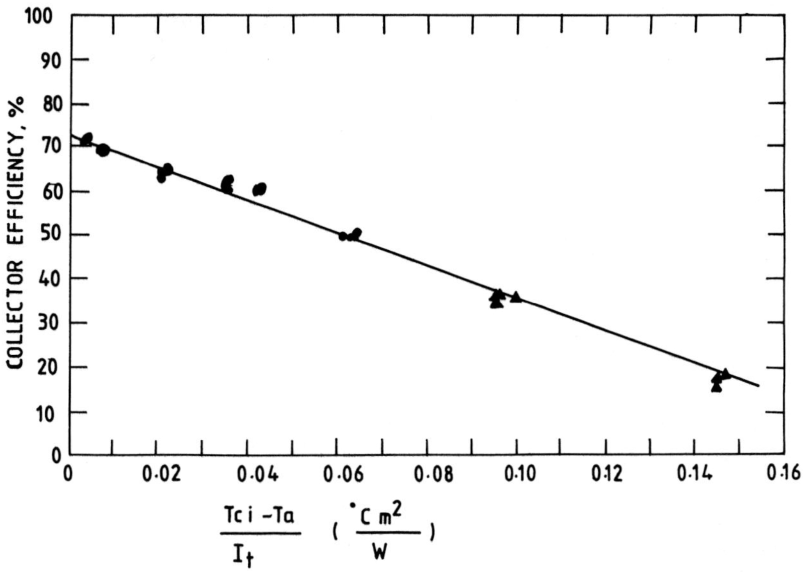

FIG.5 - NEAR-NORMAL-INCIDENCE EFFICIENCY FOR A FLAT-PLATE DOUBLE-GLAZED WATER HEATING COLLECTOR (REF. 15)

REFERENCES

1. J.A. Duffie and W.A. Beckmann, (1974), Solar Energy Thermal Processes, John Wiley and Sons, New York.
2. G.O.G. Löf, (1980), 'Flat plate and nonconcentrating collectors', Solar Energy Technology Handbook (Part A) (ed. W.C. Dickinson and P.N. Cheremisinoff), Marcel Dekker Inc., New York.
3. R.G. Nevins and P.E. McNall, (1958), 'A high flux low temperature solar collector', ASHRAE Transactions, 64, 69-82.
4. H.C. Hottel and A. Whillier, (1958), Transactions of the Conference on the use of Solar Energy, Vol. 2, Part I, Univ. Arizona Press, Arizona, U.S.A., p. 74.
5. H.C. Hottel and B.B. Woertz, (1942), 'The performance of flat-plate solar heat collectors', ASME Transactions, 64, 91.
6. S.A. Klein, (1975), 'Calculation of flat-plate collector loss coefficients', Solar Energy, 17, 79-80.
7. B.Y.H. Liu and R.C. Jordan, (1963), 'A rational procedure for predicting the long term average performance of flat-plate solar energy collectors', Solar Energy, 7(2).
8. R.W. Bliss, (1959), 'The derivation of several plate efficiency factors, useful in the design of flat-plate solar heat collectors', Solar Energy, 3(4), 55.
9. A. Whillier, (1953), 'Solar Energy Collection and its Utilisation for House Heating', D.Sc. Thesis, MIT, U.S.A.
10. A. Whillier, (1967), 'Design factors influencing collector performance, low temperature engineering application of solar energy', ASHRAE, Inc., 345 East 47th Street, New York.

11. S.A. Klein, J.A. Duffie and W.A. Beckmann, (1974), 'Transient consideration of flat-plate solar collectors', ASME J. Engineering Power, 96A, 109.
12. F.F. Simon and E.H. Buyco, (1976), 'Outdoor flat-plate collector performance prediction from solar simulator test data', NASA TMX-7107, Presented at the 10th AIAA Thermal Physics Conference, Denver, Colorado, May 27-29.
13. J.E. Hill, E.R. Streed, G.E. Kelly, J.C. Ceist and T. Kusuda, (1976), 'Development of proposed standards for testing solar collectors and thermal storage devices', NBS Technical Note 899, National Bureau of Standards, Washington, D.C., U.S.A.
14. J.E. Hill and T. Kusuda, (1974), 'Methods of testing for rating solar collectors based on thermal performance', NBSIR 74-635, National Science Foundation, Washington, U.S.A.
15. J.E. Hill, (1980), 'Standard procedures for collector performance testing', Solar Energy Technology Handbook (Part A) (ed. W.C. Dickinson and P.N. Cheremisinoff), Marcel Dekker Inc., New York.
16. ASHRAE Standard 93-77, (1977), 'Methods of testing to determine the thermal performance of solar collectors', ASHRAE, New York.
17. J.P. Jenkins and J.E. Hill, (1980), 'Testing flat-plate water heating solar collectors in accordance with the BSE and ASHRAE procedures', NBSIR 80-2087, National Bureau of Standards, Washington, U.S.A.
18. B.S.E., (1978), 'Guidelines and directions for determining the usability of solar collectors - A solar collector efficiency test', BSE, Kruppstrasse 5, 4300 Essen 1, FRG.
19. 'Standard Measurements Guide : Section on Temperature Measurements', ASHRAE Standard 41-66 (Part I), American Society of Heating, Refrigerating and Air Conditioning Engineers, Inc., 345 East 47th Street, New York, 1966.
20. 'American Standard for Temperature Measurement Thermocouples', C96.1-1964 (R1969), American National Standards Institute, 1969.
21. Instrument and Apparatus, Part II, Pressure Measurement, Supplement to the ASME Power Test Codes, Am. Soc. Mechanical Engineers, 345 East 47th Street, N.Y., 1964.
22. ASHRAE Standard 37-69, (1969), 'Methods of testing for rating unitary air conditioning and heat pump equipment' American Society of Heating, Refrigerating and Air Conditioning Engineers, Inc., 345 East 47th Street, New York.
23. F.F. Simon and Harlament, (1973), 'Flat-plate collector performance evaluation: The case for a solar simulator approach', NASA TMX-71427.

NOMENCLATURE

A_c	Cross-sectional area of solar collector, m^2
C_a	Effective heat capacity of the collector, its component and the transfer fluid, $J/^oC$
C_p	Specific heat of the transfer fluid, $J/Kg\ ^oC$
F'	Solar collector efficiency factor
F''	Solar collector flow factor
F_R	Solar collector heat removal factor
I_t	Total solar energy incident on the plane of solar collector per unit time per unit area, W/m^2
$K_{\alpha\tau}$	Incident angle modifier
$\dot{m}$	Rate of mass flow rate of the transfer fluid, Kg/s
P_{th}	Theoretical power required to move the transfer fluid through the solar collector, W
$\dot{Q}_u$	Rate of useful energy extracted from the solar collector, W
Q_L	Rate of energy loss from the solar collector, W
T_a	Ambient air temperature, ^oC
T_{ci}	Temperature of the fluid at the inlet of solar collector, ^oC
T_{co}	Temperature of the fluid at the outlet of solar collector, ^oC
$\bar{T}_f$	Average temperature of the transfer fluid, ^oC
$\bar{T}_p$	Average temperature of the absorber surface, ^oC
t	Time, S
U_L	Overall heat loss coefficient for the solar collector, $W/m^2\ ^oC$
τ_1	Time constant of the solar collector
ρ	Fluid density, Kg/m^3
η	Solar collector efficiency, %
$(\tau\alpha)_e$	Effective absorptance-transmittance factor for the solar collector
$(\tau\alpha)_{e,n}$	Effective transmittance co-absorptance product for the solar collector at normal incidence
ΔP	Pressure drop across the solar collector, N/m^2
ΔT	Temperature difference between outlet and inlet transfer fluid temperatures of the solar collector, ^oC

TECHNOECONOMICS OF SOLAR FORCED FLOW HYBRID HOT WATER SYSTEMS

Arvind Goyal, Ashvini Kumar and M.S. Sodha
Physics Department and
Centre of Energy Studies
Indian Institute of Technology
Hauz Khas, New Delhi 110 016
India

1. INTRODUCTION

Solar water-heating is a well known and proven technology in most parts of the world. An appropriate design of solar hot water systems, depends on the need, habit, resources and local climatic conditions; these considerations can be quantified by an appropriate demand pattern. Many aspects of these systems, viz. the operation under thermosyphonic (Ref.1-7), and forced convection mode (Ref.8-10), use of heat exchangers in the storage tank (Ref.10-11), and use of antifreeze solutions (Ref.12) etc., have been studied. The thermal performance of these systems have been investigated both experimentally (ref.1, 13-17) and theoretically (Ref.18-21). Several computer programs (Ref.22-27) have also been developed and validated by comparing the simulation results with the actual performance data.

In every climate, one encounters days when the incident solar radiation is very little and one needs an auxiliary energy source for meeting the hot-water demand. The augmentation of a solar system with an auxiliary system is also necessary from economic considerations because of hot-water system designed for minimum insolation conditions will be too large to be cost effective. These systems are usually known as hybrid solar water heating systems. Amongst others, Gutierrez et al. (Ref.8) studied a hybrid system and investigated the effects of auxiliary energy supply, load type and storage capacity on the performance. For optimizing the collector area and tank capacity for a given hot water demand and climate, one requires an appropriate technoeconomical model. In principle, various models developed earlier (Ref.22-23) can with some modifications be used for the purpose; however, these models require large computers and are in general complicated.

In the present paper, a simple mathematical model has been developed to evaluate the performance of a hybrid system from technical and economical points of view. The system consists of a forced flow solar hot water system and an instant electric heater as an auxiliary energy source. It may be noted here that the instant heater represents the most efficient use of the electricity. To get a numerical appreciation of the model, numerical calculations have also been carried out for the

H. P. Garg (ed.), Solar Water Heating Systems, 399–416.

climate of Delhi (28.6^{O}N), India.

The study is confined to the case of forced convection systems, when the collectors are connected in parallel to the tank.

2. THERMAL ANALYSIS

A typical forced flow solar hot water system is schematically shown in Fig.1. If the temperature of the water in the tank is lower than the desired temperature, additional energy needed to heat the water is supplied by the auxiliary energy source, viz. an instant electric heater fitted in the tap itself. In case the water temperature is above the desired temperature, the instant heater is off. Further, the water is allowed to circulate through the bank of collectors only if the difference of temperatures between the collector outlet and the top of the storage tank is more than 5°C.

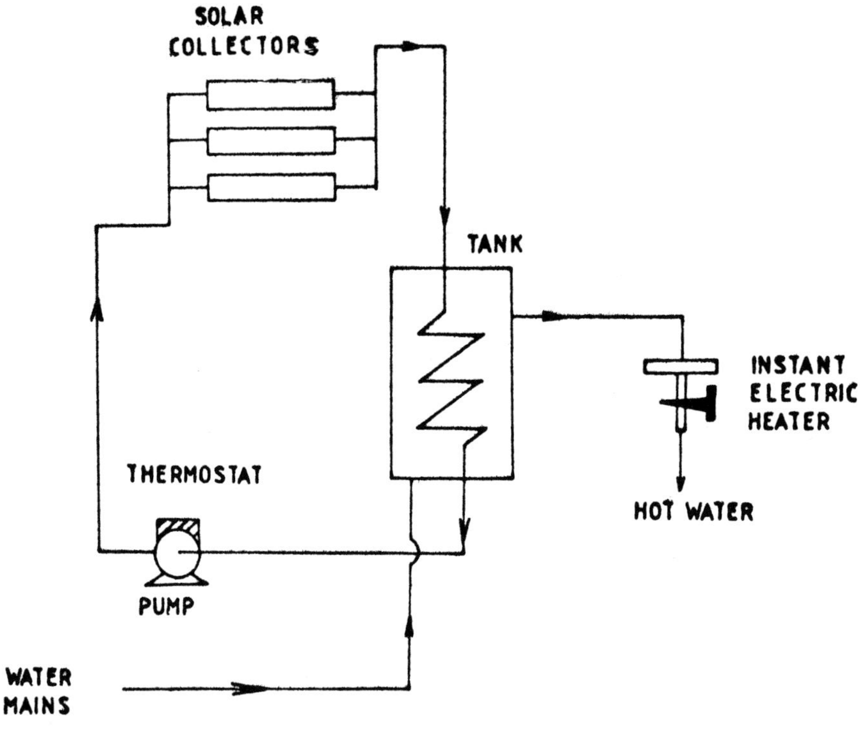

FIG.1 - SCHEMATICS OF A TYPICAL HYBRID WATER HEATING SYSTEM

Following assumptions have been made in developing the mathematical model:

(i) Collectors are connected in parallel.

(ii) Pipe losses are negligible.

(iii) Water in the storage tank is at uniform temperature.

(iv) Heat losses from the storage tank are proportional to its volume.

(v) The collectors face south and are inclined at an angle equal to the latitude from the horizontal.

If S is the amount of solar radiation per unit area, incident over the collector surface, the outlet temperature from the collectors is given by (Ref.28):

$$\frac{T_{co}-T_a-(\tau\alpha)S/U_L}{T_{ci}-T_a-(\tau\alpha)S/U_L} = \exp\left(-\frac{F'U_LA_c}{\dot{m}_cC_{wc}}\right) \quad (1)$$

The energy balance for storage tank yields the following differential equation:

$$\dot{m}_c(T_{co}-T_{ci})C_{wc} = M_wC_w\frac{dT_w}{dt} + U_tA_t(T_w-T_a) + \dot{m}(t)C_w(T_w-T_i) \quad (2)$$

where $\dot{m}(t)$ represents the time dependent withdrawal of hot water.

The energy balance equation for heat-exchanger in the storage tank may be written as

$$\dot{m}_cC_{wc}\frac{\partial T_f}{\partial x}\Delta x = -h_f(T_f-T_w)\Delta x \quad (3)$$

Solving equation (3), one gets

$$T_f(x) = T_w+(T_{co}-T_w)\exp\left(-\frac{h_fx}{\dot{m}_cC_{wc}}\right) \quad (4)$$

which gives the following expression for the collector inlet, temperature:

$$T_{ci} = T_w+(T_{co}-T_w)\exp\left(-\frac{h_fL'}{\dot{m}_cC_{wc}}\right) \quad (5)$$

Substituting for T_{ci} from equation (5) in equation (1), one gets

$$T_{co} = (1-F_1F_2)^{-1}\left[(1-F_1)F_2T_w+\left\{T_a+\frac{(\tau\alpha)S}{U_L}\right\}(1-F_2)\right] \quad (6)$$

where

$$F_1 = \exp\left(-\frac{h_fL'}{\dot{m}_cC_{wc}}\right)$$

and $$F_2 = \exp\left(-\frac{F'U_LA_c}{\dot{m}_cC_{wc}}\right)$$

Substituting T_{ci} and T_{co} from equations (5) and (6) in equation (2) and simplifying, one obtains

$$\frac{dT_w}{dt} + g(t)T_w = f(t) \tag{7}$$

where

$$g(t) = \frac{1}{M_w C_w}\left[\frac{\dot{m}_c C_{wc}(1-F_1)(1-F_2)}{(1-F_1F_2)}\left\{T_a + \frac{(\tau\alpha)S}{U_L}\right\} + U_t A_t T_a + m(t) C_w T_i\right] \tag{8}$$

and

$$f(t) = \frac{1}{M_w C_w}\left[\frac{\dot{m}_c C_{wc}(1-F_1)(1-F_2)}{(1-F_1F_2)} + U_t A_t + \dot{m}(t) C_{wc}\right] \tag{9}$$

Since the solar insolation S, the ambient temperature T_a and the temperature of water in the mains T_i are periodic function of time with a period of 24 hours, the steady state solution for $T_w(t)$ will also be periodic in nature. Thus

$$f(t) = \sum_{-\infty}^{\infty} f_n \exp(inwt) \tag{10}$$

$$g(t) = \sum_{-\infty}^{\infty} g_n \exp(inwt) \tag{11}$$

and

$$T_w(t) = \sum_{-\infty}^{\infty} T_{wn} \exp(inwt) \tag{12}$$

Substituting the expressions for f(t), g(t) and $T_w(t)$, in equation (7) and retaining the first six terms of the fourier series, the following matrix equation is obtained:

$$[A]_{13x13}\,[T_w]_{13x1} = [Y]_{13x1} \tag{13}$$

It may be noted here that the first six terms of the fourier series are usually found to be sufficient for the convergence of the series. Therefore, the remaining terms of the series have been truncated. The matrix elements may be expressed as follows:

$$A_{j,k} = g_{k-j} + (7-j)\, iw\, M_w C_w \delta_{jk}$$

where

$$\delta_{jk} = 1 \quad \text{for} \quad j = k$$
$$= 0 \quad \text{for} \quad j \neq k$$

and

$$-6 \leqslant (k-j) \leqslant 6$$

From equation (13), one gets the following expression for T_{wn}:

$$\left[T_{wn}\right] = \left[A\right]^{-1} \left[Y\right] \tag{14}$$

Substituting for T_{wn} from equation (14) in equation (12), the time variation of water temperature in the storage tank can be obtained.

The amount of heat required to raise the water temperature to a desired level may be evaluated by the expression

$$Q_a = \dot{m}(t)\, C_w\, (T_{des} - T_w)^+ \tag{15}$$

where the super script + shows that only the positive values of the expression have to be considered. This quantity of heat is provided by the instant electric heater.

Solar contribution to the total useful power may be calculated as follows:

$$\begin{aligned} Q_{us} &= \dot{m}C_w(T_w - T_i) \text{ for } T_w < T_{des} \\ &= \dot{m}C_w(T_{des} - T_i) \text{ for } T_w > T_{des} \end{aligned} \tag{16}$$

because water hotter than T_{des} is not needed. The solar fraction obtainable from the system is thus

$$S_f = \frac{\int Q_{us}\,dt}{\int Q_{us}\,dt + \int Q_a\,dt} \tag{17}$$

3. ECONOMIC ANALYSIS

Corresponding to a given initial investment, the annual cost of the total system (solar + auxiliary) may be expressed as follows (Ref.29):

$$S_c = I_s f + I_c f + M_s I_s + M_a I_c + C_a \int_o^{1\text{ year}} Q_a\,dt \tag{18}$$

where f is known as the annuity factor and is defined as

$$f = \frac{q^2(q-1)}{(q^2-1)} \tag{19}$$

where $q = (1 + \frac{Z}{100})$

and C_a = the cost of electricity (Rs/kwh).

From the prices given by Indian manufacturers it is seen that the cost of solar hot water system, including pump, controls, piping etc., is given by

$$I_s = (A_c C_c + M_w C_T)M + C_I \tag{20}$$

We have also neglected the cost of pumping energy. Substituting for I_s from equation (20) in equation (18), one gets,

$$S_c = (A_c C_c + M_w C_T)(f+M_s)M + C_I(f+M_s) + I_c(f+M_a) + MB_f \tag{21}$$

where

$$B_f = \frac{C_a}{M} \int_o^{1\ \text{year}} Q_a dt$$

The annual amount of useful energy in kwh is given by

$$Q'_{ut} = \frac{M}{3.6\text{x}10^6} \int_{\text{year}} \frac{\dot{m}}{M} C_w (T_{des} - T_i)\ dt = MQ_{ut}$$

The overall cost of useful energy (Rs/kwh) can be calculated by the following expression

$$C_u = \frac{S_c}{Q_{ut} M}$$

$$\text{or} \quad C_u = \frac{\left\{(A_c C_c + M_w C_T)(f+M) + B_f\right\}}{Q_{ut}} + \left[\frac{C_I(f+M_s) + I_c(f+M_a)}{M\ Q_{ut}}\right]$$

$$= C + C'/M \tag{22}$$

4. NUMERICAL RESULTS AND DISCUSSION

For a quantitative appreciation of the mathematical model developed here, numerical calculations have been made for the climatic data of Delhi, India.

where

and

$$-6 \leq (k-j) \leq 6$$

Values of various heat transfer coefficients have been evaluated using standard heat transfer relationships (Ref.30,31). A set of various parameters, used in calculations has been given in Table 1. Solar radiation data for horizontal surface is converted to that for inclined collector surface by using the hourly values of the monthly averaged beam and diffuse radiation (given in Tables 2 and 3) (Ref.32) and the Liu and Jordan (Ref.28) formula. Hourly values of the monthly averaged ambient air temperature have been given in Table 4. A standard daily hot water demand profile (Ref.33) used in the present calculations, has been shown in Fig.2(a).

TABLE 1 - Values of Various Parameters

Thermal analysis parameter	Cost analysis parameter
$F' = 0.84$	$C_c = 1560$ Rs/m^2
$U_L = 6$ w/m^2 oC	$C_T = 5$ Rs/Kg
$\dot{m}_c = 0.01$ Kg/sm^2 (during sunshine hours)	I_c = Rs.300
$= 0$ (during offshine hours)	$M_s = 0.05$
	$M_a = 0.0$
$\alpha = 0.9$	$C_a = 0.75$ Rs/kwh
$\tau = 0.95$	Utilization efficiency of electricity= 1.0
$T_{des} = 60^{o}C$	$Z = 10\%$
	$L = 10$ years
	C_I = Rs.3000

For the calculation of integrals, occurring in the expression for Q_a and Q_{ut}, the year is divided in twelve months and monthly averages of the hourly values of S, T_a and T_i are used, to evaluate the monthly contribution to the integral (obtained by multiplying the daily (24 hrs) integral, with the number of days in a month), the sum of monthly integrals constitutes the yearly integral as defined in the text. A typical monthly varation of the total daily hot water requirements of Delhi-like climates has also been shown in Fig.2(b). The calculations have been made for the two cases: (i) when monthly hot water demand is constant and (ii) when monthly hot water demand is variable (as shown in Fig.2b).

TABLE 2 - Monthly Averaged Hourly Values of Ambient Air Temperature (^{o}C) for New Delhi, India

MONTH	HOURS 1	2	3	4	5	6	7	8	9	10	11	12
JAN	10.9	10.4	10.0	9.6	9.2	8.9	8.6	8.7	11.0	13.8	16.3	17.0
FEB	14.6	14.1	13.5	13.0	12.7	12.3	12.0	12.4	15.2	17.9	19.2	21.9
MAR	19.5	10.9	10.4	17.7	17.3	16.0	16.6	18.0	21.1	23.6	25.7	27.1
APR	25.6	24.8	24.2	23.5	22.9	22.6	22.8	24.9	28.1	30.5	32.2	33.5
MAY	29.1	28.4	27.8	27.2	26.7	26.3	27.1	29.3	31.8	34.1	35.7	36.9
JUN	31.7	31.2	30.7	30.4	30.0	29.6	30.0	31.4	33.6	34.7	35.9	36.8
JUL	29.7	29.4	29.1	28.8	28.6	28.4	28.6	29.3	30.5	31.2	32.0	32.8
AUG	27.9	27.7	27.5	27.3	27.2	27.1	27.2	28.0	28.8	29.5	30.3	30.9
SEP	26.7	26.4	26.1	25.9	25.7	25.5	25.6	26.5	28.1	29.2	30.3	31.1
OCT	22.6	22.2	21.8	21.5	21.1	20.8	20.8	22.3	25.2	27.0	28.5	29.8
NOV	16.5	16.0	15.5	15.1	14.8	14.5	14.3	15.3	18.9	21.6	23.7	25.2
DEC	11.5	11.1	10.6	10.3	10.1	9.7	9.5	9.8	12.3	14.9	17.3	19.0

MONTH	HOURS 13	14	15	16	17	18	19	20	21	22	23	24
JAN	17.8	19.6	19.7	19.6	19.1	17.2	15.5	14.1	13.2	12.6	11.9	11.3
FEB	20.0	20.6	20.9	20.0	20.4	22.0	20.0	10.6	17.5	16.8	16.0	15.4
MAR	28.0	28.6	28.9	28.7	28.4	27.2	25.3	23.7	22.6	21.8	21.0	19.2
APR	34.0	34.0	35.1	34.9	34.7	33.7	31.9	30.2	29.0	28.0	27.4	26.5
MAY	37.7	38.3	38.5	38.7	38.0	37.5	35.9	34.1	32.6	31.5	30.7	29.9
JUN	37.7	38.3	38.5	38.4	38.2	37.7	36.7	35.4	34.3	33.6	32.9	32.3
JUL	33.3	33.6	33.7	33.7	33.4	33.0	32.4	31.5	31.0	30.5	30.1	29.9
AUG	31.1	31.5	31.6	31.5	31.4	31.0	30.3	29.6	29.1	28.8	28.4	28.1
SEP	31.0	31.9	32.0	31.8	31.5	30.7	29.5	28.7	29.2	27.7	27.3	26.9
OCT	30.0	31.2	31.3	31.1	30.6	29.0	27.2	25.9	25.0	24.4	23.5	23.0
NOV	25.9	26.2	26.4	26.0	25.2	22.9	20.9	19.6	18.7	18.0	17.3	16.8
DEC	20.1	20.7	20.9	20.7	19.7	17.2	15.9	14.7	13.9	13.2	12.5	11.9

TABLE 3 - Monthly Averaged Hourly Values of Global Solar Radiation at New Delhi, India

MONTH	HOURS													
	6	7	8	9	10	11	12	13	14	15	16	17	18	19
JAN	0	40	96	271	433	556	618	615	552	430	268	99	5	0
FEB	0	20	163	364	540	670	727	728	659	534	358	165	21	0
MAR	1	66	260	475	655	781	845	844	769	635	458	249	64	1
APR	10	130	339	554	727	848	910	903	833	699	523	318	122	10
MAY	30	177	383	581	746	865	925	920	844	719	547	352	158	27
JUN	36	164	339	510	657	757	809	813	750	639	483	318	159	36
JUL	26	130	275	425	549	640	672	671	606	518	412	273	141	32
AUG	12	101	243	392	512	607	640	659	595	510	392	255	107	13
SEP	3	79	248	431	507	698	757	747	689	571	417	244	74	3
OCT	0	36	202	406	581	701	764	758	692	569	391	191	33	0
NOV	0	9	128	319	495	620	685	680	613	490	315	125	10	0
DEC	0	3	85	259	425	543	605	605	538	418	255	86	3	0

TABLE 4 - Monthly Averaged Hourly Values of Diffuse Solar Radiation at New Delhi, India

MONTH	HOURS													
	6	7	8	9	10	11	12	13	14	15	16	17	18	19
JAN	0	3	51	103	140	160	174	172	162	137	100	51	3	0
FEB	0	13	76	126	159	183	191	190	180	160	125	75	13	0
MAR	1	43	113	159	189	210	222	226	215	197	163	113	41	1
APR	9	83	153	202	236	262	274	279	274	250	212	156	79	8
MAY	27	114	190	244	279	303	315	318	308	286	247	189	107	22
JUN	33	124	215	286	342	382	399	399	377	340	279	203	114	28
JUL	23	96	178	257	316	358	372	365	333	291	240	174	101	23
AUG	9	74	153	224	277	317	334	331	302	265	210	149	75	10
SEP	2	53	123	178	222	256	269	263	248	217	167	118	48	2
OCT	0	26	87	128	154	169	179	177	169	154	126	83	21	0
NOV	0	7	57	99	122	137	145	146	142	124	99	55	6	0
DEC	0	2	45	92	123	144	155	156	147	125	92	43	2	0

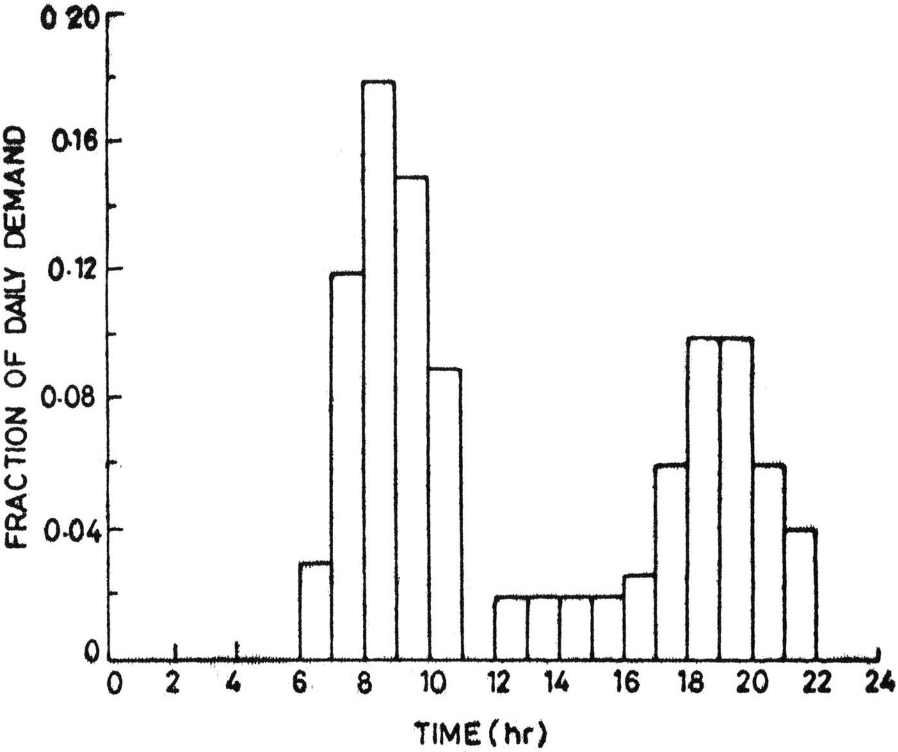

FIG.2(a) - STANDARD PROFILE OF DAILY HOT WATER REQUIREMENTS

The results of the calculations have been presented in the subsequent tables and figures. The values of collector areas and tank capacities, chosen for the calculations, range from 0.005 Mm^2 to 0.040 Mm^2 and 0.125 MKg to 1.250 MKg respectively where M represents the highest value of daily hot water demand. Tables 5 and 6 show the effect of increasing collector area on the annual averaged solar fraction ($\bar{S}_f$) as a function of tank capacity for the two above mentioned cases. In both the cases, $\bar{S}_f$ goes on increasing with increasing collector area; however, the percentage increase for large collector areas is small as compared to that for smaller collector areas. A qualitatively similar behaviour is noted for the variation of $\bar{S}_f$ with tank capacities. The effect of tank capacity on $\bar{S}_f$ is seen to be more pronounced for larger collector areas. As an illustration of the above points, it can be noted from the tables that corresponding to a tank capacity of 0.5 MKg the increase of collector area from 0.005 Mm^2 to 0.010 Mm^2 and 0.025 Mm^2 to 0.030 Mm^2 enhances $\bar{S}_f$ by 0.227 and 0.024 respectively. Similarly for collector areas for 0.005 Mm^2 and 0.030 Mm^2, a change in tank capacity from 0.5 MKg to 0.625 MKg increases $\bar{S}_f$ by 0.003 and 0.041 respectively. It may also be noted from Tables 5 and 6 that in the second case (variable monthly hot water demand), $\bar{S}_f$ assumes higher values as compared to the first case.

TABLE 5 - Variation of Annual Solar Fraction with A_c and M_w, when the Hot Water Demand is equal for all the months.

M_w (Kg) \ A_c (m^2)	0.005M	0.010M	0.015M	0.020M	0.025M	0.030M	0.035M	0.040M
0.125M	0.300	0.422	0.486	0.526	0.556	0.578	0.597	0.612
0.250M	0.342	0.510	0.591	0.637	0.666	0.687	0.702	0.713
0.375M	0.359	0.563	0.664	0.719	0.753	0.775	0.790	0.802
0.500M	0.365	0.592	0.710	0.774	0.812	0.836	0.853	0.864
0.625M	0.368	0.608	0.740	0.811	0.852	0.877	0.894	0.905
0.750M	0.369	0.617	0.760	0.836	0.879	0.905	0.921	0.932
0.875M	0.369	0.622	0.773	0.854	0.898	0.923	0.939	0.949
1.000M	0.369	0.625	0.782	0.867	0.911	0.936	0.952	0.961
1.125M	0.369	0.627	0.789	0.876	0.920	0.945	0.960	0.970
1.250M	0.369	0.628	0.793	0.882	0.928	0.953	0.967	0.975

TABLE 6 - Variation of Annual Solar Fraction with A_c and M_w, when the Hot Water Demand follows the monthly pattern.

M_w (Kg) \ A_c (m^2)	0.005M	0.010M	0.015M	0.020M	0.025M	0.030M	0.035M	0.040M
0.125M	0.368	0.492	0.551	0.586	0.610	0.629	0.644	0.656
0.250M	0.423	0.600	0.676	0.715	0.738	0.753	0.763	0.771
0.375M	0.441	0.656	0.748	0.794	0.820	0.836	0.846	0.853
0.500M	0.448	0.684	0.789	0.841	0.870	0.887	0.897	0.904
0.625M	0.450	0.699	0.814	0.870	0,900	0.917	0.927	0.934
0.750M	0.451	0.707	0.830	0.889	0.918	0.935	0.944	0.951
0.875M	0.451	0.712	0.840	0.901	0.930	0.946	0.955	0.962
1.000M	0.450	0.715	0.846	0.909	0.938	0.953	0.963	0.969
1.125M	0.449	0.716	0.851	0.914	0.943	0.959	0.968	0.974
1.250M	0.448	0.717	0.854	0.918	0.947	0.963	0.972	0.979

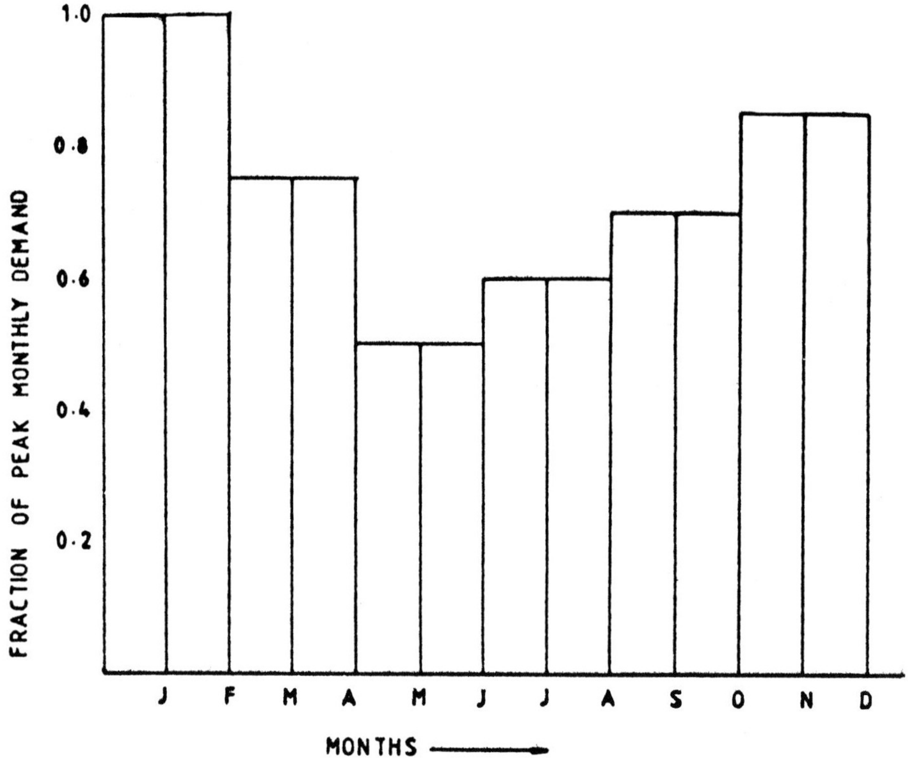

FIG.2(b) - MONTHLY DEMAND PATTERN

It is evident that the increase of collector area provides more useful energy from the sun and reduces the burden on the auxiliary energy source. However, in view of the following facts:

(i) Solar cost increases linearly with the collector area and tank capacity, and

(ii) The percentage increase in solar fraction diminishes non-linearly with the increasing collector areas, the key parameter in the optimization of collector area and tank capacity for a specific demand pattern is the cost of useful energy rather than the mean solar fraction. The cost calculations have been made for the typical cost of solar collectors in the Indian market. From equation (22), it is seen that the cost of useful energy has two components, viz. one independent of the demand and the other inversely proportional to the demand. Numerical calculations show that varying collector area for a specific tank capacity gives an optimum collector area for which the cost of useful energy is minimum. However, this optimum changes for other values of the tank capacity which indicates a need for double optimization. Figure 3 shows the variation of optimum collector areas and the corresponding cost of the useful energy for various tank capacities. The corresponding values of the annual averaged solar fraction have also been plotted in Fig.3.

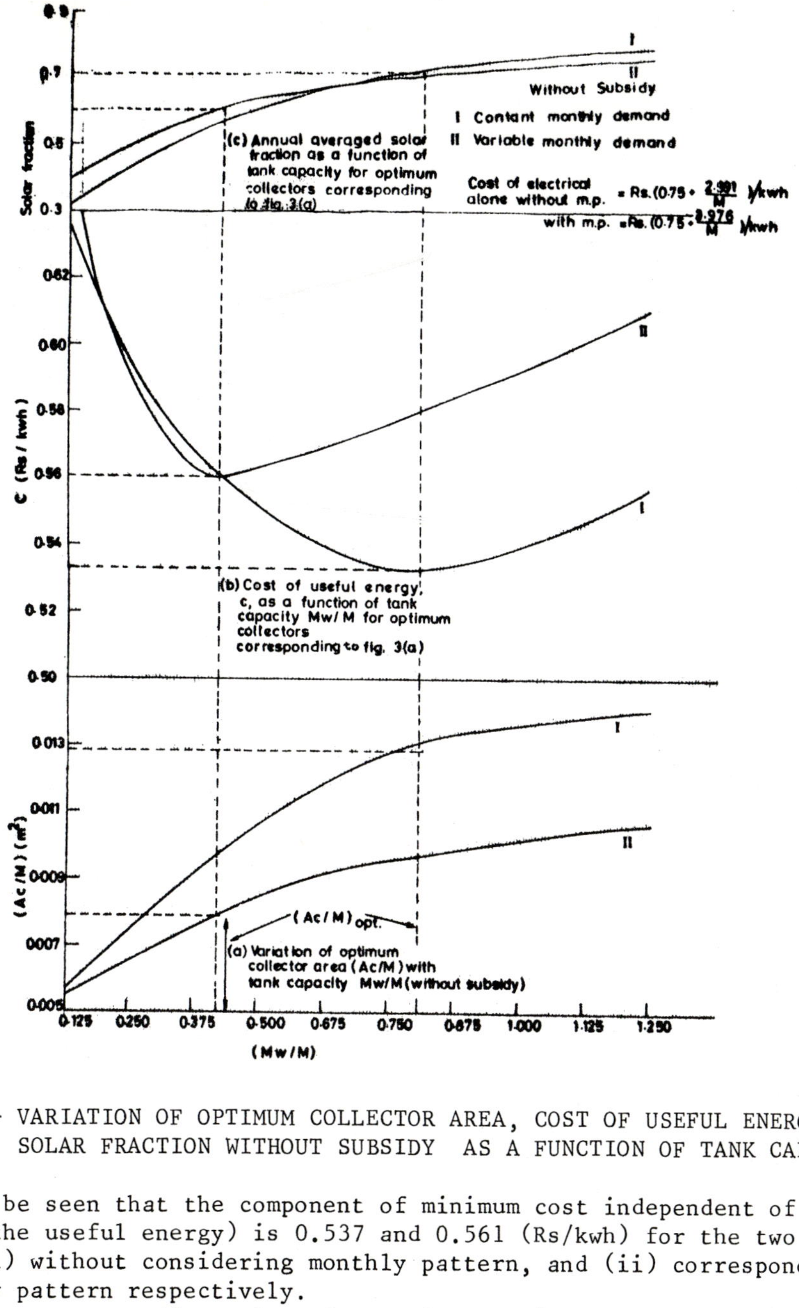

FIG.3 - VARIATION OF OPTIMUM COLLECTOR AREA, COST OF USEFUL ENERGY AND SOLAR FRACTION WITHOUT SUBSIDY AS A FUNCTION OF TANK CAPACITY

It can be seen that the component of minimum cost independent of demand C (of the useful energy) is 0.537 and 0.561 (Rs/kwh) for the two cases, viz. (i) without considering monthly pattern, and (ii) corresponding monthly pattern respectively.

The corresponding values for various design parameters have been given in Table 7.

It may, however, be noted here that the costs of useful energy in the preceding discussion (C) do not include the factor C'/M which is inversely proportional to the demand M. It is seen that

TABLE 7 - Optimized Design Parameters for a Solar Hot Water System at New Delhi, India

Parameters	Case I	Case II
(a) Without subsidy		
$(A_c/M) m^2/Kg$	0.0128	0.0079
(M_w/M)	0.8125	0.4250
$\bar{S}_f$	0.715	0.610
Cost C (Rs/kwh)	0.537	0.561
(b) With subsidy		
$(A_c/M) m^2/Kg$	0.0164	0.0135
(M_w/M)	1.106	0.875
Cost C (Rs/kwh)	0.419	0.457

$$\frac{C'}{M} = \left[\frac{C_I(f+M_s)+I_c(f+M_a)}{M.Q_{ut}}\right]$$

$$= \frac{39.11}{M} \quad \text{for case I}$$

$$= \frac{51.98}{M} \quad \text{for case II}$$

shows a significant effect on the cost of useful energy only for small values of M. Since forced circulation systems are used only for large systems ($M > 1000$ Kg/day), the cost is not appreciably effected by the magnitude of the demand in cases of practical interest.

The effect of government subsidy on the optimized values of the solar collector area and the tank capacity has been shown in Fig.4. As expected the given subsidy in India (viz. 33% of the total cost) makes the solar option more economical. This also corresponds to higher values of the optimized collector area and tank capacity.

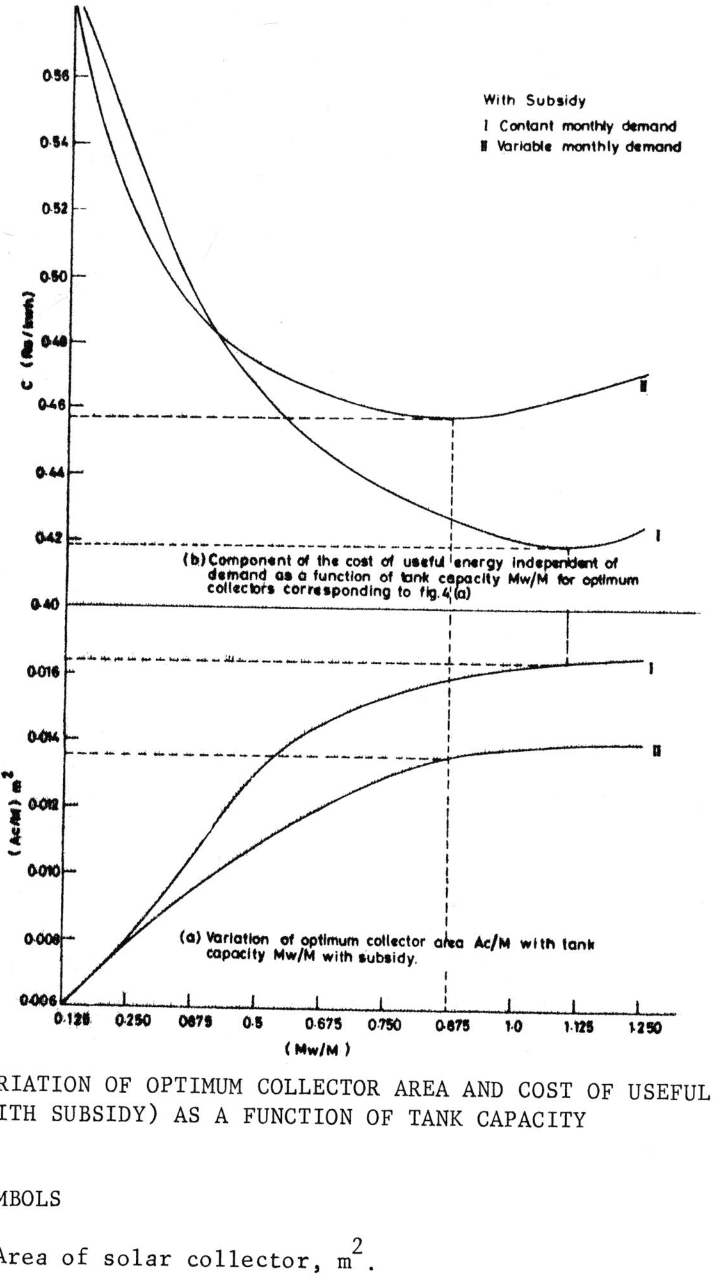

FIG.4 - VARIATION OF OPTIMUM COLLECTOR AREA AND COST OF USEFUL ENERGY (WITH SUBSIDY) AS A FUNCTION OF TANK CAPACITY

LIST OF SYMBOLS

A_c Area of solar collector, m^2.

A_t Surface area of tank, m^2.

C_c Cost of collector per unit area, Rs/m^2.

C Cost of useful energy, independent of the demand, Rs/kwh.

C'/M Cost of useful energy, depending on the demand, Rs/kwh.

C_I Installation cost, Rs.
C_w Specific heat of water in tank, J/Kg^oC.
C_T Cost of tank per unit volume, Rs/m^3.
C_{wc} Specific heat of collector fluid, J/Kg^oC.
C_u Cost of useful energy, Rs/kwh.
f Annuity factor defined in equation (19).
F' Plate efficiency factor of solar collector.
h_f Convective heat transfer coefficient.
I_c Initial investment of conventional system, Rs.
I_s Initial investment of solar system, Rs.
$\dot{m}$ Rate of hot water withdrawl, Kg/s.
$\dot{m}_c$ Rate of fluid flow in collector, Kg/s.
M Maximum value of daily hot water demand, Kg.
M_a Maintenance cost of conventional system, Rs/year.
M_s Maintenance cost of solar system, Rs per year.
M_w Storage capacity of tank, Kg.
Q_a Auxiliary power, W.
Q_{ut} Total annual useful energy, kwh.
Q_{us} Solar contribution to the total useful power, W.
S Solar radiation incident over the collectors, W/m^2.
S_c Annual cost of water heating system, Rs/year.
S_f Solar fraction obtained by the system.
$\overline{S}_f$ Averaged solar fraction as a function of tank capacity.
t Time, s.
T_a Ambient temperature, ^oC.
T_{ci} Collector fluid inlet temperature, ^oC.
T_{co} Collector fluid temperature when it enters the tank, ^oC.
T_{des} Desired water temperature, ^oC.
T_f Outlet temperature of water from the tank, ^oC.
T_i Water temperature at mains, ^oC.
T_w Water temperature in tank, ^oC.
U_L Heat loss coefficient of the solar collector, W/$m^2$$^oC$.
U_t Heat loss coefficient of tank, W/$m^2$$^oC$.
Z Percentage of interest.
α Absorptivity of collector plate.
τ Transmissivity of cover glass.

REFERENCES

1. D.J. Close, (1962), 'The performance of solar water heaters with natural circulation', Solar Energy, 6 33-40.
2. C.L. Gupta and H.P. Garg, (1968), 'System design in solar water heaters with natural circulation', Solar Energy, 12, 163-182.
3. Y. Zvirin, A. Shitzer and G. Grossman, (1977), 'The natural circulation solar heater models with linear and non-linear temperature distribution', Int. J. Heat Mass Transfer, 20, 997-999.
4. A. Shitzer, D. Kalmanoviz, Y. Zvirin and G. Grossman, (1979), 'Experimental with a flat plate solar water heating system in thermosyphonic flow', Solar Energy, 22, 27-35.
5. G.L. Morrison and D.B.J. Ranatunga, (1980), 'Transient response of thermosyphon solar collectors', Solar Energy, 24, 55-61.
6. G.L. Morrison and D.B.J. Ranatunga, (1980), Thermosyphon circulation in solar collectors', Solar Energy, 24, 191-198.
7. A. Mertol, W. Place, Webster and R. Greif, (1981), 'Detailed loop model (DLM), analysis of liquid solar thermosiphons with heat exchangers', Solar Energy, 27(5), 367-386.
8. G. Gutierrez, F. Hincapie, A. Duffie and W.A. Backman, (1974), 'Simulation of forced circulation water heaters; effects of auxiliary energy supply, load type and storage capacity', Solar Energy, 15, 287-296.
9. D. Wolf, A.I. Kudish and A.N. Simbira, (1981), 'Dynamic simulation and parametric sensitivity studies on a flat plate solar collector', Energy, 6, 333-349.
10. B.J. Brirkworth, (1975), 'Selection of design parameters for closed circuit forced circulation, solar heating systems', Solar Energy, 17, 331.
11. Francis de winter, (1975), 'Heat exchanger penalties in double loop solar water heating systems', Solar Energy, 17, 335.
12. J.M. Bradley, (1979), 'The development of a freeze-tolerant solar water heater using cross linked polyethylene as a material of construction', Presented at the 2nd Miami Int. Conf. on Alternative Energy Sources, Miami Beach, Frlorida (USA), 10-13 Dec., 1979.
13. J.T. Czarnecki, (1958), 'Performance of experimental solar water heaters in Australia', Solar Energy, 2, 2.
14. R. Farrington, D. Norcen and L.M. Murphy, (1980), 'A comparative analysis of six generic solar domestic hot water systems', Proc. System Simulation and Economic Analysis, San Diego, California, 23-25 Jan, 1980, pp 131-136.
15. A.H. Fanney, S.T. Liu and J.E. Hill, (1970), 'Experimental validation of computer programs for solar domestic hot water systems', Proc. 3rd Annual Solar Heating and Cooling,Research and Development Branch Contractors Meeting, Washington, D.C., 24-27 Sept, 1978, pp.440-444.
16. A.H. Fanney and S.T. Liu, (1979), 'Experimental system performance and comparison with computer prediction for six solar domestic hot water systems', Proc. ISES, Silver Jubilee Cong., Atlanta, Georgia, May 1979, Vol.2, pp 972-976.

17. M.S. Sodha, S.N. Shukla and G.N. Tiwari, (1983), 'Transient analysis of a natural convection solar water heater with a heat exchanger', J. Energy, 7(2), 107.
18. A. Whillier and G. Saluja, (1965), 'Thermal performance of experimental solar water heaters', Solar Energy, 9, 821.
19. D.J. Close, (1967), 'A design approach for solar processes', Solar Energy, 11, 112.
20. M.S. Sodha, G.N. Tiwari and S.N. Shukla, (1983), 'Thermal modelling of solar water systems', In Review of Renewable Energy Resources, Vol.1, Wiley Eastern Limited, New Delhi, India, pp 139-229.
21. W.E. Buckles and S.A. Klein, (1980), 'Analysis of solar domestic hot water heaters', Solar Energy, 25, 417-424.
22. S.A. Klein, W.A. Backman and J.A. Duffie, (1976), 'A design procedure for solar heating systems', Solar Energy, 18, 113-127.
23. R.G. Courtnly, (1977), 'A computer study of solar water heating', Building and Environment, 12, 73-80.
24. P.B. Howells, (1981), 'An improved computer code for the simulation of solar energy system', M.Sc. Thesis, University of Wales, Cardiff, U.K.
25. S.G. Tzafestas, A.V. Spyridonos and M.G. Koumoutsos, (1974), 'Finite difference modelling, Identification and simulation of a solar water heater', Solar Energy, 16, 25-31.
26. H. Buchberg and J.R. Roulet, (1968), 'Simulation and optimization of solar collection and storage for house heating', Solar Energy, 12, 31-50.
27. K.S. Ong, (1976), 'An improved computer program for the thermal performance of a solar water heater', Solar Energy, 18, 183-191.
28. J.A. Duffie and W.A. Backman, (1974), 'Solar Energy Thermal Processes, Wiley, New York.
29. A. Whillier and G. Saluja, (1965), 'Thermal performance of solar water heaters', Solar Energy, 9, 21.
30. H.Y. Wong, (1977), Heat Transfer for Engineers, Longman Group Ltd., London.
31. J.R. Simonson, Computing Methods in Solar Heating Design, London.
32. A. Mani, (1981), Handbook of Solar Radiation for India, Allied Publishers, New Delhi.
33. M.S. Sodha, et al., (1985), 'A technoeconomic simulation model for a hybrid solar water heating system', Int. J. of Energy Research (In press, 1985).

AUTHOR INDEX

SUBJECT INDEX